Learning Resource Centers

WITHDRAWN

D0609819

WITHDRAWN

© Learning Resource Centre

The Science and Technology of Civil Engineering Materials

PRENTICE HALL INTERNATIONAL SERIES
IN CIVIL ENGINEERING AND ENGINEERING MECHANICS

William J. Hall, Editor

The Science and Technology of Civil Engineering Materials

PRENTICE HALL INTERNATIONAL SERIES IN CIVIL ENGINEERING AND
ENGINEERING MECHANICS, WILLIAM HALL, SERIES EDITOR

J. Francis Young
Professor of Civil Engineering and Materials Science and Engineering
University of Illinois at Urbana-Champaign

Sidney Mindess
Professor of Civil Engineering and Associate Vice President Academic
University of British Columbia

Robert J. Gray
Consultant
Vancouver, British Columbia

Arnon Bentur
Professor of Civil Engineering and Vice President Research
Technion—Israel Institute of Technology

Prentice Hall
Upper Saddle River, New Jersey 07458

Library of Congress Cataloging-in-Publication Data

Young, J. Francis.
    The science and technology of civil engineering materials /
  J. Francis Young . . . [et al.].
      p. cm.   –(Prentice Hall international series in civil engineering
  –and engineering mechanics)
  Includes bibliological references and index.
  ISBN: 0-13-659749-1
  I. Materials   I. Young, J. Francis.   II Series.
TA403.S419   1998
624.I´8–DC21   97-31972
                                            CIP

Acquisitions editor: William Stenquist
Editor-in-chief: Marcia Horton
Production editor: Irwin Zucker
Managing editor: Bayani Mendoza de Leon
Director of production and manufacturing: David W. Riccardi
Copy editor: Sharyn Vitrano
Cover director: Jayne Conte
Manufacturing buyer: Julia Meehan
Editorial assistant: Margaret Weist

 ©1998 by Prentice-Hall, Inc.
Simon & Schuster/A Viacom Company
Upper Saddle River, New Jersey 07458

All rights reserved. No part of this book may be
reproduced, in any form or by any means,
without permission in writing from the publisher.

The author and publisher of this book have used their best efforts in preparing this book. These efforts include
the development, research, and testing of the theories and programs to determine their effectiveness. The author
and publisher make no warranty of any kind, expressed or implied, with regard to these programs or the
documentation contained in this book. The author and publisher shall not be liable in any event for incidental or
consequential damages in connection with, or arising out of, the furnishing, performance, or use of these
programs.

Printed in the United States of America

10 9 8 7 6 5 4 3

ISBN 0-13-659749-1

Prentice-Hall International (UK) Limited, London
Prentice-Hall of Australia Pty. Limited, Sydney
Prentice-Hall Canada Inc., Toronto
Prentice-Hall Hispanoamericana, S.A., Mexico
Prentice-Hall of India Private Limited, New Delhi
Prentice-Hall of Japan, Inc., Tokyo
Simon & Schuster Asia Pte. Ltd., Singapore
Editora Prentice-Hall do Brasil, Ltda., Rio de Janeiro

Contents

Preface

The modern civil engineer needs to deal with a variety of materials that are often integrated in the same structure, such as steel and concrete, or are used separately for construction projects, such as pavements from asphalt and portland cement concretes. Many of these construction materials have been with us for centuries, like timber, while others, like portland cement concrete and steel, are relatively new and have been used mainly during the last century. The civil engineering field is also making headway in the use of even more modern materials, such as polymers and composites. The modern principles of materials science have been applied extensively over the past three decades to construction materials, and the benefits of this approach can be seen clearly on site: The traditional construction materials used at present are far superior to those of the past (achieving, for example, concrete strength levels greater by an order of magnitude), and there is increased use of synthetic and composite materials that are specially formulated for civil engineering applications.

As a result of these changes and the expected dynamic developments in this field, there is a clear trend in the industry to move from the empirical-technological approach of the past to one which incorporates both the technology and materials science concepts. In view of this modern trend, there is a need for a revision in the materials education of the civil engineer. Traditionally, materials science and construction materials have been taught almost as separate entities. Materials science teaching was based mainly on texts developed for courses for engineering areas in which metals are of the greatest interest, with some reference to other materials, such as polymers and ceramics. Construction materials were taught thereafter independently, giving greater attention to their technology and much less to their science. As a result, civil engineers were limited in their overall view of construction materials and were lacking some of the concepts of materials science, such as surface properties, which are of prime importance in construction materials but receive hardly any attention in the traditional materials science texts.

This book offers a new approach, in which the science and technology are integrated. It is divided into four parts; the first two provide the general concepts of materials, referring to their fundamental structure and mechanical properties (Part I is titled "The Fundamentals of Materials," and Part II is "Behavior of Materials under Stress"). The other two parts of the book deal with specific construction materials (the titles are as follows: Part III, "Particulate Composites: Portland Cement and Asphalt Concretes"; Part IV, "Steel, Wood, Polymers, and Composites"). The parts of this book dealing with general materials science concepts are presented in an approach which is directed toward civil engineering needs and emphasizes surface properties and amorphous structures. The parts of this book dealing with the actual construction materials are written with the view of combining the materials science and engineering approaches with an emphasis on materials characteristics of particular interest for civil engineering applications.

This book is designed primarily for use at the undergraduate level, but it can also serve as a guide for the professional engineer. Thus it includes reference to standards and specifications. It is intended to serve as a basis for a two-semester course. However, it is designed to be flexible enough to be adjusted for shorter courses. Such a course could be based on all of Part I, three of the chapters in Part II (Chapter 5, "Response of Materials to Stress," Chapter 6, "Failure and Fracture," and Chapter 7, "Rheology of Fluids and Solids"), and selected chapters dealing with specific construction materials, in view of the intended scope of the shortened course.

1

Atomic Bonding

1.1 INTRODUCTION

Introductory courses in chemistry have discussed atomic structure and the way in which chemical bonds serve to ensure that atoms achieve stable electron configurations by adding, removing, or sharing electrons. In this chapter we will simply review the characteristics of the various types of bonds that can form in materials. These are summarized in Table 1.1 and Fig. 1.1 and can be divided into two major categories: the strong (primary) bonds between atoms (ionic, covalent, and metallic) and the weak (secondary) van der Waals bonds between molecules. The position of an element in the periodic table determines the type of chemical bonds it can form.

1.2 IONIC BONDS

Elements in Groups I and II readily lose electrons to form cations (i.e., they are strongly electropositive), while at the other end of the periodic table, elements in Groups VI and VII readily gain electrons (they are strongly electronegative). Thus, when these elements are brought together there will be an exchange of electrons to form ionic compounds containing M^+ and X^-, where M is a Group I element and X is a Group VII element; or M^{2+} and X^{2-}, where M is a Group II element and X is a Group VI element.

The interaction energy between a pair of ions is proportional to $(z^+z^-e^2)/r$, where z is the ionic charge and r is the distance between ions. However, we seldom find discrete ion pairs; rather ions of a given charge try to be surrounded by as many ions of the opposite charge as possible (Fig. 1.1b). In the crystalline state

TABLE 1.1 Summary of bond types.

Bond Type	Bond Energies (kJ · mol⁻¹)	Typical Materials	Typical Elements	Remarks
Ionic	500–1200[a]	Ceramic Oxides Gypsum Rock salt Calcite	Compounds of Gp I, Gp II	All exist as crystalline solids.
Covalent	150–750[a, b]	Diamond Glasses Silicon carbide	Gp IV, Gp V, Gp VI	States of matter at room temperature depend on intermolecular attraction.
Metallic	50–850[a]	Metals	Elements of Gp I–III. Transition metals. Heavy elements of Gp IV and V.	May be liquid or solid depending on binding energies.
Hydrogen	10–30[c]	Water	F, O, N	Can be considered weak ionic or strong van der Waals. Strongly influences material behavior.
van der Waals	0.05–5	Thermoplastic polymers	Compounds of all elements	Primarily intermolecular bonds. Dominate the behavior and microstructure of construction materials, such as concrete and asphalt.

[a] Lattice energies of crystal.
[b] Isolated multiple covalent bonds (as formed in N_2, for example) can be as strong as 950 kJ · mol⁻¹.
[c] Single hydrogen bond is about 2 kJ · mol⁻¹.

Figure 1.1
The principal types of crystalline binding. In (a) neutral atoms with closed electron shells are bound together weakly by the van der Waals forces associated with fluctuations in the charge distributions. In (b) electrons are transferred from the alkali atoms to the halogen atoms, and the resulting ions are held together by attractive electrostatic forces between the positive and negative ions. In (c) the valence electrons are removed from each alkali atom to form a community electron sea in which the positive ions are dispersed. In (d) the neutral atoms appear to be bound together by the overlapping parts of their electron distributions.

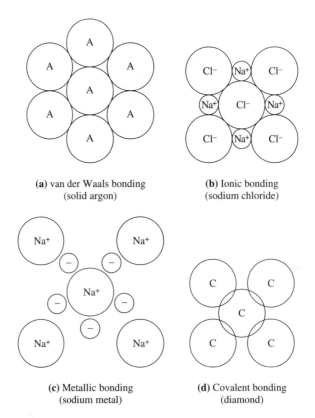

(a) van der Waals bonding (solid argon)

(b) Ionic bonding (sodium chloride)

(c) Metallic bonding (sodium metal)

(d) Covalent bonding (diamond)

Part I

THE FUNDAMENTALS OF MATERIALS

(see Chapter 2), they take up specific arrangements to maximize the interactions between ions of opposite charge; but in the gaseous or liquid state these ions are free to move about. Hence, ionic solids will not conduct electricity unless they are either in the molten state or dissolved in water, where the ions are free to move under electric gradients.

1.3 COVALENT BONDS

For most elements, the need to lose or gain electrons will not be sufficient to form ions (or the ions will be unstable), and valency requirements are satisfied by sharing electrons. The simplest situation is the sharing of one pair of electrons between two elements, as in hydrogen (H_2) or methane (CH_4), where there are four covalent C—H bonds. Two atomic orbitals, which each contain one electron, overlap and combine to form a single molecular orbital lying between the two atomic nuclei (Fig. 1.1d). When more than one covalent bond is formed by an element, the atomic orbitals may combine to adopt certain directional arrangements to increase their degree of overlap with other atomic orbitals and hence the strength of the covalent bond. This process is called hybridization. More than one atomic orbital from each atom may be involved, leading to multiple bonds between atoms.

Another complication is that electrons are seldom shared equally between two dissimilar atoms, but usually the electrons spend more time near the more electronegative atom (i.e., the atom with a greater tendency to attract electrons). There is thus a statistical separation of charge so that the bond has a permanent *dipole*. Such a bond is said to have partial ionic character, since an ionic bond implies complete separation of charge. Conversely, an ionic bond can be said to have some covalent character; in this case, the atomic orbitals of the ions are distorted, leading to a distortion of the ideal packing of ions in the crystal toward more directional arrangements that favor this distortion.

Covalent bonds thus lead to the formation of a specific grouping of atoms (molecules) in which all the atoms achieve stable electron configurations. Only a few materials are bound together principally by covalent forces acting between atoms in all directions; natural diamond and synthetic silicon carbide are common examples. Most covalent materials are composed of covalently bonded molecules; whereas the bonds between the atoms within the molecules are strong, the bonds between atoms in adjacent molecules are generally much weaker and involve van der Waals forces (see discussion in Sec. 1.5). Covalent molecules may range from simple molecules, like H_2, to the very complex macromolecules, such as organic polymers, which may contain many thousands or millions of atoms in a single molecule.

1.4 METALLIC BONDS

Atoms of electronegative elements (e.g., chlorine, oxygen, or sulfur) can satisfy their electron needs through covalent bonding. But this possibility is not open to the electropositive elements since these elements wish to lose electrons while covalent bonding effectively adds electrons to an atom. This problem is solved by the metallic bond: All atoms give up electrons to a "common pool," becoming positive ions with a stable electron configuration. The free electrons occupy extended delocalized orbitals lying between the positive metal ions (Fig. 1.1c) so that the electrons are independent of any particular ion. These electron "clouds" bind the ions together but allow the electrons freedom of movement so that metals can conduct electricity and can move rapidly to effect transfer of thermal energy (high thermal conductivity).

1.5 VAN DER WAALS BONDING

Weak attractions can occur between molecules with nonpolar bonds, or between single atoms. The latter are sometimes called dispersion forces and were first postulated to explain the nonideal behavior of gases like helium, argon, hydrogen, and nitrogen. As shown in Fig. 1.2a, dispersion forces are considered to arise from *fluctuating dipoles*. Although over an extended time the centers of positive charges (protons in the nucleus) and negative charges (electrons in the orbitals) coincide, momentarily they are misplaced due to the continuous movement of the electrons, thus creating a transitory dipole. This dipole induces a transitory dipole in an adjacent atom (or molecule), and weak bonding is obtained. Stronger attractions can arise from interactions between adjacent permanent dipoles in polar molecules when suitably oriented (Fig. 1.2b). Intermediate van der Waals attractions, between permanent and fluctuating dipoles, are obtained.

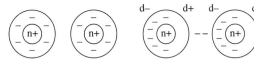

(I) No attraction (symmetrical charge distribution)

(II) Attraction by fluctuating charge distribution

(a) Dispersion forces arising from fluctuating dipoles

(b) Intermolecular interactions through permanent dipoles (in polymer molecules)

Figure 1.2
Schematic representation of van der Waals forces.

Relative contributions from the various effects to the total van der Waals attractions are compared in Table 1.2 for different covalent molecules. All these attractions fall off rapidly as the distance between the atoms increases ($\alpha\ 1/r^6$) so that their influence extends only about one-tenth of a nanometer in space. It can be seen

TABLE 1.2   Contributions to van der Waals bonding energy in solids (kJ · mol^{-1})[a]

Molecule	Dipole--Dipole	Dipole-Induced Dipole	Dispersion	Total	Boiling Point(°C)
Argon (A)	0.00	0.00	2.03	2.03	−197
Carbon Monoxide (CO)	0.00	0.008	2.09	2.10	−192
Hydrogen Chloride (HCl)	0.79	0.24	4.02	5.05	−85
Ammonia (NH$_3$)	3.18	0.37	3.53	7.07	−33
Water (H$_2$O)	8.69	0.46	2.15	11.30	100

[a] Calculated from dipole moments and polarizability (the ease with which orbitals can be distorted) of the atoms and bonds involved.

that the strength of the van der Waals bonds between molecules strongly influences the boiling point of the compounds. The high dipole-dipole interactions in ammonia and water are due to the special case of hydrogen bonding discussed next.

Although the effects of van der Waals attractions are most noticeable between covalently bonded molecules, they also exist in the extended atomic arrays of ionic and metallic solids. The effects are not obvious, but calculations have shown that van der Waals forces may account for up to 10% of the total binding energy in some crystals, although the figure is generally much less. However, van der Waals attractions are extremely important in controlling the properties of surfaces (see Chapter 4).

1.5.1 Hydrogen Bonding

A special case of van der Waals bonding arises from strong electrostatic interactions between the hydrogen atom and O, F, or N atoms in molecules containing highly polar C—O, H—F or H—N bonds. We consider the case for water because it is of the most interest to us. Because the O—H bond is highly polar, the H atom has an appreciable net positive charge while the oxygen atom is negative. The small size of the hydrogen atom allows it to approach an oxygen atom on an adjacent molecule closely so that a strong electroactive interaction is set up between the two (Fig. 1.3a). These attractions are strong and lead to the anomalous behavior of water that is vital to our existence. For example, if it were not for hydrogen bonding, water would freeze at much lower temperatures, and its boiling point would be lower, too. In the absence of hydrogen bonding, the boiling point of water should be about $-100°C$—lower than that of hydrogen sulfide, which cannot form hydrogen bonds because the sulfur atom is too large. Hydrogen bonding is also responsible for the minimum density of water being at 4°C and for the fact that ice floats (both are vital properties for aquatic life in cold climates). Hydrogen bonding in water contributes about $10 \text{ kJ} \cdot \text{mol}^{-1}$ of energy. Water will form hydrogen bonds with hydroxylated surfaces, such as cellulose (wood) or hydrous metal oxides (hydrated cement). Hydrogen bonding requires a suitable molecular shape (stereochemistry) that allows the two atoms to come close together (see Fig. 1.3b). Such bonding contributes to the high mechanical performance and heat resistance of some modern polymers (e.g., Kevlar®, nylon).

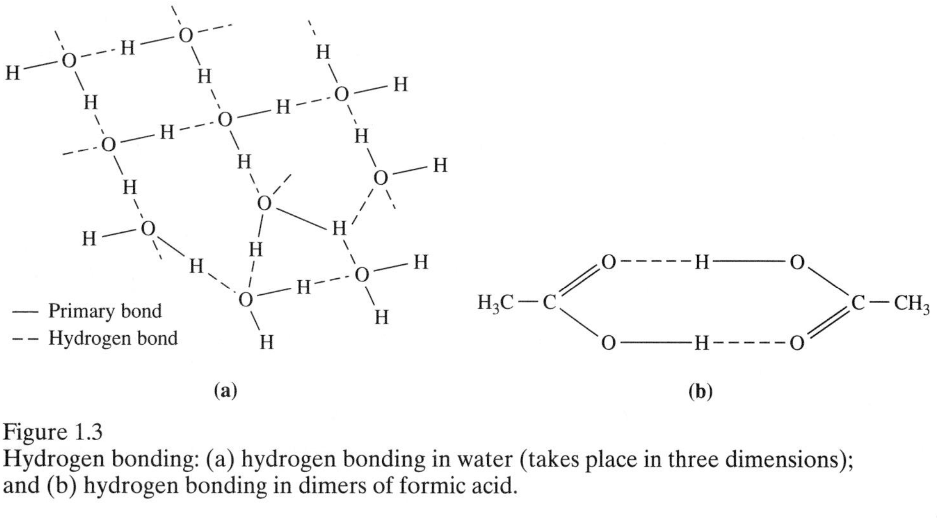

Figure 1.3
Hydrogen bonding: (a) hydrogen bonding in water (takes place in three dimensions); and (b) hydrogen bonding in dimers of formic acid.

The cohesive forces that hold a solid together are directly related to the interatomic bonding within the solid. These attractive forces are not uniform over distance since repulsive forces are also involved between atoms: As the atoms or ions approach each other, repulsion is generated by the electrons in the external orbitals. To evaluate interatomic forces, it is necessary to calculate the net attractive forces involving all atoms, and this can be done most simply for an ionic crystal. The attractive force between a pair of ions is given simply by Coulomb's law:

$$F_{attr} = \frac{(z_1 \cdot z_2)e^2}{r^2} \tag{1.1a}$$

and the potential energy involved in this attraction (i.e., the energy associated with bringing the two charges from an infinite distance to a distance r) is

$$U_{attr} = -\frac{(z_1 \cdot z_2)e^2}{r}, \tag{1.1b}$$

where z_1 and z_2 are the charge associated with each ion, e is the electron charge, and r is the distance between them. The negative sign in Eq. 1.1b denotes a decrease in potential energy. Equation 1.1 represents two isolated ions, but they actually exist in a crystal lattice surrounded by ions in a definite geometrical relationship. Thus for 1 mol (N atoms) the expression becomes

$$U_{attr} = \frac{-N \cdot A \cdot (z_1 \cdot z_2)e^2}{r} \quad kJ \cdot mol^{-1}, \tag{1.2}$$

where A is the Mandelung constant, which depends solely on the geometry of the crystal (for a cubic crystal A is 1.75 and it can be calculated for other geometries as well). Equation 1.2 indicates increasing attraction as r gets smaller (i.e., as the ions get closer together), but eventually the electron orbitals on each ion begin to overlap and cause repulsion, according to the relationship

$$U_{rep} = \frac{NB}{r^a} \quad kJ \cdot mol^{-1}, \tag{1.3}$$

where B and a are constants ($a \sim 9$). Thus the net attractive energy is the algebraic sum of these two quantities:

$$U_{cryst} = U_{attr} + U_{rep} = \frac{-N(z_1 \cdot z_2)A \cdot e^2}{r} + \frac{NB}{r^a} \quad kJ \cdot mol^{-1}. \tag{1.4}$$

A plot of U_{cryst} as a function of r gives the curve shown in Fig. 1.4a, which is known as the *Condon-Morse diagram*. The crystal has maximum stability when U_{cryst} is a minimum. The depth of the potential energy "well" indicates the strength of the cohesive forces within the crystal. The value of r which corresponds to U_{min} is r_0 and represents the equilibrium distance between ions in the crystal (the interatomic distance). The way the energy varies with r tells us how the crystal will respond to various conditions.

Although this calculation was carried out for ionic solids, the same general treatment can be applied to all materials. For completely covalently bonded materials (e.g., diamond) and metals, the Condon-Morse diagram will have a deep energy well, similar to that in Fig. 1.4a. However, many covalent molecules are actually bonded together by van der Waals forces (e.g., plastics). In these cases, Eq. 1.4 becomes

$$U_{cryst} = \frac{-NP(\mu_1^2 \cdot \mu_1^2)}{r^6} + \frac{-Q \cdot \alpha \cdot \mu_2}{r^6} + \frac{-R \cdot \alpha_1 \cdot \alpha_2}{r^6} + \frac{N \cdot B}{r^9} \quad kJ \cdot mol^{-1}, \tag{1.5}$$

$$\text{(dipole-dipole)} \qquad \text{(induced dipole)} \qquad \text{(dispersion)} \qquad \text{(repulsion)}$$

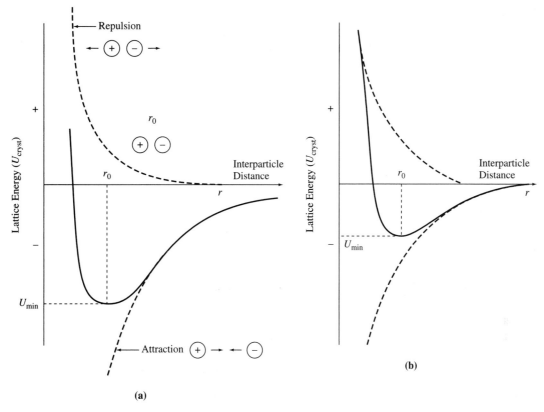

Figure 1.4
Condon-Morse diagrams for: (a) strongly bonded solids; (b) weakly bonded solids
(van der Waals attraction).

where μ and α are dipole moments and polarizabilities, respectively, and P, Q, and R are constants. The U_{attr} term is the different van der Waals forces summed between all the molecules. Since this term is proportional to $1/r^6$ (Eq. 1.5) rather than $1/r$ (Eq. 1.4), the effects fall off much more rapidly so that the Condon-Morse curve has a much shallower well (Fig. 1.4b) and r_0 will be larger. The depth of the well depends on the exact magnitude of U_{attr}: It is very sensitive to the extent to which each of the forces contribute, which is strongly dependent on molecular structure. This is reflected in a comparison of the melting points and boiling points (which is a measure of the cohesive energy) of various molecules (Table 1.3). The number of atoms involved in the summation (i.e., molecular size) is also important, as is strikingly illustrated in comparing the alkane (paraffin) series of nonpolar hydrocarbons (where only dispersion forces are involved) in Table 1.4.

1.7 THERMAL PROPERTIES OF SOLIDS

Several properties of solids can be determined directly from their Condon-Morse diagrams, and this emphasizes some interesting relationships between properties. Only at $0°K$ ($-273°C$) will $U_{solid} = U_{min}$ and the atoms occupy equilibrium spacings. At higher temperature, thermal energy imparts motion to the atoms so that part of the potential energy is converted to kinetic energy. The atoms thus vibrate about the value of r_0, as indicated in Fig. 1.5a. The vibrational energy, like all energies associated

Compound	Melting Point (°C)	Boiling Point (°C)	Bonding Forces
Helium	−272.2	−268.6	Atomic dispersion forces only
Hydrogen	−259.1	−252.5	Also bond dispersion forces
Nitrogen	−209.9	−195.8	Bond dispersion forces involving multiple bonds with more electrons
Carbon dioxide	−56.6	−78.5	Two polar bonds per molecule
Hydrogen sulfide	−85.5	−60.7	Strongly polar bonds
Water	0	100	Strong hydrogen bonding

TABLE 1.4 Properties of selected alkanes

Molecule	Formula	Melting Point (°C)	Boiling Point (°C)	Use
Methane	CH_4	−183	−162	Gaseous fuel
n-Propane	C_3H_8	−187	−42	Gaseous fuel
n-Octane	C_8H_{18}	−57	120	Liquid fuel (gasoline)
n-Hexadecane	$C_{12}H_{34}$	18	280	Liquid fuel (kerosene)
n-Tetracosane	$C_{24}H_{50}$	51	Decomposes	Grease (vaseline)
n-Pentacontane	$C_{50}H_{102}$	92	Decomposes	Solid wax
Polyethylene	$(CH_2)_n$	120	Decomposes	Plastic

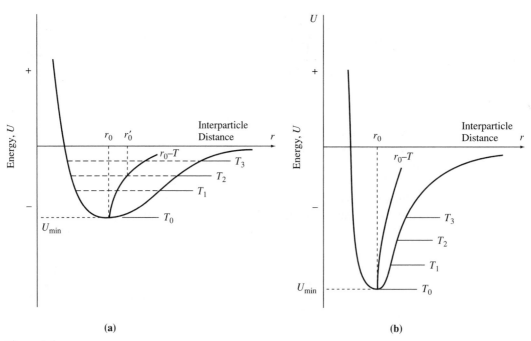

(a) (b)

Figure 1.5
Effect of temperature on interatomic distances: (a) shallow energy well;
(b) deep energy well.

with atoms and molecules, is quantized. Only discrete energy levels can be attained. A quantum of vibrational energy is called a phonon, analogous to the photon, which is a quantum of light energy. Phonons thus can be regarded as waves that exist within the lattice, which can be excited by temperature; as the temperature rises, the atoms are promoted to higher vibrational energy states until eventually complete dissociation of the solid occurs. The boiling point is thus proportional to the depth of the potential energy well (i.e., a high boiling point indicates that strong, cohesive forces hold the solid together).

It can be seen from Fig. 1.5a that as atomic vibrations increase, the mean interatomic distance increases from r_0 to r_0', because the potential energy well is not symmetric about r_0. The difference between r_0 and r_0' as a function of temperature (i.e., the slope of the $r_0 - T$ line) is proportional to the coefficient of thermal expansion (α). Since a deep potential energy well is much narrower, the coefficient is roughly proportional to $1/U_{min}$ so that strongly bonded solids have lower thermal expansions. Because the Condon-Morse diagram is asymmetric, the slope of the $r_0 - T$ line changes with increasing temperature, and α is constant only over a restricted temperature range. Eventually the increase in r_0 makes the packing of the atoms, ions, or molecules unstable and phase changes occur: either to a new form of packing (polymorphism) or to the liquid state. The melting point will thus also bear some relationship to U_{min}.

Other important thermal properties of materials can be related back to their basic structure. The heat capacity of a solid represents the energy required to excite phonons (i.e., raise the lattice to a higher vibrational energy state). For all monatomic crystalline solids (ionic crystals and metals) the heat capacity is about $3R$, which is 25 J $\cdot$ mol^{-1} $\cdot$ K^{-1} at room temperature. The value is higher for solids which have polyatomic molecules or ions because now additional intermolecular vibrational modes are possible about the mean atomic position. In liquids rotational energies of the molecule are also involved, and thus the specific heat is considerably higher (see Table 1.5).

The thermal conductivity of a material is a measure of the thermal energy transmitted across a unit area per unit time:

$$\frac{dQ}{dt} = -K\frac{dT}{dx}, \tag{1.6}$$

where dQ/dt is the heat flux and K is the thermal conductivity. It is a random process depending on the temperature *gradient* rather than temperature *difference*. Hence the specific heat influences the thermal conductivity. The process of heat transfer is analogous to the kinetic theory of ideal gases. In a gas molecular motion is proportional to temperature; molecules heated at one end of a cylinder gradually transfer their excess kinetic energy to all other molecules so that the temperature is equalized. In solids the analogous process occurs through phonons being excited thermally and gradually transferring their excess energy through random interactions to phonons in other parts of the material. Thermal conductivities are thus

TABLE 1.5   Contributions to specific heat

Material	n^a	Energy Contributions (J $\cdot$ mol^{-1} $\cdot$ K^{-1})		
		Vibrational	Rotational	Total
Solid Fe	1	25	—	25
Liquid CS_2	3	35	25	60
Liquid CCl_4	5	70	38	108

[a] n = number of atoms in molecule.

Sec. 1.7   Thermal Properties Of Solids

strongly affected by impurities and lattice imperfections because of the mobile phonons associated with the electron "gas." Amorphous solids thus have very low thermal conductivities. Comparisons of thermal properties of some materials are given in Table 1.6.

TABLE 1.6   Comparison of thermal properties of materials.[a]

Compound	Phase	Coefficient of Thermal Expansion $10^{-6} \cdot K^{-1}$	Specific Heat		Thermal Conductivity $W \cdot m^{-1} \cdot K^{-1}$
			$(J \cdot mol^{-1} \cdot K^{-1})$	$J \cdot kg^{-1} \cdot K^{-1}$	
Nitrogen	Gas	—	28.4	1050	0.03
Water	Liquid	200	76	4200	0.50
Aluminum	Monatomic solid	24	25	460	350
Iron	Monatomic solid	12	25	460	120
Granite	Polycrystalline material	7–9	—	800	3
Silica glass	Amorphous inorganic solid	0.5	—	800	1
Polystyrene	Amorphous organic solid	~ 150	—	1200	0.15

[a]Values at room temperature. All properties are temperature dependent.

1.8   BONDING FORCES

Differentiating Eq. 1.4 with respect to r gives the force of attraction between adjacent atoms:

$$F = \frac{dU}{dr} = \frac{-NA(z^+ \cdot z^-)e^2}{r} + \frac{aNB}{r^{a+1}} \qquad (1.7)$$

and $dU/dr = 0$ when $r = r_0$ since U is a minimum at this value. The curve of F versus r is given in Fig. 1.6 for strongly and weakly bound materials. The slope of the curve about r_0 is approximately linear and gives a measure of the restoring force that acts on the atoms for small displacement from the equilibrium position. These displacements are elastic, so the slope gives a measure of Young's modulus of elasticity:

$$E \propto \left(\frac{dU}{dr}\right)_{r=r_0} \qquad (1.8)$$

and hence E is related to U_{min}. Therefore, both thermal and mechanical properties relate back to the binding energy and should relate to one another. That this is so can be seen in Table 1.7 over a wide range of binding energies.

The force-distance diagram also gives information about the theoretical strength of a solid. The applied stress required to stretch the bonds is given by the force-distance curve (Fig. 1.6). The fracture stress (strength) (σ_f) is the maximum force per unit area required to continue stretching the bonds so that fracture will occur as the atoms separate without further stress applied. Thus the relation between the applied stress and the interatomic distance can be presented as a sine function:

$$\sigma(r) = \sigma_f \cdot \sin\left(\frac{2\pi r}{\lambda}\right). \qquad (1.9)$$

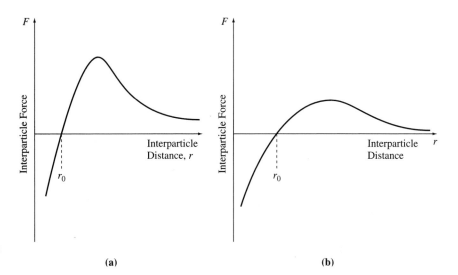

Figure 1.6
Force-distance curve for
strongly (a) and weakly (b)
bound solids.

(a)   **(b)**

TABLE 1.7   Relationships between metal properties that are
dependent on binding energies

Element	Boiling Point (°C)	Modulus of Elasticity (GPa)	Coefficient of Thermal Expansion ($10^{-6} \cdot °K^{-1}$)
Li	1332	25	56
Al	2056	70	25
Fe	2998	210	12
W	5927	500	4

The work required for fracture (W_f) is the area under the curve, which can be approximated as half a sine wave:

$$W_f = \int_0^{\lambda/2} \sigma(r) \cdot dr = \int_0^{\lambda/2} \sigma_f \cdot \sin\left(\frac{2\pi r}{\lambda}\right) dr. \tag{1.10}$$

That is,

$$W_f = \left(\frac{\lambda \cdot \sigma_f}{\pi}\right). \tag{1.11}$$

Since two surfaces are formed when fracture is complete, the work (W_s) required to do this (see Chapter 4) is given by

$$W_s = 2\gamma_s. \tag{1.12}$$

Since $W_f = W_s$,

$$2\gamma_s = \left(\frac{\lambda \cdot \sigma_f}{\pi}\right) \tag{1.13}$$

or

$$\sigma_f = \left(\frac{2\pi \cdot \gamma_s}{\lambda}\right). \tag{1.14}$$

Since for elastic deformation

$$\sigma(r) = \frac{E \cdot r}{r_0} = \sigma_f \cdot \sin\left(\frac{2\pi r}{\lambda}\right) \tag{1.15}$$

and for small strains

$$\sin\left(\frac{2\pi r}{\lambda}\right) \approx \frac{2\pi r}{\lambda}$$

thus

$$\sigma_f \approx \frac{E \cdot r}{r_0} \cdot \frac{\lambda}{2\pi r} \approx \frac{E \cdot \lambda}{2\pi r_0}. \tag{1.16}$$

Equation 1.15 indicates that the theoretical cohesive fracture strength of a material, σ_f, is a function of E, λ, and r_0, all of which depend on its cohesive energy. Thus, from Eqs. 1.16 and 1.14,

$$\sigma_f^2 \approx \frac{2\pi \cdot \gamma_s}{\lambda} \cdot \frac{E \cdot \lambda}{2\pi r_0} \tag{1.17}$$

or

$$\sigma_f = \left(\frac{E \cdot \gamma_s}{r_0}\right)^{1/2}. \tag{1.18}$$

Equation 1.18 indicates that the theoretical fracture strength, σ_f, of a material is a function of E, γ_s, and r_0, whose values reflect its cohesive energy. It will be shown later (Chapter 6) that in practice the theoretical strength predicted by Eq. 1.17 is not achieved due to defects and flaws in the material.

BIBLIOGRAPHY

Any standard freshman chemistry text. E.g.

STEVEN S. ZUMDAHL, *Chemistry,* 3rd Ed., Heath & Co., Lexington, MA, 1993, 1123 pp.

DONALD A. MCQUARRIE and PETER A. ROOD, *General Chemistry,* Freeman, N.Y., 1987, 876 pp.

Any standard introductory text in materials science and engineering. E.g.,

WILLIAM F. SMITH, *Foundations of Materials Science and Engineering,* 2nd Ed., McGraw-Hill Inc., 1993, 882 pp.

MILTON OHRING, *Engineering Materials Science,* Academic Press, 1995, 827 pp.

PROBLEMS

1.1. What would be the expected dominant bonding type for the following materials: (i) Boron nitride (BN); (ii) silicon nitride (Si_3N_4); (iii) zirconia (ZrO_2); (iv) nickel (Ni); (v) fluorspar (CaF_2)?

1.2. Many thermoplastic polymers have low moduli of elasticity and low melting points. What kind of bonding will be responsible for these characteristics?

1.3. Provide an expected order of melting points for the polymer series $(CHX)_n$ where X = OH, F, CI, CH_3. Justify.

2

The Architecture Of Solids

Now that we have discussed the nature of bonding, we need to consider how atoms and molecules form the solid materials that we use in everyday life. This depends on the nature of the chemical bond of the molecular structures and on the assembly of different phases in a composite structure. The possibilities which will be discussed in this chapter are shown in Table 2.1.

2.1 THE CRYSTALLINE STATE

2.1.1 Metallic Crystals

We will discuss metals first because they are the simplest examples of solid structure. The metallic bond is an array of cations bonded by delocalized electrons. To maximize the binding energy, the cations are placed in a close-packed array. The possible arrangements are determined by considering the atoms as a collection of solid spheres of equal radius (the ionic radius). Three possible packings exist:

FCC = face-centered cubic packing
HCP = hexagonal close-packing
BCC = body-centered cubic packing.

These names derive from the unit cells, which are the smallest volumes that describe the structures.

The packing in HCP and FCC structures is shown in Fig. 2.1. The HCP packing is an array of spheres packed in layers with an ABABAB . . . sequence; that is, there are two layers arranged with the spheres in one layer fitting in the interstices of the lower (and upper) layer. The FCC packing is an ABCABCABC . . . sequence. In this

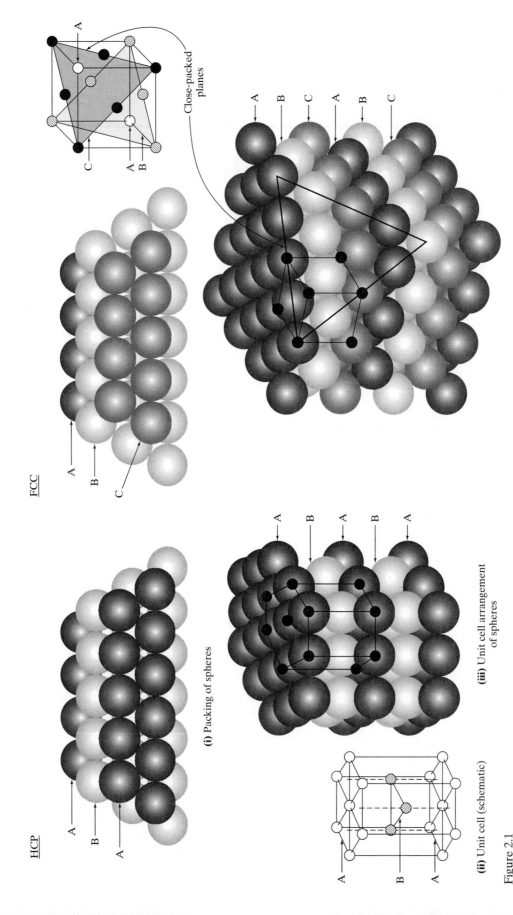

FCC

A
B
C

(i) Packing of spheres

A
B
C
A
B
C

Close-packed
planes

C
A
B
A

HCP

A
B
A

A
B
A
B
A

**(iii) Unit cell arrangement
of spheres**

A
B
A

(ii) Unit cell (schematic)

Figure 2.1
Packing of unit cells of HCP and FCC structures (after Moffat, Pearsall, and Wulf,
The Structure and Properties of Materials, Vol. I: Structures, Wiley, 1964, Figs. 3.2–3.4,
pp. 48, 50–51).

TABLE 2.1 Classification of Solid Structures

Type	Description	Examples
Ionic crystals	Close-packed array of ions (alternate M^{n+} and X^{n-})	E.g., crystalline salts, common salt (NaCl), fluorite (CaF_2)
Metallic crystals	Close-packed array of metal cations	Iron, copper
Covalent crystals	Packing of atoms to satisfy directional covalent bonds	Diamond, silica, clay minerals
Molecular crystals	Packing of specific molecules held together by van der Waals bonds	Crystalline polyethylene, ice
Amorphous materials	Irregular packings of ions, covalently bonded atoms, or distinct molecules	Metglas, soda glass, amorphous polymers
Composite materials	Particles or fibers dispersed in a continuous matrix	Fiber-reinforced plastics, portland cement concrete

case, the close-packed layer is represented by an inclined plane in the unit cell. The BCC structure is also an arrangement of planes stacked in an ABABAB . . . sequence, but unlike HCP, the close-packed planes are not parallel to a side of the unit cell but are diagonal planes. The BCC structure can be visualized as intersecting primitive cubes (see Fig. 2.2).

The HCP and FCC structures represent the closest packing of spheres, since each sphere is surrounded by 12 touching neighbors. About two-thirds of the metallic elements crystallize in these arrangements, as do the rare gas elements when solidified at low temperatures. The BCC structure is less dense; each sphere is surrounded by eight touching neighbors. Most of the remaining one-third of metals have BCC structures. Many of these are transition metals (Fe, Cr, etc.), which are known to be partly covalent and hence to have directional bonds. Alkali metals (Na, K, etc.) are BCC at high temperatures, but transform to FCC or HCP at very low temperatures. It appears that the relatively large thermal vibrations of the atoms at high temperatures (see Chapter 1) favor the more open structure.[1] Changes in crystal structure without a change in composition are called *polymorphism* or *allotropy*.

2.1.2 Ionic Crystals

Ions of the Same Charge

When cations and anions are to be packed into a crystal lattice, the problem becomes one of packing spheres of unequal sizes, since the cation and anion radii will not be equal. In most instances the cations are smaller than anions because they have fewer electrons and hence fewer filled atomic orbitals. We can describe ionic crystal structures as packed arrays of anions, with the cations located in holes within the array. If the cation is bigger than the holes, the anions will be pushed apart to accommodate the cations, and eventually an alternate anion arrangement will be favored to give a denser packing. If the cation is too small for the holes, the packing will be unstable and must be rearranged to allow the anions and cations to touch each other (see Fig. 2.3). The geometries depend, therefore, on the radius ratio R, where $R = r_c/r_a$. There is a critical value of R, R_c, below which a different, less dense

---

[1]This is a consequence of the fact that the atoms (and ions) do not behave as "hard spheres," which is the model we have been using because it is easy to visualize and treat mathematically. Atoms and ions really behave as "soft spheres," which can be deformed by directional bonding and thermal vibrations.

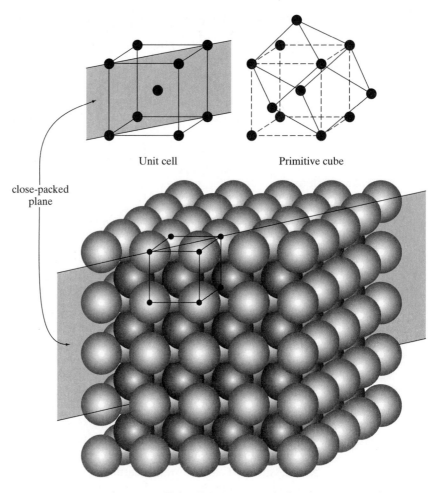

Figure 2.2
Packing of unit cell of BCC structure (after Moffat, Pearsall, and Wulf, *The Structure and Properties of Materials, Vol. I: Structures,* Wiley, 1964, Figs. 3.2, 3.4, pp. 48, 51).

Unit cell

Primitive cube

close-packed plane

Unit cell arrangement of spheres

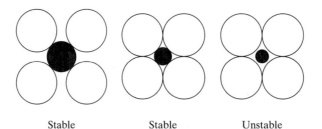

Figure 2.3
Effect of cation size on anion packing.

Stable   Stable   Unstable

packing of the anions becomes stable. The different stable packings are presented in Table 2.2, and the most common geometries are shown in Fig. 2.4.

The most stable crystal structures are those in which the ions are surrounded by as many neighbors of the opposite charge as possible, since this increases the total binding energy (i.e., deepens the potential energy well). This extra energy, obtained by packing the ions together, is known as the *lattice energy*. The size of the ions, as well as their packing, affects the lattice energy. The highest lattice energies are found in crystals containing close-packed arrangements of small ions (e.g., LiF). The extremely high energy of LiF makes it insoluble in contrast to the fluorides of neighboring elements.

TABLE 2.2   Stable Packing of Spheres of Different Sizes

Coordination Number[a]	R_c [b]	Packing Geometry	Examples of Ionic Crystals
12	1.0	HCP	—
12	1.0	FCC	—
8	0.73	cubic[c]	CsCl
6	0.414	octahedral	NaCl, MgO
4	0.225	tetrahedral	ZnS, SiO$_2$
3	0.155	trigonal	B$_2$O$_3$
2	< 0.155	linear	—

[a] The number of nearest touching neighbors.

[b] E.g., below R_c = 0.73 the cubic packing is no longer stable because the cation is too small. The stable packing is now octahedral, which remains stable until R_c = 0.414, when tetrahedral becomes stable. Above R_c = 0.74 octahedral packing will be stable, but less stable than cubic packing since the anions will be farther apart.

[c] The anions are packed in a simple cubic array (at the corners of a cube) with the cation in a body-centered position.

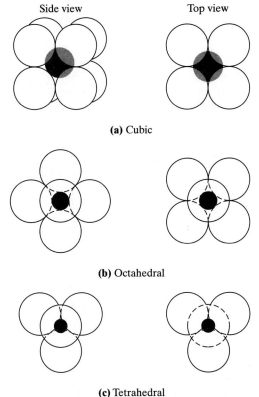

Figure 2.4
Common cation-anion geometries.

The relationships of these packing geometries to crystal structure are shown in Fig. 2.5. The structure can be considered as interpenetrating lattices of cations and anions. Alternatively, we can think of the cations and anions collectively occupying a single lattice site.

Ions of Different Charge

The preceding discussion has been concerned with the effects of differing atomic size, while each anion and cation has the same charge. When the charges are not the

Figure 2.5
Ionic arrangements for (a) eight-fold coordination in CsCl, projected on a plane. This would look the same in all directions and can be described as interpenetrating cubic lattices of Cl^- and Cs^+ ions. (b) six-fold coordination in NaCl, which can be viewed as interpenetrating FCC lattices of Cl^- and Na^+ ions, and (c) four-fold coordination in ZnS (wurzite). The ions are drawn smaller for clarity showing an FCC arrangement of S^{2-} ions with Zn^{2+} ions in the interstices between the S^{2-} ions, so that each is in tetrahedral array around the other.

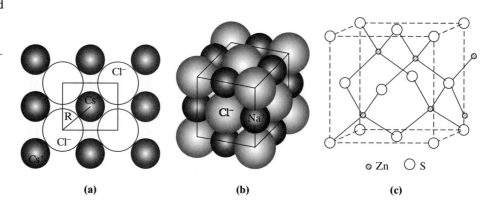

(a)   (b)   (c)

same, the crystal structure must be modified to accept more ions of lower charge to maintain electroneutrality. Consider the simplest case of calcium fluoride (CaF_2), where there must be twice as many F^- ions as Ca^{2+} ions. Calcium fluoride has a packing with a coordination number of 8 since $R_c = 0.8$ (Fig. 2.6). The simple cubic packing arrangement is modified to accommodate the extra fluoride ions. The fluoride anions are cubic close packed, but only half the cation sites are filled.

Another example is calcium hydroxide [$Ca(OH)_2$], which is present in hardened concrete. It has a coordination number of 6 in an octahedral arrangement, in which the hydroxide ions are in a close-packed hexagonal array with the calcium ions in the octahedral holes (Fig. 2.7). However, every second layer of octahedral holes is unoccupied to maintain electroneutrality. As a result, the hydroxide ions on either side of this plane are not held together by strong ionic forces, but only weak van der Waals secondary attractions. The crystal is therefore readily cleaved along this plane, as can be seen in fracture surfaces of portland cement binder. This is an extreme example of crystal anisotropy: Most crystals show a degree of anisotropy

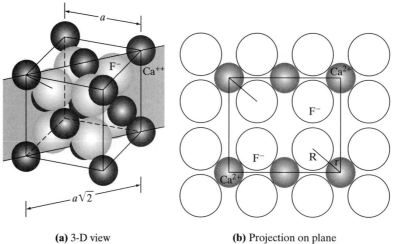

(a) 3-D view   (b) Projection on plane
                           indicated in (a)

Figure 2.6
Crystal structure of CaF_2.

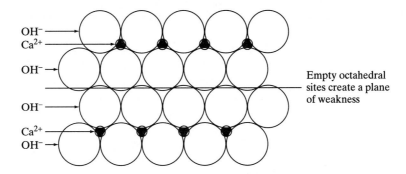

Figure 2.7
Crystal structure of
Ca(OH)$_2$.

OH$^-$
Ca^{2+}
OH$^-$
OH$^-$
Ca^{2+}
OH$^-$

Empty octahedral
sites create a plane
of weakness

because different planes within the crystal have denser atomic packing (and thus stronger bonding) than others. Anisotropy is observed in ionic crystals but is most marked in covalent crystals, where directional bonds dominate.

2.1.3 Covalent Crystals

In close-packed arrays of anions, the spaces between can be either octahedral or tetrahedral (see Fig. 2.8a). Therefore, an alternate view of a crystal structure is to view it as a packed coordination polyhedra (Figs. 2.8b and 2.8c), which shares edges and faces. The anions are at the vertices of the polyhedral and the cation at the

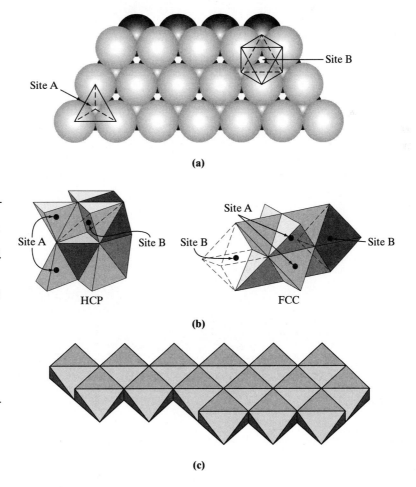

Figure 2.8
Arrangements of tetrahedral (a) and octahedral (b) sites: (a) location in a close-packed array of spheres (after Moffat, Pearsall, and Wulf, *The Structure and Properties of Materials, Vol. I: Structures,* Wiley, 1964, Fig. 3.10, p. 58,); (b) representation of HCP and FCC packings as arrays of coordination polyhedra (after Moffat, Pearsall, and Wulf, *The Structure and Properties of Materials, Vol. I: Structures,* Wiley, 1964, Figs. 3.11b, 3.12b, pp. 58–59,); and (c) representation of HCP as an array of octahedra (after H. F. W. Taylor, *Chemisrty of Cement,* Vol. I, Academic Press, 1964, Fig. 3, p. 174).

Site A
Site B

(a)

Site A
Site B Site B
HCP

Site A
Site B
Site B
FCC

(b)

(c)

centers. This type of visualization becomes important at low coordination numbers (e.g., the zinc blende structure, Fig. 2.5c) and when dealing with the directional nature of bonding typical of covalent structures. We will illustrate the influence of covalency on crystal structure by considering various silicate structures since these have relevance in portland cements, clays, rock minerals and ceramics.

The basic unit is the silicon-oxygen tetrahedron (SiO_4), which can be linked together in various 1-D, 2-D, and 3-D arrangements (Fig. 2.9). Orthosilicates have discrete tetrahedral ions (SiO_4^{4-}) replacing simple anions, as shown in Fig. 2.10. The oxygen atoms form a distorted HCP structure, with the silicon atoms filling one-eighth of the tetrahedral sites and calcium ions occupying one-half of the octahedral cation sites. A similar type of structure occurs for carbonates, sulfates, etc. Pyrosilicates (or disilicates) have the $Si_2O_7^{6-}$ ion held together by cations, while in the extreme case infinite SiO_3^{2-} linear chains are present, as in wollastonite (Fig. 2.11).

Fully two-dimensional silicate structures are sheet silicates, which are typified by clay minerals. Consider kaolinite, which is an HCP array of oxygen and hydroxide ions (Fig. 2.12a). We can visualize kaolinite as having sheets of $[Si_2O_5^{2-}]$ composition fitting together with sheets of $[Al_2(OH)_4]^{2+}$ (Fig. 2.12b). The charge and geometry of the two sheets match perfectly, and each polar sheet combination is bound to the

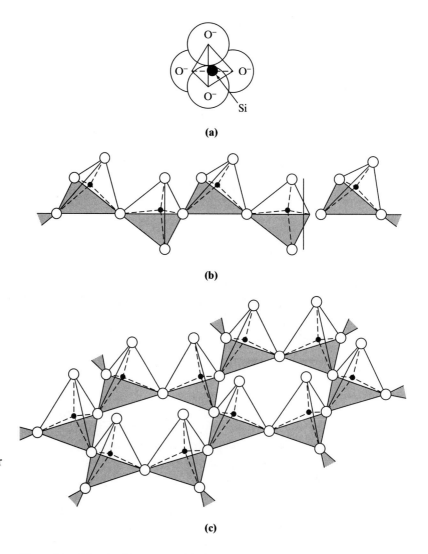

Figure 2.9
Silicon-oxygen (silicate) structure: (a) the basic unit of SiO_4^{4-} tetrahedron. Four oxygen atoms surround a silicon atom. (b) Silicate chain structure (1-D). (c) Silicate sheet structure (2-D).

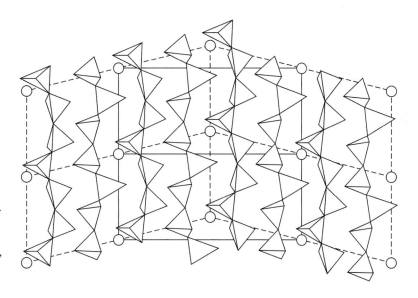

Figure 2.10
Crystal structure projection of γ—Ca_2SiO_4 (the olivine structure). The oxygens (large circles) are in HCP packing; the shaded circles lie between or above the open ones. The silicon atoms (not shown) lie at the center of the tetrahedra. The calcium atoms (small circles) are at octahedral sites; the open ones lie in a plane above or below the filled ones (after H. F. W. Taylor, *Chemistry of Cement,* Vol. I, Academic Press, 1964, Fig. 6, p. 146).

Oxygen

Calcium

Figure 2.11
Structure of chain silicates: wollastonite showing linear chains of the silica tetrahedra (after H. F. W. Taylor, *Chemistry of Cement* Vol. I, Academic Press, 1964, Fig. 4, p. 142).

others only by weak van der Waals forces. Thus, kaolinite crystallizes as flat sheets and is isotropic only in two dimensions: Kaolinite typifies a "two-layer" clay (Fig. 2.12b).

This molecular structure has several consequences regarding material properties. Clays crystallize as flat plates which cleave easily along the weakly bonded plane. Mica is well known in this regard in that it forms very large crystals. Mica can readily be cleaved into very thin sheets, which were often used as electrical insulation and as a window glazing, although now mica has been replaced by modern materials. Dry clays have a slippery feel due to the plates sliding across one another and can be used as a solid lubricant. Talc is the most familiar example of this. Most clays

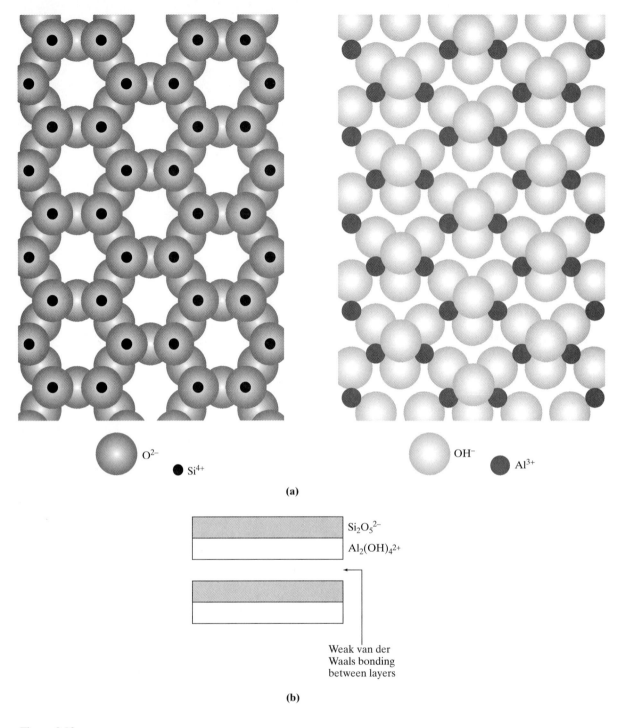

O^{2-}

Si^{4+}

OH^-

Al^{3+}

(a)

$Si_2O_5^{2-}$

$Al_2(OH)_4^{2+}$

Weak van der
Waals bonding
between layers

(b)

Figure 2.12
Structure of kaolinite showing condensation of silicate and aluminate sheets (a) Silicate sheets formed by linked chains of many minerals (left). Each silicon atom (black)
is surrounded by four oxygens; each tetrahedron shares three of its oxygens with three
other tetrahedrons. Notice the hexagonal pattern of "holes" in the sheet. An aluminate sheet (right) consists of aluminum ions (hatched) and hydroxide ions (speckled).
The top layer of hydroxides has a hexagonal pattern. If the two sheets are superposed
(as if they were facing pages of a book) they mesh, forming kaolinite. (b) Schematic
description of the two-layer kaolinite structure.

cannot be used for this purpose because they readily adsorb water between the layers. Water adsorption reduces the strength of a clay and in some cases will be accompanied by a pronounced swelling as water pushes the layers apart. However, many clays have metal cations located between the layers, which increase the bonding between them and reduce swelling potentials. These cations balance the net negative charges on the layers caused by impurities in the layer structures.

Finally, silica tetrahedra can be fully bonded in three dimensions to four other tetrahedra. Several different packings are possible. Cristoballite (Fig. 2.13) has the diamond cubic structure, which is an FCC packing modified to accommodate tetrahedral coordination.[2] We can visualize the FCC cube as composed of eight smaller cubes having an oxygen atom at each corner. Alternate small cubes contain silicon atoms at the corner of a tetrahedron, and these are additional sites in the unit cell.

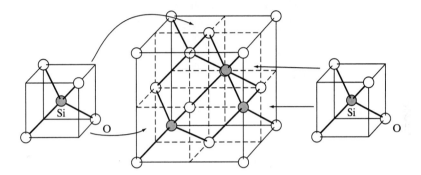

Figure 2.13
Arrangement of silica
tetrahedra in the diamond
cube lattice unit cell
(cristoballite structure).

Cristoballite and quartz are examples of two different forms of the same compound (SiO_2) that have different crystal structures determined by two different atomic packings. These forms are called *polymorphs*. Many crystalline compounds exhibit polymorphism. The different packings have different binding energies, and therefore one polymorph is energetically favored for a particular set of conditions (e.g., temperature and pressure), while under another set of conditions another polymorph will be the most stable.

2.1.4 Crystals and Unit Cells

It has been shown that a diversity of chemical bonding leads to a wide variety of crystal structures. All structures can be described by a unit cell, which is the smallest volume that represents the structure. Any crystal is made up of unit cells stacked three dimensionally. A unit cell is an arrangement of points in space (a space lattice) where atoms are located. The HCP, FCC, and BCC are just three examples of a total of 14 distinct space lattices that describe different 3-D arrangements of lattice points. These are called *Bravais space lattices* and can be found in any basic materials science text. Since most civil engineers will not need to study crystal structures in depth, we will not show all the Bravais lattices here.

Atomic planes within crystals or directions within crystals are identified by Miller indexes. Such information is only needed when determining crystal structure, describing an atomic plane, or identifying directions of dislocation movements. In this text we will discuss dislocations qualitatively without reference to Miller indexes. An introduction to Miller indexes can be found in any basic materials science text.

It must be remembered that although it is possible to grow large single crystals in favorable circumstances, all crystalline structural materials are polycrystalline. As

[2]The zinc blende structure (Fig. 2.5c) is very similar.

the solid phases form (see Chapter 3), many separate crystals nucleate and grow. The crystals grow until they impinge on neighboring crystals. Both metals and ceramics are composed of relatively large crystals that are irregular in shape and size (Fig. 2.14) and randomly oriented with regard to their atomic lattices.

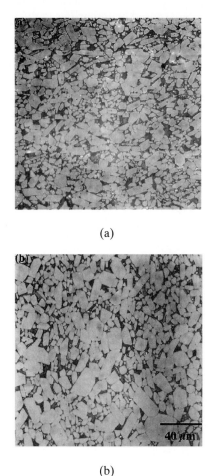

(a)

(b)

Figure 2.14
Microstructure of (a) nickel (× 170) and (b) alumina (× 350), showing large crystals with boundaries between them.

2.2 DEFECTS AND ATOMIC MOVEMENTS IN CRYSTALLINE SOLIDS

2.2.1 Defects in Crystals

We have been discussing ideal atomic arrangements in crystals, but the perfect crystal does not occur in nature with any more frequency than does the ideal gas. Perfection may be approached, but is never attained: Most crystals of interest do not even come close. Thus, the responses of materials of interest to stress (and other properties) are determined by how imperfections respond. Table 2.3 lists the type of defects that can occur in crystals, most of which will be discussed in this section.

Point defects are points in the crystal lattice which either have an atom missing (a *vacancy*), an extra atom present (an *interstitial atom*), or an atom of a different size replacing the expected atom (a *substitutional atom*). These defects are shown in Fig. 2.15, along with the way in which the surrounding crystal lattice is distorted. The additional strain caused by the distortion raises the energy of the crystal (i.e., lowers the lattice binding energy) and limits the number of point defects that can occur

TABLE 2.3   Classification of Crystal Defects

Type	Specific Examples	Influence on Material Properties
Point	Interstitial Substitutional	Solid solution strengthening Swelling of clays
Line	Dislocations	Ductility in metals Work hardening
Surface	Grain boundaries	Grain size strengthening
Volume	Pores	Site of stress concentrations and crack initiation
	Inclusions	Precipitation hardening

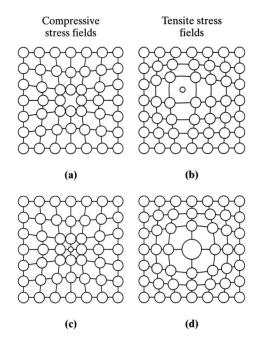

Figure 2.15
Point defects in crystals: (a)
vacancy; (b) interstitial; (c)
substitutional (small), and
(d) substitutional (large).

before the crystal becomes unstable. Thus, calcium ions (ionic radius 0.1 nm) cannot be replaced by too many magnesium ions (ionic radius 0.078 nm) before increasing lattice strains will cause the mode of crystal symmetry to change and thus limit magnesium substitution.

However, point defects inhibit the movement of dislocations, leading to solid-state strengthening, as is found in simple alloys (see Fig. 2.16 and the discussion in this section). Point defects can also lead to charge imbalance, most commonly where an ion of one charge substitutes for an ion of a different charge. For example, aluminum has an ionic radius of 0.051 nm, which is not too different in size from that of silicon (ionic radius 0.041 nm), so aluminum can occupy the same tetrahedral sites. But the difference in ionic charge ($+3$ versus $+4$) can have important consequences. Consider the clay mineral kaolin (Fig. 2.14b). The SiO_2 and $Al_2(OH)_4$ sheets combine to give polar, but electroneutral, layers. Partial substitution of aluminum in the silicate sheets leaves the layer with a net negative charge, because Al^{3+} replaces Si^{4+}. External cations must thus be located between the layers to balance the net charge. The size and charge of the interlayer cations determines the distance between the layers and the extent to which water can enter between the layers to cause swelling. This is shown in Fig. 2.17. [Note that silicon does not substitute in the aluminum sheets because silicon does not fit well in octahedral sites.]

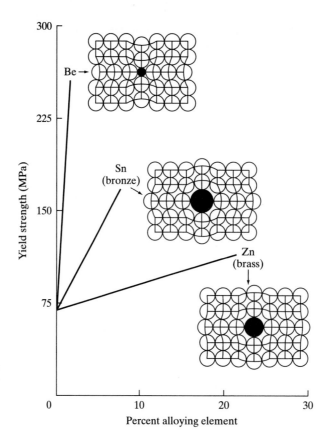

Figure 2.16
Strengthening of pure copper by alloying with substitutional atoms. The greater the mismatch in atomic size, the greater the strengthening effect.

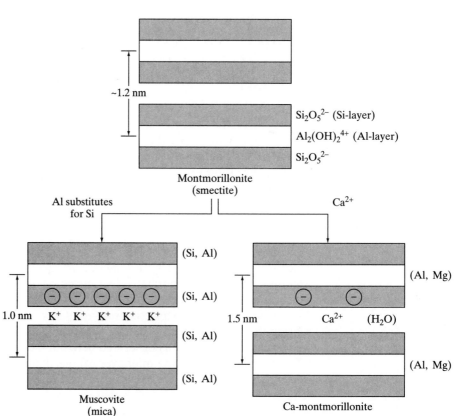

Figure 2.17
Effect of ionic substitution on the structure of clay minerals.

Frequently, however, point defects occur in pairs, to avoid disruption of charge balances. The two most common are a *Frenkel* defect (Fig. 2.18a), which is a vacancy-interstitial pair formed when an ion moves from a normal lattice point to an interstitial site; and a *Schottky* defect (Fig. 2.18b), which is a pair of vacancies created by the omission of both a cation and an anion. A point defect pair also tends to minimize the strain within the crystal lattice caused by the defects. Point defects allow atoms to diffuse more readily through a crystal lattice; atomic diffusion is an important process in the heat treatment of metals.

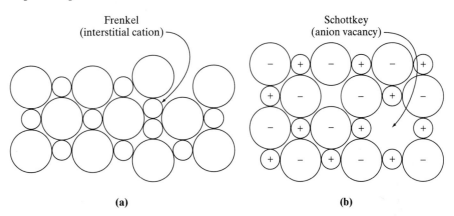

Figure 2.18
Schematic illustration of paired point defects.

Frenkel
(interstitial cation)

Schottkey
(anion vacancy)

(a)　　　　　**(b)**

Line defects are imperfections in a crystal lattice in which a line of atoms become mismatched with their surroundings. Line defects are commonly known as *dislocations*. There are two distinct types: *edge* dislocations and *screw* dislocations. An edge dislocation is formed by the addition of an extra partial plane of atoms in the lattice. The dislocation thus forms a line through the crystal and is denoted by the symbol ⊥. The distortion induced in the lattice by the edge dislocation leads to zones of compressive and tensile strain (Fig. 2.19). The distortion is measured by the

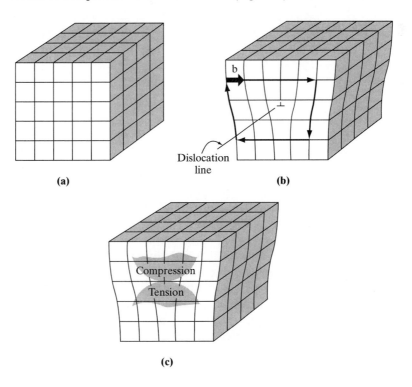

Figure 2.19
Distortions caused by edge dislocation: (a) undistorted lattice, (b) distortion measured by the Burgers vector, and (c) zones of compression and tension.

(a)　　　　　**(b)**

Dislocation line

Compression

Tension

(c)

Burgers vector (Fig. 2.19b), which is obtained by traversing a path around the dislocation and comparing it to an ideal path in a perfect lattice. The Burgers vector has a length which is an integral number of atomic spacings, while its direction is perpendicular to the dislocation line.

A screw dislocation is illustrated in Fig. 2.20. Consider a plane cutting through a crystal. If the atoms immediately adjacent to the plane slip past each other one atomic distance, then a screw dislocation is created. The Burgers vector is parallel to the dislocation line, and atomic movement can be regarded as a spiral movement along the line. The formation of the screw dislocation makes it easier for crystals to grow from solution. The growth of a crystal requires new planes of atoms to be laid down in the required sequence of packing (e.g., ABAB . . . or ABCABC . . .). A screw dislocation creates B or C sites in what would normally be a plane of only A sites. In a perfect crystal, a new plane has to form by adsorbed atoms at the surface aggregating to form B or C sites (a much more difficult process).

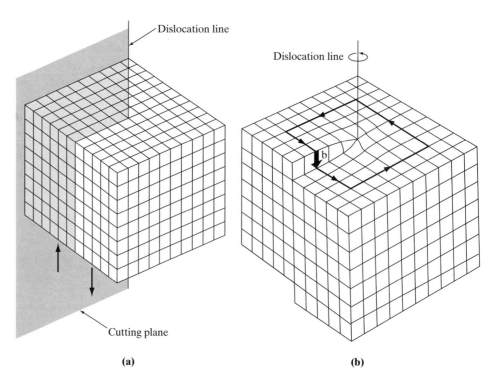

Figure 2.20
Creation of a screw dislocation: (a) cutting plane along which atoms are displaced, and (b) displacement of atoms creating the screw dislocation and its Burgers vector.

(a)   **(b)**

Most of the dislocations show a mixed character, which can be resolved as edge and screw components. These are known as dislocation loops since they are curved rather than linear defects and may form a continuous loop within the crystal (Fig. 2.21). A complete loop is made up of edge and screw dislocations of opposite signs.

The interfacial zone between different crystals in a polycrystalline solid represents an area of defect concentrations, called *plane defects,* within the solid. This occurs because the atoms are packed in one crystal (or grain) in one orientation, and must change to a different orientation in the adjacent crystal. As can be seen in Fig. 2.22a, this mismatch can be accommodated by an array of point defects. The array actually encompasses a volume, but since it extends over only a few atomic diameters, it can be considered as a plane at the micron level. Since it represents the

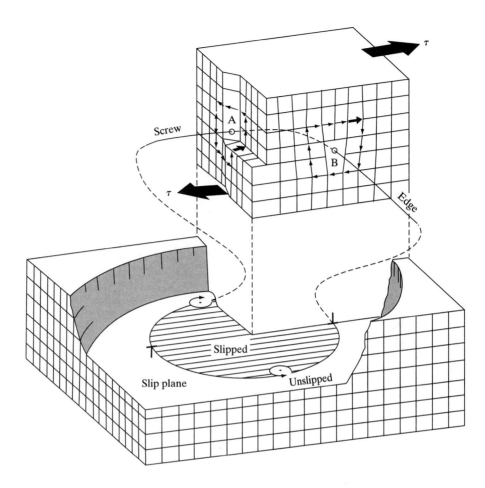

Figure 2.21
Part of a dislocation loop in a crystal, showing up as a screw dislocation as it enters the crystal on the left (*A*) and edge dislocation as it leaves to the right (*B*) (after H.W. Hayden, W. G. Moffatt and J. Wulff, *The Structure and Properties of Materials, Vol. 3, Mechanical Behavior,* Wiley, 1965, p. 65).

boundary separating a marked change in crystal orientation, it is not unexpectedly called a *grain boundary*. The localized atomic mismatch at the boundary tends to hinder atomic diffusion and so is often the place where atomic impurities concentrate during processing. Grain boundary segregation may result in second phases (often amorphous) developing at crystal interfaces. Another kind of grain boundary is formed by line defects. This is the *low-angle grain boundary* (Fig. 2.22a), which is formed by the alignment of edge dislocations (Fig. 2.22b). Such an alignment represents a minimum in the interaction forces between dislocations.

Three-dimensional regions of defects, called *volume defects,* are *voids* and *inclusions*; they are considered to be regions external to the crystal structure. Voids are generally small pores left by incomplete filling of the available volume during processing, but in some cases they are cracks caused by internal stresses. Vacancies moving under such stresses may coalesce at certain points. Inclusions are regions of a second phase contained wholly within a single grain. Inclusions will most often arise from the uneven distribution of impurities in the starting materials used to form the solid matrix. Also, during thermal processing, rapid cooling may prevent impurity atoms from diffusing to grain boundaries, so that a second phase forms within the crystal. When crystals are growing from a melt or solution, the interruption of crystal growth at one point by an obstacle of some kind may distort crystal growth patterns and result in the inclusion of part of the liquid in the growing crystal.

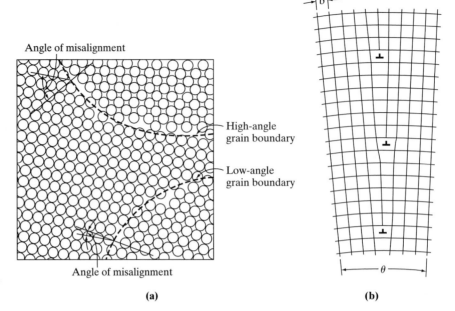

Figure 2.22
(a) Schematic representation of low- and high-angle grain boundaries and the arrangement of atoms in their vicinity (b) Alignment of edge dislocations forming a small-angle grain boundary (after W. D. Callister, *Materials Science and Engineering,* Wiley, p. 78).

Angle of misalignment

High-angle grain boundary

Low-angle grain boundary

Angle of misalignment

(a)

(b)

2.2.2 Atomic Movements

In an ideal crystal, movement of atoms is very difficult because of high energy barriers to motion through the lattice; defects provide low-energy paths for atomic motion. Thus, defects are responsible for the mobility of atoms under various external stimuli. Movements of point defects are the main mechanism for the migration of impurities in solids, and such movements influence sintering processes. Movements of dislocations underlie plastic deformation.

Atomic Diffusion

If an atom is to diffuse through a crystal lattice, it must move from its equilibrium position through a point where there are strong repulsive forces from its nearest neighbors. In addition, the lattice will be distorted in the immediate vicinity of the displaced atom. Both of these changes will raise the energy of the moving atom so that there is a high energy barrier to such motion (q), as illustrated in Fig. 2.23. The energy required to promote atomic motion comes from the thermal energy of the random atomic fluctuations, which increase with temperature. The magnitude of such energy is kT. The ease with which an atom can move is proportional to the ratio q/kT. This is embodied mathematically in a rate equation:

$$\frac{ds}{dt} = V \exp\left(\frac{-q}{kT}\right) \tag{2.1}$$

where ds/dt is the rate of self-diffusion, V is a term related to the frequency of atomic motions, the exponential term reflects the probability that an atom will attain enough energy to cross the barrier q, k is the Boltzman constant ($0.138\mu J\ K^{-1}$), and T is the temperature in K. Equation 2.1 describes a typical *thermally activated rate process,* which was a concept first proposed for homogeneous chemical reactions (the Arrhenius equation). For an assembly of N atoms (1 mol), Eq. 2.1 becomes

$$\frac{ds}{dt} = V \exp\left(\frac{-E_a}{RT}\right) \tag{2.2}$$

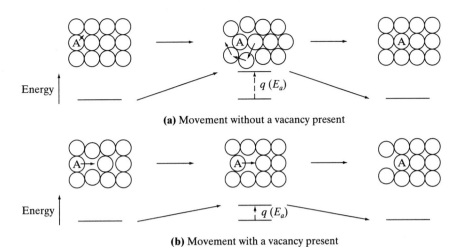

Figure 2.23
Atomic movement (diffusion) within a crystal:
(a) movement without vacancy present;
(b) movement with vacancy present.

(a) Movement without a vacancy present

(b) Movement with a vacancy present

where E_a is the activation energy of self-diffusion, and R is the gas constant (1.987 cal/mol K). In the presence of a vacancy, the amount of local lattice distortion and concomitant movement of adjacent atoms is much less and hence considerably reduces the energy barrier to diffusion. In effect, the vacancy provides a space in the lattice that an atom can move into; one can think of the vacancy diffusing in the opposing direction. On the other hand, interstitial atoms will impede self-diffusion since an occupied interstitial site will restrict the ability of an atom to be displaced permanently from its equilibrium position in the lattice. In contrast, the interstitial atoms themselves will be able to diffuse readily through the lattice if they are smaller than the host atoms, since the energy barrier between adjacent sites will be lower.

Since Eq. 2.1 and 2.2 contain temperature in the exponential term, the rate of diffusion is strongly dependent on temperature. At high temperatures atomic diffusion can be rapid. Self-diffusion is of importance in sintering reactions whereby adjoining particles are welded together (see Chapter 3), while diffusion of impurities may lead to concentrations at grain boundaries (where high defect concentrations can more readily accommodate impurity ions) or to the segregation of a second phase. The driving force for segregation is the minimization of the energy of the system (the same driving force that underlies all chemical changes).

Crystal Slip

Plastic deformation in metals has been shown to be due to slip along atomic planes under the influence of shear forces induced by an applied load as shown in Figure 2.24a. This is shown diagrammatically in Fig. 2.24c and is discussed in more detail in Chapter 5. The extent to which slip will occur will depend on (1) the magnitude of the shear stress, which will depend on the orientations of the atomic planes (remember that crystals are oriented randomly in a solid) and the magnitude of the applied stress; and (2) the bonding between the atomic planes, which is a function of the intrinsic bond strengths and the distance between the planes. The planes along which slip can occur most easily can be determined from the crystal structure; they are the planes on which the atoms are most closely packed. There are 12 in the FCC structure and 3 in the HCP structure. If the bond strengths are known, the theoretical shear strength can be calculated for these planes. This is the force required to move the

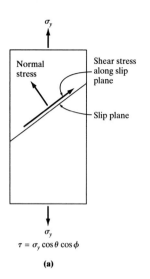

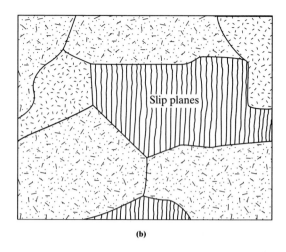

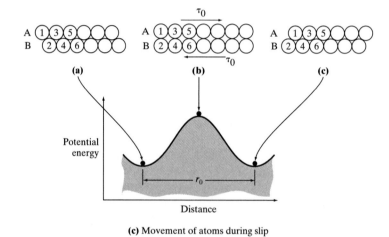

Figure 2.24
Slip along atomic planes
under an applied stress:
(a) stresses, (b) slip planes
in actual material (after R.
A. Higgins, *Properties of
Engineering Materials,*
Krieger, 1977, Fig. 5.1,
p. 52); (c) movement of
atoms during slip.

close-packed atoms past one another through the intermediate position of high energy (Fig. 2.24c). It can be shown that

$$\tau_0 = \frac{Gb}{2\pi a_0} \tag{2.3}$$

where τ_0 is the theoretical shear strength, G is the shear modulus, b is the interatomic spacing on a crystallographic plane of close-packed arrangement, and a_0 is the separation distance between the sheared planes. The shear modulus (G) is proportional to the bond strength of the material. The observed shear strength (Table 2.4) is two to three orders of magnitude smaller than that calculated from Eq. 2.3. This difference strikingly illustrates just how much easier atomic movements are in the vicinity of dislocations. Slip can occur effectively by the movement of the dislocation in the opposite direction, as shown in Fig. 2.25. Only a relatively few atoms are involved in the slip motion (those which are adjacent to the dislocation). It can be seen from Fig. 2.25a and b that an edge dislocation moves in the direction of shear, while a screw dislocation moves normal to the shear stress but in the same plane. The net result in each case is a shear deformation equal to the Burgers vector. Dislocation loops cause slip by movement of both edge and screw components (Fig.

TABLE 2.4 Comparison of Calculated and Theoretical Shear Strengths

| Material | Shear Strength (MPa) | | Shear Modulus (MPa) |
	Theoretical	Observed	
Copper	0.8×10^5	2.2×10^3	4.9×10^5
Iron	1.3×10^5	3.2×10^3	7.7×10^5
SiC	3.3×10^5	1.8×10^3(tensile)	20.6×10^5

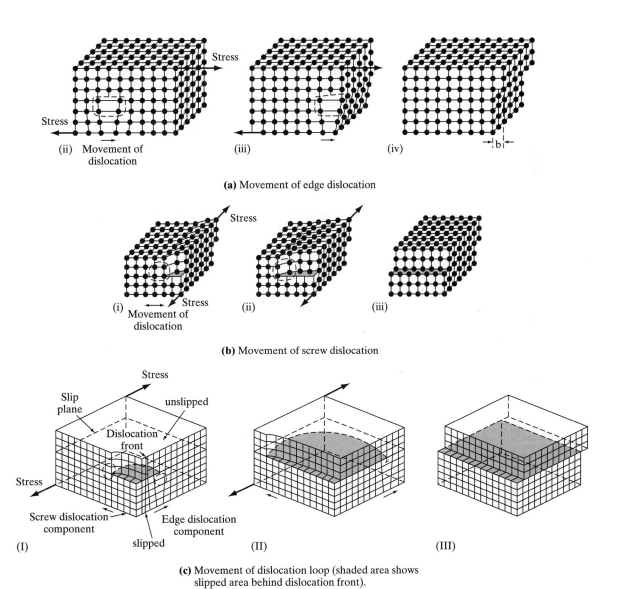

(a) Movement of edge dislocation

(b) Movement of screw dislocation

(c) Movement of dislocation loop (shaded area shows slipped area behind dislocation front).

Figure 2.25
Crystal slip by movement of dislocations: (a) movement of edge dislocation; (b) movement of screw dislocation; and (c) movement of dislocation loop (after R. A. Higgins, *Properties of Engineering Materials,* Krieger, 1977, Figs. 5.9, 5.11, 5.12, pp. 77–79).

2.25c). A simple analogy of slip by deformation movement is the removal of a wrinkle in the middle of a carpet. Trying to smooth out the carpet by pulling at one edge is difficult because of the large frictional forces that must be overcome when moving a large area of the carpet. Smoothing out the wrinkle by pushing it to the end moves only a small fraction of the carpet in the vicinity of the wrinkle, so that frictional forces are relatively slight.

The movement of a single dislocation causes slip of one atomic distance: a movement that can only be seen under high-resolution electron microscopy. However, slip deformation can be readily seen under low resolution ($\times 200$) by optical microscopy, which requires slip of several hundred atoms. Although crystals contain about 10^6 cm of dislocation lines per cm^3 of crystal, this is not a high enough energy density to account for the resolvable deformation on a slip plane. Thus, there must be a mechanism by which dislocations are generated under an applied stress. Such a mechanism has been observed to occur and is known as a *Frank-Read* source, after its discoverers. There are sites within a crystal where a dislocation is pinned; pins being points at which mobility is prevented or restricted. Such sites can be a cluster of point defects, intersecting dislocations, or particles of a second phase. When a dislocation becomes pinned, the shear force moving the dislocation will force it to bow between the pinning sites (Fig. 2.26a). When the shear stress is greater than $2Gb/l$ (where G = shear modulus, b = the Burgers vector, and l = the dislocation length), the dislocation will move spontaneously, creating a dislocation loop without destroying the original dislocation. Thus, many dislocations can be created from a single source.

On the other hand, dislocations may also be decreased by *annihilation processes*. Annihilation is responsible for the effects of annealing, which reduce the energy of the crystal by removing the energy associated with the dislocations. As shown in Fig. 2.26b, adjacent dislocations of opposite sign (i.e., Burgers vectors in opposite directions) can combine to form a perfect crystal, a row of vacancies, or a row of interstitial atoms. On the other hand, adjacent dislocations of the same sign will repel each other.

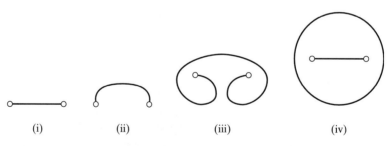

(i) (ii) (iii) (iv)

(a) Frank-Read source of dislocation generation

Figure 2.26
Examples of dislocation generation and annihilation: (a) Frank-Read source of dislocation generation; (after D. R. Askeland, *The Science and Engineering of Materials,* PWS-Kent Publishing Company, 1985, p. 127). (b) Annihilation of dislocations through the attraction between two edge dislocation of opposite signs lying on the same slip plane.

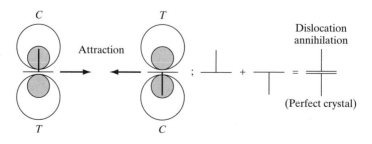

(b) Attraction between two edge dislocations of opposite signs lying on the same slip plane, leading to annihilation

The movement of dislocations through the crystal during the slip process may be impeded as the dislocation crosses imperfections, such as an impurity atom, precipitated particle, vacancy, interstitial atom, or another dislocation. The barrier at the intersection may affect the whole dislocation, and additional stress is required to move the dislocation forward. This mechanism is responsible for the phenomenon of "strain hardening," which is the increase in stress required to maintain the continuation of a slip process after it has started. Strain (work) hardening is often the result of interactions between dislocations, which in the actual material exist in various planes. When a dislocation moves in a slip plane and crosses another dislocation not in the plane, a local displacement will occur, and the result could be the formation of a line of vacancies or interstitials, or part of the dislocation may be displaced to a slip plane parallel to the original one. This displacement in the dislocation line is called a *jog* and energy is consumed in its production (so that it acts as a barrier to the movement of the dislocation).

Dislocations have the ability to avoid obstacles in their path by changing their slip planes. An edge dislocation does this by a process called *climb*. This requires the presence of vacancies adjacent to the obstacle and allows the atoms forming the dislocations to move, thereby raising the slip plane one atomic distance and so bypassing the obstacle. Thus, climb is a process that is only common at higher temperatures, which favor the formation and movement of vacancies.

Dislocation motion in ceramic materials is generally much more difficult than it is in metals. The ionic structure makes the Burgers vector much larger; it is equal to the anion-anion distance rather than the atom-atom distance. Therefore, greater energy (i.e., stress) is required to move the dislocations. Brittle failure often occurs before this happens. Ceramics, therefore, show little ductility at room temperature but are more ductile at high temperatures, where vacancy motion can promote dislocation movement.

2.3   THE AMORPHOUS STATE

When the regularity of atomic packing is completely absent, a solid is said to be non-crystalline or amorphous. The most familiar kind of amorphous solids are glasses. These form from melts when the rate of cooling is too fast to allow nucleation and growth of ordered crystals. For a crystal to grow, enough atoms must come together in the right packing to create an energy minimum, which is the lattice binding energy; until this happens, energy will have to be used to overcome the mutual repulsions of the atoms. This situation is shown schematically in Fig. 2.27 and results in a nucleation energy barrier that must be overcome before crystal growth can occur. Thermal energy can help atoms to overcome the nucleation barrier when the temperature provides a comparable energy (given by RT). If the temperature is too low,

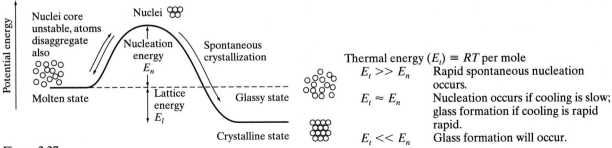

Figure 2.27
Schematics of the nucleation of crystals.

nuclei of the critical size needed for growth will not be able to form and hence crystals will not be able to grow. During cooling there is a dynamic situation; if the rate of cooling is faster than the time required to achieve critical nucleation, an amorphous structure will form. Amorphous solids are frequently called glasses, after the first human-made amorphous material, silica glass, which was commonly used for glazing.

Because the atoms in a glass are irregularly arranged, the packing is not as dense as in crystals and therefore lattice energies are not as high. Glasses are, therefore, *metastable* with respect to the crystalline forms that can occur. Some glasses will spontaneously transform to a lower-energy crystalline phase if the nucleation barrier is not too high. This phenomenon is called *devitrification*. An otherwise stable glass can be induced to devitrify by raising the temperature (so that the available thermal energy more closely matches the nucleation energy; i.e., $E_t = E_n$, in Fig. 2.27), or by adding nucleating agents that lower the height of the barrier. This is the principle of glass ceramics. An inorganic glass is first formed with the addition of atoms that will later act as nucleating agents. The glass is then annealed at a temperature well below its melting point but sufficiently high enough that $E_t = E_n$. Controlled crystal growth will then occur, forming a dense, strong body.

In theory, any chemical composition can be made to form a glass provided its cooling rate is fast enough. Even metals, which are the most easily crystallized of all solids, can be induced to form a glass when cooled at rates grater than $10^{6\circ}$C/min. Ceramic materials form glasses more readily since the more complicated packings needed to accommodate specific molecular geometries require longer times to develop nuclei of the critical size. Many polymer melts crystallize with difficulty because of the greater molecular complexity (Sec. 2.4) and the low melting temperatures. Molecular motions are too slow to allow the large molecules to orient themselves to form critical nuclei, and in many cases, the molecules are too bulky to pack together in a regular manner.

Although they have random packings, glasses can be considered in terms of ideal, defect-free structures, as shown in Fig. 2.28. Fused silica has a random arrangement of silica tetrahedra, with each tetrahedron linked to four others. In soda glass, this arrangement is disrupted by the presence of sodium ions: Some Si—O—Si bonds are replaced by Si—O—Na$^+$ bonds. Soda glass, therefore, has lower binding energy and melts at a much lower temperature than fused silica ($\sim$600°C versus 1200°C). Fused silica devitrifies more readily (to cristoballite), but soda glass is more chemically reactive and is readily attacked by alkalis, for example. Granulated blast furnace slag and fly ash are aluminosilicate glasses that chemically react with alkalis in cementitious materials.

Another kind of amorphous material is that formed by the rapid precipitation of solids from solution. The principles embodied in Fig. 2.28 apply here also, except

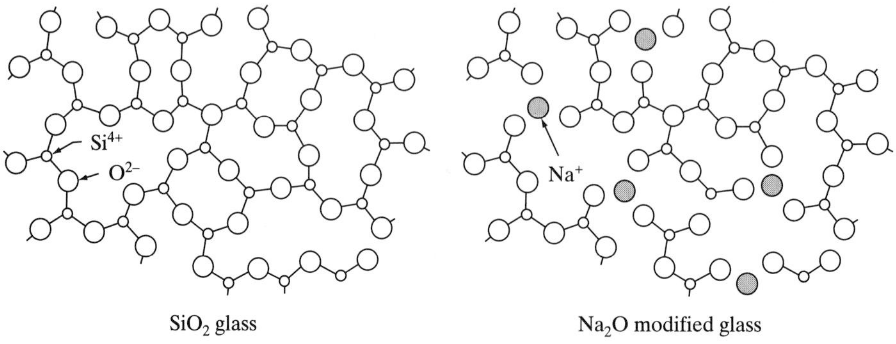

Figure 2.28
Amorphous structure of silicate glasses: (a) ideal structure of amorphous silicate glass, and (b) amorphous structure of soda glass, where the Na$^+$ ions interfere with the continuity of the silica network (after D. R. Askeland, *The Science and Engineering of Materials,* PWS-Kent Publishing Company, 1985, p. 310).

Si^{4+}
O^{2-}

Na^+

SiO$_2$ glass

Na$_2$O modified glass

that ionic concentrations replace temperature as the means to cross and overcome the nucleation barrier. Unlike glasses, which form as a homogeneous phase, amorphous precipitates are a collection of individual particles (which may, however, strongly agglomerate). Many solids which are initially amorphous when first precipitated will spontaneously crystallize when kept in contact with the solution, reflecting the more rapid diffusion of ions within a liquid phase and which can be enhanced by spontaneous dissolution-precipitation processes. If molecular structures are complex, however, this may not happen. A good example is the major binding component in concrete, which is an amorphous precipitate of a hydrated calcium silicate formed from the reaction of portland cement with water. (Even in this case, there exist very small domains of nanometer dimensions with ordered molecular packings.) The precipitate forms a solid matrix with particles packing in a manner analogous to crystals in a polycrystalline solid. But these microstructures are severely defected, with high concentrations of planar and volumetric defects. Solid microstructures based on amorphous materials only form from systems that have a high solids content. In cases where the solids content is much lower, precipitation processes cannot lead to a strong body since the low concentration prevents strong attractions between particles. Nevertheless, the high surface areas of such amorphous precipitates lead to significant van der Waals interactions between adjacent particles, which can cause gelling. This is the basis of colloidal behavior, which is further discussed in Chapter 4. A third kind of amorphous material forms from reactions in the gas phase and generally results in very small individual amorphous particles, or loose arrays of agglomerated particles.

2.4   THE POLYMERIC STATE

Plastics are a class of materials made from large molecules (polymers), which are composed of a large number of repeating units (monomers). The monomers react chemically with each other to form extended molecular chains containing several hundred to several thousand monomer units. Most monomers are organic compounds, and the typical polymer is therefore characterized by a carbon chain backbone (Fig. 2.29), which may be linear or branched. The molecular structure of the unit that makes up the very large molecules controls the properties of the material. The rigidity of the chains, density and regularity of packing (i.e., crystallinity) within the solids, and interactions between the molecular chains can be altered and thus change the bulk properties of the elastic. Two major classes of polymers are used: *thermoplastics* and *thermosets*. The main difference between the two is the nature of bonds between the chains: secondary van der Waals in the thermoplastics and chemical crosslinks in the thermosets (Fig. 2.29). This gives rise to considerable differences in properties. Despite this, there are sufficient similarities that make it useful scientifically to consider the two classes together as a single class of materials. The properties of polymers with respect to their application in the field of civil engineering will be discussed in Chapter 15. The nature of bonds and structure of the polymers will be described in this section. The production of polymeric materials, which involves polymerizing monomers and forming plastic bodies, is beyond the scope of this book.

2.4.1 The Polymeric Molecule

An important characteristic of the carbon backbone is the tetrahedral bonding configuration (Fig. 2.30a), which allows largely free rotation of one segment of the carbon backbone relative to another (Figs. 2.30b and 2.30c). This characteristic is due to

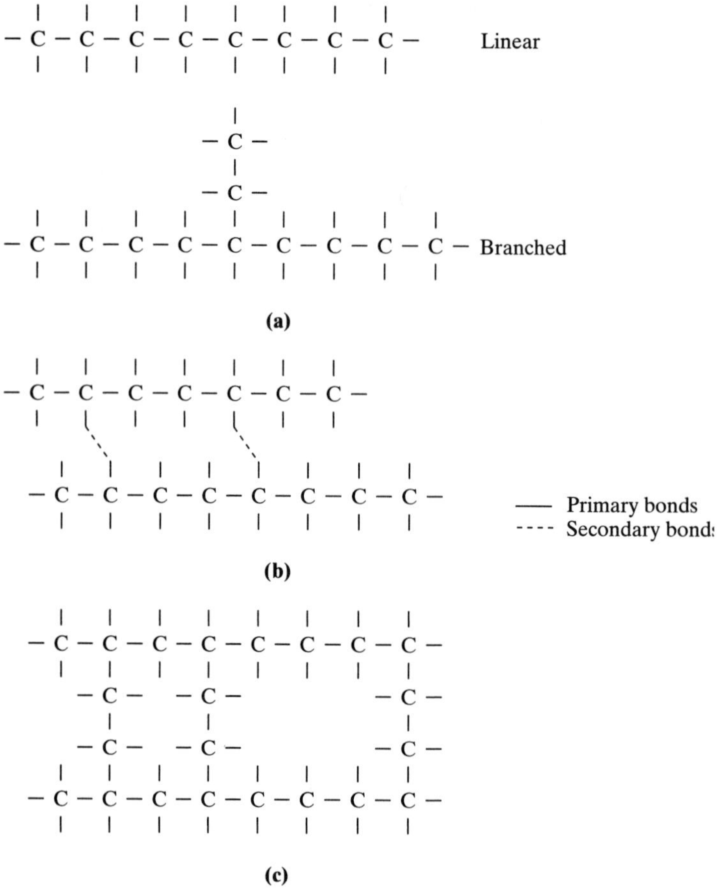

Linear

$$- C - C - C - C - C - C - C - C -$$

Branched

$$- C - C - C - C - C - C - C - C - C -$$
(with a branch of $-C-$, $-C-$ above)

(a)

$$- C - C - C - C - C - C - C -$$
$$- C - C - C - C - C - C - C - C -$$

—— Primary bonds
---- Secondary bonds

(b)

$$- C - C - C - C - C - C - C - C -$$
$$- C -   - C -   - C -$$
$$- C -   - C -   - C -$$
$$- C - C - C - C - C - C - C - C -$$

(c)

Figure 2.29
Schematic representation of the structure of polymers: (a) individual chain of carbon backbone, linear or branched; (b) thermoplastic polymer with secondary bonds between polymeric molecules; and (c) thermoset polymer with chemical bonds crosslinking between polymeric molecules.

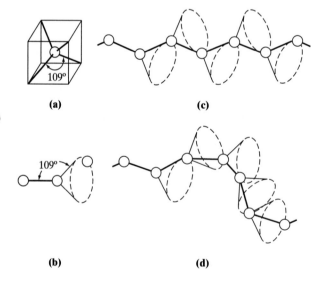

Figure 2.30.
The geometrical shape of the polymeric molecule: (a) tetrahedral structure of the bonds; (b) rotation between adjacent segments; (c) straight, extended conformation; and (d) coiled conformation (after D. R. Askeland, *The Science and Engineering of Materials,* PWS-Kent Publishing Company, 1985, pp. 336, 340).

the fact that the rotation does not involve any stressing of the C—C bond, as the distance and bond angle between the neighboring atoms remain the same. Thus, the molecule can assume different *conformations* in which the chain becomes twisted (Fig. 2.30d). The two extremes would be an extended long chain and a very coiled one (Fig. 2.31). In practice, the material is made up of an assembly of chains with

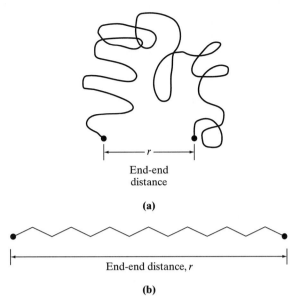

Figure 2.31
Schematic presentation of
polymer chain in the ex-
tended (a) and coiled conformations (b) with the
end-to-end distance, r.

different coiling configurations (think of a plate of spaghetti, a "nest of vipers," or tangled strings) with some distribution around a mean conformation. The molecules are not static, and they change their conformation continuously.

The coiling nature of the polymeric molecule is one of its most important characteristics. This is the basis for elastomeric materials, which can be extended considerably and maintain continuity (i.e., extension and coiling of the chain), in contrast to conventional solids, where deformation is limited. The difference is in the fact that in conventional solids, primary chemical bonds must be stressed, whereas in the elastomeric polymers, the free rotation allows for large changes in the size of the molecule (end-to-end distance) with only a modest load, as long as the rotation between the segments is largely unhindered. On removal of the load, the extended molecules coil back to the mean conformation.

In the discussion of the behavior of the individual polymer chain, it was assumed that segments between carbon atoms can rotate almost without any hindrance. However, this is not always the case, and the flexibility of the chain can be impaired by several factors, the most important of which are as follows:

1. Steric limitations due to bulk or polar sidegroups along the chain which interact with each other during the rotation of segments
2. van der Waals bonds between neighboring chains
3. Chemical bonds between chains (crosslinking)
4. Change in the bonding within the chains.

Thus, in practice, the flexibility of the individual chain may not be obtained and the polymeric material will then behave more like a rigid solid.

2.4.2 Thermoplastic Polymers

In a thermoplastic material, the hindrance to polymer flexibility can arise from steric limitations and van der Waals bonding. To maintain the flexibility, the segments along the polymer chain will need to acquire sufficient intrinsic energy to overcome these barriers to rotation. This might be achieved by thermal activation following the concepts discussed for thermally activated processes (i.e., Eq. 2.2). Thus, the overall behavior of the thermoplastic polymer will depend on the balance between these influences. If the temperature is sufficiently high, there is enough thermal energy to enable unhindered rotation of the individual chains. This may result in two effects: flexibility of the chain itself, as well as easy mobility of one chain past the other as the van der Waals bonds are easily broken by the vibrant bond rotation of the chain segments. Under such conditions, the polymer behaves as a viscous liquid.

The individual van der Waals bonds between neighboring atoms located on different chains can be weak. Yet because the polymer chains are long, with an enormous number of atoms, the accumulating effect of individual van der Waals attractions can provide a strong intermolecular bond, enough to inhibit rotation and flexibility of the chains at room temperature and create a solid material. Waxes and greases are "semisolids": polymer precursors that have relatively short chains and thus develop insufficiently strong intermolecular bonding to form solids with useful mechanical properties. Even with the longer polymer chains, the intermolecular bonds will not be sufficient to fully hinder rotation and flexibility, resulting in a polymer which will soften at temperatures in the range of 70 to 200°C.

The dependency of the behavior of polymers on temperature, as discussed previously, can be quantitatively described in terms of two transition temperatures: *melting temperature*, T_m, and *glass transition temperature*, T_g. Above T_m, the polymer behaves like a viscous liquid as the chains are flexible and able to coil back and forth and to slide past each other. As the temperature is reduced below T_m, the van der Waals attraction becomes sufficiently large to hinder large-scale coiling and slipping of molecules, and the thermodynamics is such as to favor crystallization (i.e., arrangement of the molecules in an orderly structure). For this to occur, the molecules must be linear and have a regular structure to allow orderly packing. Also, crystallization is enhanced when larger than usual intermolecular attractions are generated, such as in molecules where hydrogen bonding can be developed. The orderly structure maximizes the intermolecular bonding and, as a result, rotation is eliminated and a rigid solid is obtained, as seen schematically in Fig. 2.32. If the polymer chains are irregular in shape, they cannot be packed into a regular structure and the cooling below T_m does not lead to any drastic change in the arrangement of the chains (they remain in the coiled and entangled structure without any repeatable order; i.e., they will be amorphous). Although the polymer becomes much stiffer, it cannot yet be considered a rigid solid; the chain flexibility is reduced but local chain rotation is still feasible. The material behaves like an undercooled liquid of very high viscosity. When cooled below T_g, the material shows a drastic change in properties since its structure is now "frozen" in the amorphous state (Fig. 2.32a) and the chain segments have lost their ability to rotate. It behaves as a relatively brittle, rigid solid. These changes in structure and properties can be readily quantified by monitoring changes in specific volume (Fig. 2.32b) and modulus of elasticity (Chapter 15) as a function of temperature.

Since the glass transition involves only a change from a highly deformable state to one that is more rigid, it is a second-order transition that does not involve a change in state (i.e., amorphous to crystalline structure, as in T_m), but rather a change in the mobility of the molecules. Monitoring T_g is an excellent means of assessing molecular mobility and the factors which affect it. The lack of chain mobility below T_g accounts for the performance of the material as a brittle solid, but since the structure remains amorphous, it can be considered as a glass. The preservation of the amorphous structure accounts for the observation that glass transition involves only a rate of change of specific volume with temperature, whereas in crystallization a marked decrease in the specific volume occurs (Fig. 2.32b). Parameters that reduce the chain flexibility and enhance interchain bonding will lead to an increase in T_g.

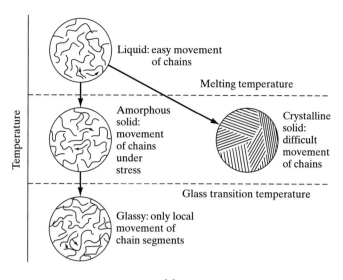

(a)

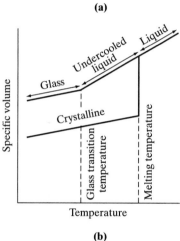

(b)

Figure 2.32
The effect of temperature on the structure and behavior of thermoplastic polymer (a) structure and (b) specific volume (after D. R. Askeland, *The Science and Engineering of Materials,* PWS-Kent Publishing Company, 1985, pp. 345, 347).

Thus, the presence of bulky and electrically charged substituents will lead to higher values of T_g, as seen in Table 2.5 for polyvinyl chloride and polystyrene (which has the highest glass transition among the vinyl polymers).

Crystallinity in polymeric materials is different from that in low-molecular-weight compounds, both in the structure of the crystals and in the degree of crystallinity that can be achieved. Due to the large size of the polymeric molecules, considerable movement is required for the chains, as they are being cooled through T_m, to be arranged in a regular crystalline structure. As a result, the crystal structure is formed by regular folding of the chains into a lamellar morphology (Fig. 2.33); a portion of the polymer chains always remains in an amorphous state separating the lamellar crystallites. The lamellar crystal thickness corresponds to the fold period

Figure 2.33
Crystalline structure of a polymer, showing the lamellar morphology of the crystalline phase and the amorphous nature of the chain segments between the crystallites.

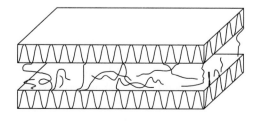

TABLE 2.5 Composition and Transition Temperatures of Different Thermoplastic Polymers

Class	Composition		T_m (°C)	T_g (°C)	Remarks
Vinyl with one substituent	$\left[\begin{array}{c} H \quad R \\ -C-C- \\ H \quad H \end{array}\right]_n$				
	R = H	Polyethylene	135	−120	~80% Crystalline
			115	−120	~50% Crystalline
	R = CH$_3$	Polypropylene	176	−27	CH$_3$ group inhibits crystallization
	R = OH	Polyvinylalcohol	—	85	Hydrogen bonding between chains
	R = Cl	Polyvinylchloride	212	87	PVC
	R = C$_6$H$_5$	Polystyrene	—	100	Benzene ring prevents crystallization
Vinyl with two substituents	$\left[\begin{array}{c} H \quad R_1 \\ -C-C- \\ H \quad R_2 \end{array}\right]_n$				
	R_1 = CH$_3$ R_2 = vinyl	Polyisoprene	28 28	−73 −73	*cis*-Natural rubber *trans*-Gutta Perch
	R_1 = CH$_3$ R_2 = CO$_2$CH$_3$	Polymethyl-methacrylate		100	
Fully substituted vinyl	$\left[\begin{array}{c} R \quad R \\ -C-C- \\ R \quad R \end{array}\right]_n$	Polytetrafluoro-ethylene (PTFE) R = F	327		E.g., Teflon, all hydrogen atoms replaced by fluorine
Polyamide	$\left[\begin{array}{c} H \quad O \\ -R_2-N-C-R_1- \end{array}\right]_n$		265	50	Nylon 66 $R_1 = R_2 = C_6H_{12}$
Polyester	$\left[-R_2-\overset{H}{\underset{H}{C}}-\overset{O}{C}-\bigcirc-\overset{O}{C}-O-\overset{H}{\underset{H}{C}}-\overset{H}{\underset{H}{C}}-O-R_2-\right]_n$		270	80	Dacron
Polycarbonate	$\left[-R_1-\bigcirc-\overset{CH_3}{\underset{CH_3}{C}}-\bigcirc-O-\overset{O}{C}-O-R_2-\right]_n$		—	150	Polycarbonate

and is about 10 nm. Therefore, a crystalline polymer consist of an intimate mixture of crystalline and amorphous phases, which can assume a variety of morphologies at a wide range of crystallinity (in the range of ~20 to 80%). Materials of this kind can show both crystalline (melting) and glass transitions, as demonstrated in Table 2.5 for several polymers.

Crystalline polymers usually have superior mechanical properties and chemical stability in comparison with their amorphous counterparts. This is a reflection of the higher intermolecular bonds which can be developed in the ordered structure, where the chain segments in the crystal can be packed closely together. The ease of crystallization depends on the regularity of the molecular structure and the bulkiness of the substituting groups along the molecular chains. In vinyl-based polymers, where the repeating unit is

R is a substituting group (substitutes for H), and more than one of the H atoms can be substituted by similar or different groups. The nature of this substitution is important. For example, polypropylene $\{R = CH_3\}$ is more readily crystallized than polyisobutylene $\{R = (CH_3)_2\}$; while polyvinyl chloride (PVC) $\{R = Cl\}$ is always amorphous. The size of the latter substituent makes it more difficult to achieve regular packing of the chains. These characteristics can be clearly seen in the differences in the transition temperature values in Table 2.5. However, for the substituent groups to have a positive effect on crystallization, they should be regularly arranged with respect to the backbone (isotactic structure) rather than randomly arranged (atactic structure).

The presence of functional groups which can form hydrogen bonding can also enhance crystallization, if they are regularly placed; examples are esters and amides (Chapter 15). Therefore, polyesters and polyamides have high values of T_m (Table 2.5) and are used to produce fibers in which superior properties are generated by the presence of stable, oriented crystals. Another requirement for crystallization is the linearity of the polymer chain. Branched chains cannot be packed into a crystalline and ordered structure because of the interference of the side branches. A notable example is polyethylene: When it is polymerized as a linear chain, it can be crystalline, whereas branched polyethylene remains amorphous. Thus, there are two forms of polyethylene which are widely used: (1) high-density polyethylene, which is crystalline and is a stiff solid with good thermal stability; and (2) low-density polyethylene, which is amorphous and thus soft and flexible.

2.4.3 Elastomeric Polymers

Elastomeric materials may be considered as a special category of polymers, although their structure and behavior is closely linked with the concepts described for the thermoplastic polymers. A thermoplastic polymer with structural characteristics which eliminate crystallization will exhibit above T_g a behavior which combines the elastomeric nature of the individual polymer chains (reversible coiling) and a viscous characteristic associated with the ability of one chain to slip over the other (Fig. 2.34). To obtain truly elastomeric behavior, the slipping should be prevented while maintaining the freedom of chain coiling. This can be achieved by very light crosslinking, where the crosslinks prevent slipping but are sufficiently spaced apart that there is enough chain length that can exhibit coiling. Natural and synthetic elastomers consist, therefore, of thermoplastic polymers which cannot crystallize and whose T_g is below room temperature. Natural rubber has sterically asymmetric isoprene units that reduce van der Waals bonding and promote chain coiling (Fig. 2.35). Sulphur is used to create crosslinks in a process known as *vulcanization*.

2.4.4 Thermosetting Polymers

Thermosetting polymers are characterized by dense crosslinking between polymeric molecules. Strictly speaking, the thermoset material consists of a single three-dimensional polymeric molecule. This structure results in a hard, rigid, inflexible

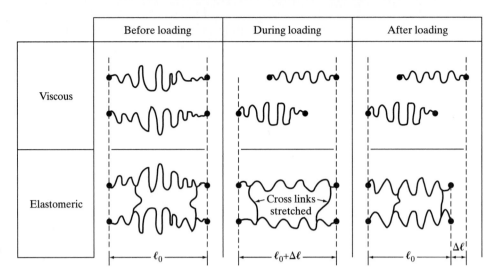

Figure 2.34
Schematic description of elastomeric and viscous behavior.

	Before loading	During loading	After loading
Viscous			
Elastomeric		Cross links stretched	

ℓ_0 — $\ell_0+\Delta\ell$ — ℓ_0 — $\Delta\ell$

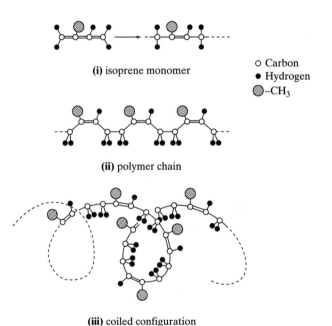

(i) isoprene monomer

○ Carbon
● Hydrogen
◉ –CH₃

(ii) polymer chain

Figure 2.35
The molecular configuration and conformations of polyisoprene (after R. A. Higgins, *Properties of Engineering Materials,* Krieger, 1977, Figs. 13.14 and 13.15, pp. 290–291).

(iii) coiled configuration

solid which generally starts to decompose before melting. Hence these materials do not show the variety of response to temperature which was observed in thermoplastics. However, even though the bonding between chains is strong, there is still a place for coiling on a small scale of chain segments, even if dense crosslinking is present. This does not allow for any chain slippage but leads to a modulus of elasticity which is smaller than in conventional materials, such as steel and ceramics.

2.4.5 Rigid Rod Polymers

Polymers which have multiple bonds within the backbone (eg. $C = O$) lose their ability to freely rotate and so behave as rigid rods with relatively limited flexibility. When the temperature is lowered, the chains align and form strong molecular bonding. This structure is usually crystalline. Liquid crystals are a familiar example of a rigid rod polymer. Polyamides, such as Kevlar, are rigid rod polymers characterized by a high modulus of elasticity and high tensile strengths as a result of their crystalline structures.

2.5 THE COMPOSITE STRUCTURE

Engineering materials used in practice are rarely made up of only a single phase of atoms or molecules grouped together in a structure, as described in the previous section for the crystalline, amorphous, and polymeric states. Although the control of the structure at these basic states can provide a means for generating materials of a variety of properties, this is still not sufficient for many engineering purposes. Therefore, in practice, many engineering materials consist of a composite structure (i.e., a structure made up of a heterogeneous mixture of two or more homogeneous phases bonded together). The different phases may not be distinguishable to the naked eye, and thus the material might be referred to as a homogeneous material (although on a microscopic level it can be seen to consist of different phases). Examples include steel, which is an alloy consisting of different phases (see Chapters 3 and 13), and plastics, which consist of a polymeric phase and dispersed fillers (see Chapter 15). Composites which are clearly heterogeneous include portland cement and asphalt concretes, which consist of a binding phase (portland cement paste or asphalt) in which aggregates of sizes up to about 25 mm are dispersed (see Chapters 9, 11, 12); fiber-reinforced composites, in which fibers are dispersed in a cement or polymer matrix (see Chapter 16); and laminated composites made of a layered structure.

There is no generally accepted classification of composites according to their structure. Composites used for construction may be described in terms of the matrix, which is the continuous phase, and inclusions, which are embedded in the matrix. In the field of construction we are usually interested in inclusions which strengthen the matrix (reinforcing inclusions), but we must also consider other influences, such as fillers (which are added to polymeric matrices to enhance their wear resistance, fire resistance, and durability to external weathering conditions). A broad classification may be based on the geometry of the inclusion (particulate versus fibrous) and its dispersion in the matrix (oriented, laminated, random), as seen in Fig. 2.36 and Table 2.6.

TABLE 2.6   Different Types of Composites

Composite Type	Components	Examples
Particulate (micro)	Hard particles Ductile matrix	Duraluminum alloy Bearing metals
Particulate (macro)	Hard particles Brittle matrix	Cermets; Concretes
	Hard particles Ductile matrix	Filled plastics
Fiber reinforced	Strong brittle fibers Ductile matrix	Fiberglas Wood
	Strong fibers Ductile matrix	Fiber-reinforced metals
	Strong fibers (brittle or ductile) Brittle matrix	Fiber-reinforced cement and concrete
Laminates (micro)		Perlitic steel
Laminates (macro)		Plywood Filament-wound epoxy tubes

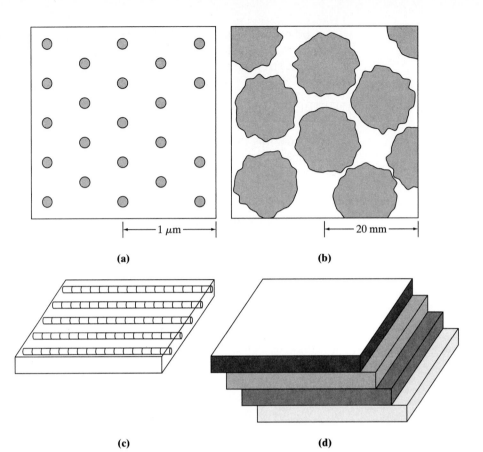

Figure 2.36
Schematic description of the structure of various composites: (a) composite with dispersed phase of extremely small particles (~10 nm–100 μm); (b) particulate composite with dispersed particles of large size (~100 μm–50 mm); (c) fibrous composite; and (d) laminate composite.

(a)   (b)

(c)   (d)

The definition and description of the composite structure as just outlined is far reaching. In practice, the term *composite* is limited to materials that are human made by combining the different phases together in a production process and in which the content of the dispersed phase in the matrix is substantially large. This more strict definition excludes materials such as steel alloys (where the content of the dispersed phase is usually small) and naturally occurring materials such as wood, which are not human made.

BIBLIOGRAPHY

Any standard introductory text in materials science and engineering. E.g.,

A. K. GALWAY, *Chemistry of Solids,* Chapman & Hall, 1967, 210 pp.

WILLIAM F. SMITH, *Foundations of Materials Science,* 2nd Ed. McGraw-Hill Inc., 1993, 882 pp.

MILTON OHRING, *Engineering Materials Science,* Academic Press, 1995, 827 pp.

PROBLEMS

2.1. In scanning electron micrographs of fracture surfaces of hardened portland cement paste, large $Ca(OH)_2$ crystals frequently show cleavage surfaces at the micron level. Why do $Ca(OH)_2$ crystals tend to cleave on fracture?

2.2. Carbon has two different polymorphs: diamond and graphite. Diamond is hard and strong while graphite shows lubricating qualities similar to talc. Based on these observations, what kind of bonds are causing these properties in each polymorph?

2.3. Rank the order of importance of defects (volume, plane and point) in controlling mechanical properties of the following materials: Copper; sintered alumina, polypropylene. Give reasons for your rankings.

2.4. Predict the packing geometries for the following compounds: LiF, KCl, KI, RbBr, CsI. (Ionic radii in nm: $Li^+ = 0.06$; $K^+ = 0.133$, $Rb^+ = 0.148$; $Cs^+ = 0.169$; $F^- = 0.136$, $Cl^- = 0.181$, $Br^- = 0.195$; $I^- = 0.216$).

2.5. **1.** Given $E_a = 20$ kcal/mole and $V = 8 \times 10^{-3}$ cm^2/sec calculate the diffusion of carbon in iron at (1) 500°C, (2) 550°C, (3) 1050°C.

2. Make the same calculations for nickel in iron where $V = 0.5$ cm^2/sec and $E_a = 66$ kcal/mole.

3. Compare the two sets of values. Why does carbon diffuse more readily than nickel?

2.6 Explain the difference between an elastomer, a thermoplastic, a thermoset, and a liquid crystal.

3

Development of Microstructure

3.1 INTRODUCTION

We have now discussed atomic bonding and molecular structures and have considered their effects on compound properties. Another important aspect is the way in which the ionic, metallic, and covalent compounds form macroscopic solids. The details of a solid's structure can usually only be seen at high magnifications under a microscope; thus we call such details the *microstructure*. In some cases a material's structure may be visible with the naked eye, in which case we would call it a *macrostructure*. Concrete has an identifiable macrostructure: Gravel particles and sand grains can be seen embedded in hardened cement paste. Each of these components has its own microstructure. Microstructures are infinite in their variety, and a given material may form different microstructures under different conditions. This will be discussed in detail in this chapter and during the in-depth discussion of each material.

The purpose of this chapter is to discuss, in general terms, the processes by which the microstructures of materials are formed and to examine the general principles guiding the relationships between microstructure and bulk properties. Table 3.1 lists the most common processes by which microstructures are developed. We will only consider in detail crystallization from melts or solution and sintering. We will also introduce phase diagrams which are important to predicting the behavior of various material systems during the development of microstructures.

3.2 SOLIDIFICATION

Metals, polymers, and glass are generally formed by solidification from melts. Solidification of polymers and glasses, however, does not usually involve crystallization because in these cases the mixtures of atoms (in inorganic glasses, like silicate glass) or large molecules (in polymers) are not able to rearrange to an ordered structure

TABLE 3.1 Formation of Different Kinds of Microstructures

Conditions	Material	Remarks
Crystallization from melts	Metals Igneous rocks	
Precipitation or crystallization from solution	Hardened cement paste; Clays	Amorphous and crystalline components
Solidification without crystallization	Silicate glass Thermoplastics	Limited crystallization for some plastics
Solidification from gas phase	Silica fume Carbon black	Amorphous powders
Sintering of inorganic powders	Ceramics Bricks	May involve reaction between two kinds of powders, and/or partial melting (liquid phase sintering)

before solidification is complete. Most substances, even metals, can form a glassy structure (i.e., solidify without atomic ordering) if the cooling rates are fast enough. The process of rapid cooling to form a glassy structure (glass) is called *vitrification*. Cooling for glass formation must be extremely rapid in the case of metals and ionic salts. The formation of "metglass," for example, may require cooling rates approaching 10^{-6}°C.s^{-1}. On the other hand, glasses may start to crystallize (*devitrify*) spontaneously at some time after cooling is finished; controlled devitrification is used to produce glass ceramics. Some polymers may be induced to crystallize partially under very slow cooling rates by allowing their molecules to become aligned over localized regions. Crystallization of polymers is more readily obtained by the application of directional stresses.

In contrast, cementitious materials develop microstructures by solidification from solutions formed as the cement particles dissolve in water. The products may be largely amorphous (as in the case of portland cement) or crystalline (e.g., calcium aluminate cement). Most cements react with pure water, but other liquids may be used; some refractories are formed by bonding oxides through reaction with phosphoric acid. A cement is really the combination of a reactive solid and liquid, although if the liquid is water it is usual to refer to the solid alone as the cement.

3.2.1 Crystallization from Melts

When crystals form from a molten mass, the time-temperature curve (Fig. 3.1) shows a plateau which lasts for the duration of solidification. This plateau is caused by the liberation of heat during crystallization, the latent heat of fusion, due to the low potential energy of the crystalline state (see Chapter 1). Often a certain amount of supercooling[1] (or undercooling) occurs because a critical number of atoms must come together to form a critical nucleus which can grow spontaneously as a crystal. The formation of critical nuclei represents an energy barrier that must be overcome due to the energy required to form an interface (Fig. 3.2). Supercooling depends on the rate of cooling and can be avoided by "seeding" the melt with crystals or providing foreign matter, such as dust particles, to act as nucleation sites. In either case, the energy barrier is eliminated by introducing preformed surfaces.

[1]A melt can often be cooled below the temperature at which thermodynamics predicts spontaneous crystallization without crystals forming. When this happens, a melt is said to be supercooled and is the analogy of supersaturation in homogeneous solutions.

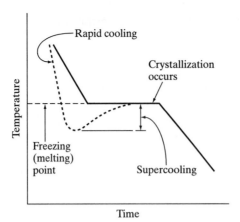

Figure 3.1
Time-temperature curve
during solidification of a
melt.

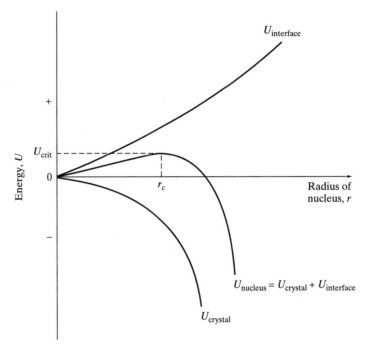

Figure 3.2
Energy associated with the
formation of a nucleus.
$U_{\text{interface}}$ is the energy required to form the nucleus
(creation of a surface), and
U_{crystal} is its cohesive
energy:

$$U_{\text{nucleus}} = U_{\text{crystal}} + U_{\text{interface}}.$$

Once nucleation and growth of crystals begin in the melt, the concomitant liberation of latent heat has implications for continuing crystal growth. Growth will
occur preferentially in the direction in which supercooling first occurs [Fig. 3.3a(i)],
which is the direction in which cooling is taking place most readily. The release of latent heat will exceed the rate of cooling at the growth site, thereby raising the local
temperature and reducing the degree of supercooling. Hence the rate of crystal
growth in this direction is greatly diminished. Now the zone of supercooling extends
behind rather than *ahead* of the growing crystal face. The direction of maximum
crystal growth now takes place laterally and encourages the rapid growth of side
branches normal to the initial growth direction [Fig. 3.3a(ii)]. Several side branches
may develop at points on the crystal most favorable for growth in the new direction;
their length depends on the degree of supercooling at that point. This phenomenon
is known as *dendritic growth*.

The release of latent heat accompanying this lateral growth will, in turn, reduce the degree of supercooling in this direction, while cooling of the melt as a

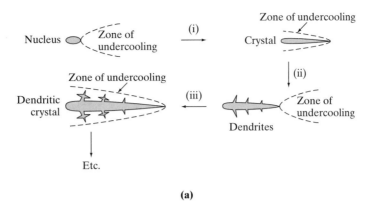

Direction of greatest heat loss

(a)

Figure 3.3
(a) Schematic representation of dendritic crystal growth; (b) dendritic crystals.

(b)

whole restores the supercooling in the original direction. Further directional growth occurs, as shown in Fig. 3.3a(iii). In some cases, additional subsidiary branches may develop from the first dendritic branches. The growth cycles (i)–(iii) will continue until the dendrites begin to impinge on their neighbors. The subsidiary branches then grow gradually until each dendrite becomes a solid grain (Fig. 3.4). The original dendritic structure can seldom be seen in the final polycrystalline solid. However, impurities in the melt sometimes precipitate along the dendrite surfaces during cooling, thereby providing an outline of the initial structure.

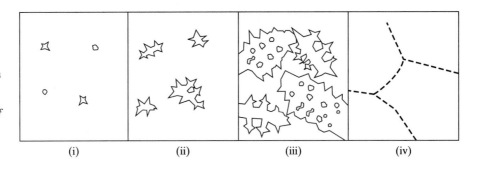

Figure 3.4
Dendritic solidification of a metal, from nuclei to complete grain growth (after R. A. Higgins, *Properties of Engineering Materials*, Krieger, 1977, Fig. 3.20, p. 49).

| (i) | (ii) | (iii) | (iv) |

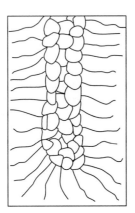

Figure 3.5
Schematic of effect of directional cooling from the sides of a metal ingot. Columnar crystals on the outside, equi-axed crystals due to slow cooling at the center.

Dendritic growth is most pronounced when directional cooling occurs at intermediate rates of cooling. The final crystal shape tends to reflect the directional nature of cooling, being columnar in morphology with the long axis in the direction of cooling (Fig. 3.5). However, columnar or fibrous crystals may also form if the configurations of molecules packed in the crystal are elongated in one direction. (This can be seen on a macrosopic level in basalt lava flows; slow directional cooling gives rise to the characteristic columnar structure seen, for example, in eastern Washington State.)

If cooling is very rapid, a high degree of supercooling develops. This promotes the formation of many critical nuclei and the growth of a large number of very small crystals without a highly developed dendritic structure. If cooling is very slow, only a few nuclei develop into crystals, which grow slowly and uniformly in all directions because heat flow tends to even out directional supercooling. These crystals are equi-axed (i.e., their physical dimensions are similar along all three orthogonal axes).

3.2.2 Crystallization from Solution

Crystallization may also occur from solution. In this case, similar considerations apply, except that nucleation and growth are now controlled by the degree of supersaturation. Saturation occurs when the crystal is in equilibrium with its solution, so that the rate of dissolution equals the rate of crystallization. Ion fluxes (i.e., flow of ions in a particular direction), which are analogous to temperature gradients, will be controlled by the rate of dissolution of a reactive solid or the rate of mixing of two liquids. Temperature may play a role in that it can change the level of saturation; but once the crystals have formed, their rate of growth depends on the diffusion of ions to the growing faces. Rapid precipitation is likely to form amorphous powders which may later crystallize, but certain compositions may form amorphous solids even when precipitation is slow. These solid masses are analogous to glasses.

3.3   PHASE CHANGES ON HEATING AND COOLING

We saw in Chapter 1 that the cohesive (bonding) energy of solids can be calculated from the attractions between atoms and ions integrated over the whole crystal structure. The arrangement of atoms is thus important. We also noted that the cohesive energy is decreased by temperature as atomic vibrations increase. It is possible to calculate the cohesive energy for different crystalline structures that are theoretically possible (considering only geometrical limitations in packing), and these can, in principle, be determined for a range of temperatures. It is thus possible to construct a diagram, as shown in Fig. 3.6, where α and β represent two different packings of atoms in a crystal with the same composition; α has the lower energy at high temperatures and will thus be the stable phase, while at lower temperatures β is energetically favored. At the transition temperature the two energies are equal. However, the phase may persist below the transition temperature if there is an energy barrier for nucleation of β. In this case, α is *metastable* and may be induced to change to the stable β phase at a later time by the application of additional energy (thermal or mechanical) or by the addition of seed crystals of β.

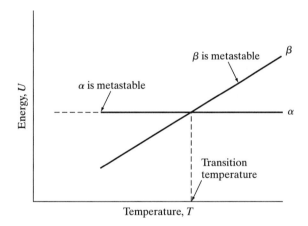

Figure 3.6
Energy diagram for two phases α and β of the same composition.

3.4   PHASE DIAGRAMS

3.4.1 One-component Systems

This treatment also applies to different states of matter; for example, α can be the liquid phase[2] and β the solid phase, with the transition temperature being the melting point. In most cases, it is not possible to calculate or measure the cohesive energy of a phase. Figure 3.6 is thus usually determined by experiment and represented as a temperature-pressure plot: a *phase diagram*. Figure 3.7 is the schematic phase diagram for a pure substance—a one-component system—covering the whole range of

[2]A *phase* is defined as a homogeneous region in a system on which the properties of the system are uniform. It has a specific composition and atomic structure and can be physically distinguished from other phases. It can consist of a single component or multiple components.

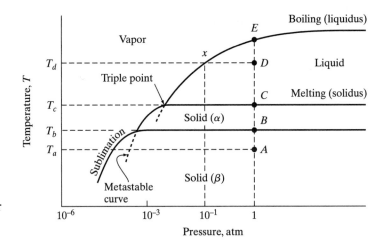

Figure 3.7
Schematic temperature-pressure phase diagram for a single-component material.

possible phase changes. The diagram is constructed by determining the phases present after equilibrium is achieved under different combinations of temperature and pressure. The interpretation of this diagram is relatively simple. Point A corresponds to ambient temperature and pressure, indicating that this substance exists as a solid with certain packing (β) in the crystal. Raising the temperature over T_b while keeping the pressure constant makes the packing of β less stable thermodynamically than α, and thus we pass into the α field at point B, which determines the polymorphic transformation temperature. At point C melting of the α polymorph occurs, while at point E (the boiling point) liquid spontaneously turns to vapor: At this point the vapor pressure equals the ambient pressure. At point D, the vapor pressure is much lower than 1 atm; if the external pressure is reduced while keeping the temperature constant, the liquidus is crossed at point X, which means the liquid will start to boil. The pressure at which this occurs equals the vapor pressure at point D($\sim$0.1 atm). The liquidus is the vapor pressure curve for the substance.

Similarly, reducing the pressure at point A crosses the sublimation curve at Y. The β solid will vaporize spontaneously at this point without melting; its vapor pressure corresponds to $\sim$10^{-4} atm. Other points to note are that the phase transformations involving solids and liquids are not greatly affected by pressure, and the *triple point* is the point at which solid, liquid, and vapor exist simultaneously. It is important to remember that the phase diagram predicts thermodynamic *equilibrium* and does not predict the *rate* at which the changes will occur. As discussed earlier, problems in nucleating a phase may make the rate of change infinitely slow and allow the system to move along the metastable extensions of the phase boundaries.

3.4.2 Two-component Systems

In many instances, the systems of interest involve two substances (components). Thus, a third parameter—that of relative concentration—must be added, and the phase diagram should be drawn as a three-dimensional figure. Since this is hard to do within the two-dimensional limitations of the printed page, it is customary to represent the phase diagram by a slice taken at constant pressure (usually 1 atm). Figure 3.8 shows the simplest form of a two-component phase diagram which describes many binary alloy systems. In this case, components A and B are completely miscible with each other.

This simple diagram can be used to illustrate two important concepts: the *phase rule* and the *lever rule*. The phase rule determines the number of *degrees of freedom* (i.e., the number of independent variables) needed to describe the system:

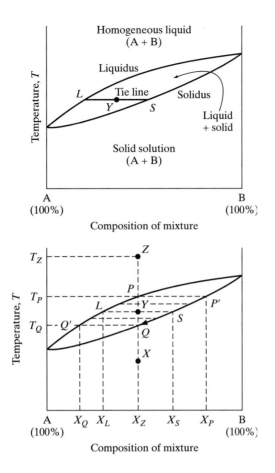

Figure 3.8
Phase diagram of two completely miscible components.

$$F + P = N + 1, \tag{3.1}$$

where F is the number of degrees of freedom, P is the number of phases present, and N is the number of components. Consider point X in Fig. 3.8. Here $P = 1$ and $N = 2$, so $F = 2$. This means that we can alter both variables—temperature and composition—without altering the number of phases present. The solid solution area in which X is located is said to be *bivariant*. Point Y, on the other hand, lies in a two-phase area (liquid + solid), and the phase rule tells us that it is *univariant* (i.e., only one experimental variable can be altered without changing the phase composition of the system). We can use the lever rule to calculate the amount of each phase present at Y:

$$\frac{\text{weight of liquid}}{\text{weight of solid}} = \frac{YS}{YL}. \tag{3.2}$$

The liquid has the composition X_L and the solid X_S.

Now consider cooling composition X_Z from temperature T_Z (point Z). The system remains liquid until we reach point P on the liquidus, at which time the solid of composition X_P begins to appear. As we lower the temperature to T_Q, application of the lever rule tells us that the amount of solid formed continues to increase while its composition moves from X_P to X_Z. At Q, the whole system solidifies to a solid of composition X_Z. The complete miscibility of the two components allows the composition of the solid to change continuously. For this to occur, sufficient time must be allowed for atoms of B in the initially formed solid to diffuse out and thus to achieve the thermodynamically required conditions. The phase diagram represents the *equilibrium* conditions; if equilibrium is not reached rapidly, a *metastable* phase assembly may result. Metastable systems can persist indefinitely, while others may slowly revert to the equilibrium state. Slow transformations tend to have adverse effects on materials performance.

Figure 3.8 also tells us that this alloy does not have a definite melting point, as does a pure substance (A or B), but rather it melts (or solidifies) over a range of temperatures—in this case T_P to T_Q (i.e., unlike A or B, where solid and liquid can only coexist at one temperature, called an *invariant point*). Thus, the alloy can have solid and liquid together over a wide range of temperatures, although in differing amounts.

3.4.3 Systems with Partial Immiscibility

A more common two-component phase diagram is one in which the components exhibit limited mutual solubility. A typical phase diagram, as shown in Fig. 3.9, has two regions of partial miscibility: one where solid A separates out from the liquid, and one where solid B separates. Consider cooling from point Z (temperature T_Z) on the B-rich side of the diagram. At this temperature, the system is a single liquid, but on cooling to temperature T_P (point P), separation of pure solid B commences. B continues to separate out, with the composition of the liquid moving from X_Z to X_L along the liquidus line ($P \rightarrow L$). At point Y the lever rule tells us that

$$\frac{\text{weight of solid B}}{\text{weight of liquid}} = \frac{YL}{BY}. \tag{3.3}$$

At point S, the liquid has changed to composition X_E (point E). Further cooling will solidify the remaining liquid without any more compositional change. The solidified liquid will consist of an intimate mix of solids A and B. Considering the same changes on the A-rich side, it can be seen that at temperature T_Z (point Z'), some solid has already separated out. Further cooling to T_Y causes the separation of more A so that

$$\frac{\text{weight of solid A}}{\text{weight of liquid}} = \frac{Y'L'}{AY'}. \tag{3.4}$$

The composition of the liquid has moved from X_N to $X_{L'}$. At point S the liquid has once again reached the composition X_E, and further cooling will solidify the remainder with no additional change.

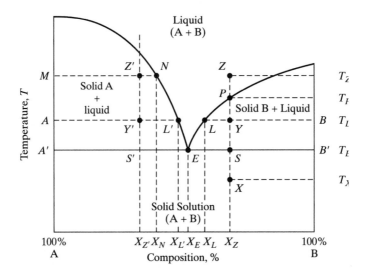

Figure 3.9
Phase diagram of two partially miscible components.

Point E is called the *eutectic point,* and the solid phase that freezes out is the *eutectic.* The eutectic is not a true solid solution, but is rather an intimate mixture of solids A and B that creates a unique microstructure. At E, three phases exist simultaneously: solids A, B, and eutectic. Application of the phase rule gives

$$F + 3 = 2 + 1; \ (\text{i.e., } F = 0).$$

Thus, there are no degrees of freedom at E. This means that no change in temperature, composition, or pressure can take place without a change in the system. E is thus an *invariant point.*

Another invariant point is the *peritectic point.* Whereas at the eutectic point a liquid crystallizes as a mixture of two solids, at the peritectic point a solid (S) and liquid crystallize to a new solid (N). Both the eutectic and peritectic involve a liquid as one phase; if only solid phases are involved, the invariant points are referred to as *eutectoid* and *peritectoid,* respectively.

More complex binary phase diagrams may contain several eutectic and peritectic points due to the formation of several different solid phases (including true solid solutions). The iron-carbon phase diagram, which underlies the heat treatment of steels, is a good example of this, and its interpretation will be discussed in Chapter 14.

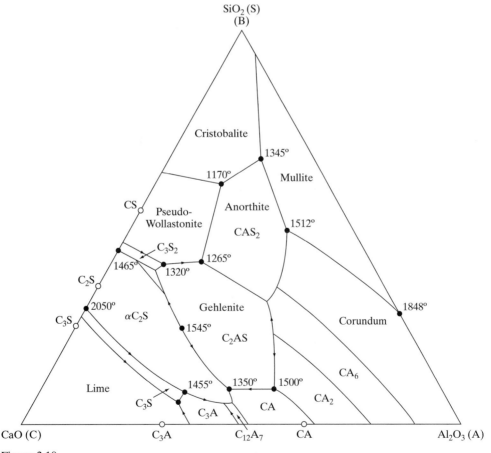

Figure 3.10
Ternary diagram for CaO—Al$_2$O$_3$—SiO$_2$. Numbers are temperatures in °C; arrows show temperature gradients along some phase boundaries.

3.4.4 Three and Four-component Systems

A three-component (tertiary) phase diagram requires the composition of the three components as well as temperature to describe the system at constant pressure. This can only be done using a three-dimensional figure. The system is usually represented as a projection onto the compositional plane, as shown in Fig. 3.10. The effect of temperature is handled by giving the temperatures at which invariant points occur and using arrows to indicate rising or falling temperatures along phase boundaries. Often temperature contours (lines of equal temperatures) are given. The interpretation of a tertiary phase diagram is no different from a binary one except that it is complicated by the absence of a specific temperature axis. However, the use of such diagrams is not needed in this book, and it will not be treated in detail. Similarly, we will not discuss quarternary diagrams. Suffice it to say that these must be broken down into tertiary subsections of the complete system (which would need a 4-D diagram). The components of these subsections can be specific compounds.

3.5   SINTERING

Another major approach to forming microstructures is the sintering of compacted powders. A solid bridge or "neck" is formed between adjacent particles (Fig. 3.11) that are in close contact when they are heated at a temperature well below the melting point. Bulk shrinkage may accompany sintering, resulting in a density increase as the spaces between the particles are eliminated. Sintering occurs at temperatures above about half the melting temperature (T_m, measured in K) and involves several processes, illustrated schematically in Fig. 3.12.

1. Surface tension: loss of surface energy as surfaces merge
2. Local deformations under bulk stresses
3. Bulk diffusion
4. Surface diffusion.

For example, bulk diffusion in copper is negligible at 300 K, while at $T = 0.75\ T_m$, the atoms can diffuse at about 45 nm s^{-1}. At this temperature, the average lifetime of an atom on the surface is 1 s, while at 300 K it is 10^{37} s. Thus, it is clear that melting is not necessary for appreciable solid transport to occur.

The formation of necks between particles is proportional to ct^m, where t is time, m is characteristic of the particular type of mass transport (see Table 3.2), and c is a constant. It may be possible to enhance these slow, diffusion-controlled, mass transport processes by the judicious use of foreign materials. For example, a small amount of nickel will enhance grain boundary diffusion in tungsten, while the addition of a small amount of water promotes an evaporation-condensation mechanism.

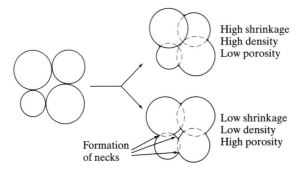

High shrinkage
High density
Low porosity

Low shrinkage
Low density
High porosity

Formation
of necks

Figure 3.11
Sintering to low and high porosities.

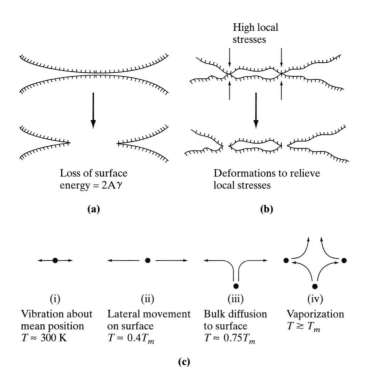

Figure 3.12
Schematic representations
of different sintering
processes.

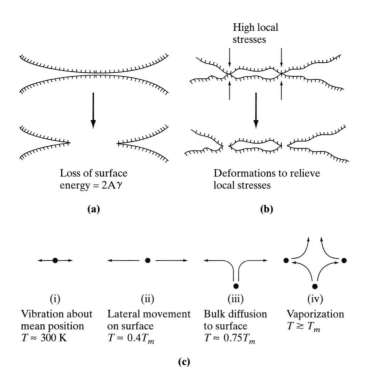

Loss of surface
energy = 2Aγ

(a)

High local
stresses

Deformations to relieve
local stresses

(b)

(i)	(ii)	(iii)	(iv)
Vibration about mean position $T \approx 300$ K	Lateral movement on surface $T \approx 0.4T_m$	Bulk diffusion to surface $T \approx 0.75T_m$	Vaporization $T \gtrsim T_m$

(c)

TABLE 3.2   Sintering Mechanisms

Process	Material	Shrinkage	m^c
Viscous flow (plastic deformation)	Glass	High	1/2
Volume diffusion	Copper	Moderate	1/5
Surface diffusion	Ice	None	1/7
Grain boundary diffusion[a]	Tungsten	Moderate	1/4–1/6
Evaporation-condensation[b]	Sodium chloride (common salt)	None	1/3

[a] This is a surface diffusion process from interparticle contact area to the neck (see Fig. 3.11), while surface diffusion is from a free surface.
[b] Evaporation from particle surface and condensation in the neck.
[c] See Section 3.5 for definition of m.

On the other hand, sintering may be interrupted by the presence of impurities at the surface (for example, elimination of adsorbed volatiles, which may separate particles and create porosity).

Sintering can be used to make complex shapes and need not involve the formation of a new phase at all (as in the case of powder metallurgy or the formation of alumina bodies). In many ceramic processes, however, sintering and reaction occur together, and in some cases a liquid phase may partially form. This will lead to highly effective densification and is the basis of *liquid-phase sintering*. Here only a small fraction of the material becomes liquid, but this takes place at the grain boundaries, thus drawing the solid particles together by surface tension forces; moreover, diffusion of ions is much more rapid through a liquid than through a solid. Sintering also involves *grain growth* (i.e., small particles are gradually adsorbed into larger ones). This also occurs in crystallization processes either in solution (by dissolution-reprecipitation processes) or in solids (by diffusion mechanisms). The driving force for grain growth is the annihilation of high-energy surfaces.

3.6 MICROSTRUCTURE

This aspect will be discussed in more detail in the chapters devoted to each material, but it is appropriate to draw attention to a few universal principles at this stage. Material properties can be dependent on porosity, grain size, and composite structure.

3.6.1 Porosity

Internal voids strongly affect the strength of all materials; they are particularly important for brittle materials, such as ceramics, concrete, or cast iron. A good estimate of the total void space (porosity) can be obtained from density measurements. This estimate is used, for example, to estimate the sintering efficiency in ceramic bodies.

Several different expressions have been used to relate strength to porosity. Three of the more commonly used ones are given in Eqs. 3.5 to 3.7:

$$\sigma_u = \sigma_0 (1 - P)^n \tag{3.5}$$

$$\sigma_u = \sigma_0 \exp(-kP) \tag{3.6}$$

$$\sigma_u = k' \ln(P_0/P), \tag{3.7}$$

where σ_u and P represent strength and porosity; σ_0 is the strength at zero porosity (intrinsic strength); P_0 is the porosity at zero strength (critical porosity); and k and k' are constants. These equations are all empirical, so none is intrinsically better than the others (although one may provide a better fit for a particular set of data) (See Fig. 3.13). All three predict a very strong influence of porosity on strength, and similar equations can be used to predict other material properties, such as modulus of elasticity. The potentially important secondary characteristics, such as pore shape, size, distribution, and orientation, are not taken into consideration.

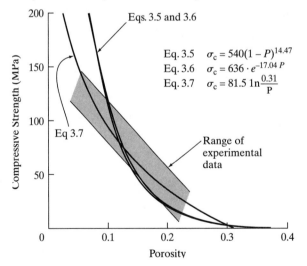

Figure 3.13
Strength-porosity relationships for hardened cement paste.

3.6.2 Grain Size

Correlations of strength with grain (crystal) size are also to be found; a semilogarithmic relationship is the most common (Fig. 3.14). It can be seen that materials with a fine-grained crystallinity are generally stronger than those with coarse crystals. This is

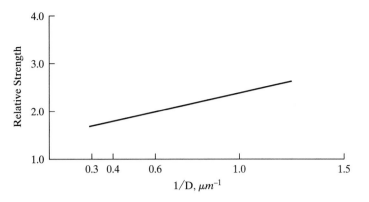

Figure 3.14
Relative strength of hot-pressed silicon nitrides as a function of mean grain size, D (after M. M. Eisenstadt, *Introduction to Mechanical Properties of Materials,* Macmillan, 1971, Fig. 4.9, adapted from Spriggs and Vasilar, *J. Amer. Ceram. Soc.*).

attributed to the disruption of crack propagation or plastic deformation processes by the atomic discontinuities at grain boundaries.

3.6.3 Composite Microstructures

Heat treatment of alloys leads to composite microstructure when limited solubility or eutectic compositions occur. The example most frequently cited is the pearlitic structure (Fig. 3.15), which is the eutectic in carbon steels. Layers of cementite (Fe_3C) are present in a ferrite (BCC iron) matrix forming a laminated structure. The hard but brittle cementite imparts increasing strength and hardness to the soft, ductile ferrite matrix, but at the expense of ductility. The composite structure has properties that lie between those of the two components. There are a number of other alloys that develop laminated eutectic structures.

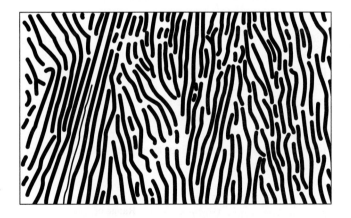

Figure 3.15
Pearlitic structure of eutectoid steel.

The other major type of composite microstructure is the dispersion of a second phase within a matrix in the form of very small particles (Fig. 3.16). When these microscopic particles are harder than the matrix, they impede the movement of dislocations (see Chapter 2). As a dislocation encounters an array of these particles, it is "pinned" at each particle site. The dislocation front has to bow around the particles, leaving a dislocation loop around each particle. These loops act as additional barriers to subsequent dislocation movement. A greater resistance to dislocation movement raises the yield point of a metal and generally increases its hardness. The composite also tends to retain these properties since resistance to dislocation movement also hinders recovery. Many dispersion hardened metals contain 0.1 to 1.0 μm metal oxide particles as the dispersed phase. Examples are sintered aluminum powder, which contains 6% Al_2O_3 in an aluminum matrix, and TD-nickel, with ThO_2

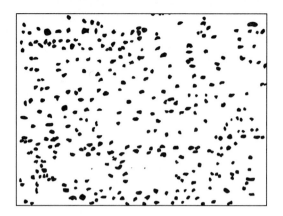

Figure 3.16
Dispersion of fine particles in a matrix.

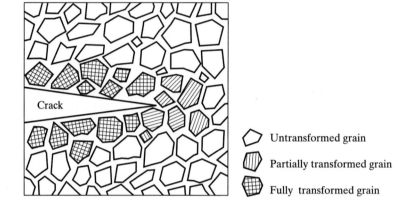

Figure 3.17
Partially stabilized zirconia-crack arrest mechanism.

Untransformed grain

Partially transformed grain

Fully transformed grain

dispersed in a nickel matrix. Other metals are more commonly dispersion hardened by precipitation of an intermetallic phase from a solid solution during cooling. These precipitates are much smaller (~10 nm), and such materials are sometimes referred to as precipitation hardened. Common examples are bearing metals, where intermetallic precipates provide hardness while still maintaining ductility and an alloy of high toughness results. Duraluminum alloys develop small, coherent precipitates within the crystals of α-aluminum which strengthen the aluminum matrix. At high temperatures the small, dispersed precipitates tend to grow into large, noncoherent precipitates which do not impede dislocation movement effectively; hence strength decreases.

A different kind of dispersion hardening is found in PSZ (partially stabilized zirconia) ceramics. When a zirconia body is cooled, the cubic polymorph is the stable phase at room temperature. However, some small crystals can be induced to remain as the high-temperature monoclinic polymorph. This metastable phase transforms to the stable one when a crack propagates in its vicinity (see Fig. 3.17). The energy involved in this transformation serves as a barrier to the movement of cracks and can thus serve to increase the strength and the fracture energy of the material.

BIBLIOGRAPHY

R. A. Higgens, "Properties of Engineering Materials," Krieger, 1977, 441 pp.

C. G. Bergeron and S. H. Risbud, "Introduction to Phase Equilibria in Ceramics," American Ceramic Society, 1984, 158 pp.

PROBLEMS

3.1 Discuss the differences and similarities between supercooling and supersaturation.

3.2 Given the binary system Ge-Si determine the following:
  (a) the temperature at which the solid solution attains an equilibrium composition of 60% Si-40% Ge.
  (b) the composition of the equilibrium liquid.
  (c) the proportion of solid to liquid if the original composition of the mixture was 50 wt% Si.
  (d) the temperature at which the original composition in (c) becomes completely molten.

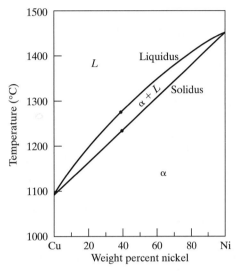

3.3 Given the binary system A-B determine the following:
  (a) The temperature at which 50% of solid B is in equilibrium with a liquid composed of 55% B and 45% A.
  (b) The composition of the initial mixture for (a).
  (c) The temperature at which solid B first appears.
  (d) The composition of the eutectic.

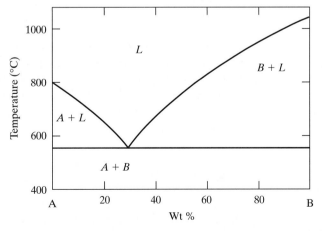

3.4 Why is excessive grain growth detrimental to strength? What processes enhance grain growth?

4

Surface Properties

4.1   SURFACE ENERGY AND SURFACE TENSION

We saw in Chapter 2 that the potential energy of an atom in a phase is determined by all the atoms which act as its "nearest neighbors." An atom in the bulk of the phase (solid or liquid) sees an equal interaction energy in all directions. But when an atom is at or near the surface, it will see an asymmetric force field because it has no neighbors (or fewer) on one side of it (see Fig. 4.1b). As a result, these atoms have a higher potential energy, and their position on the Condon-Morse diagram will be about one-third up the potential energy well (Fig. 4.1c). It can be seen that the equilibrium distance between surface atoms (r_0') is also greater than between those in the bulk phase.

If the material can reach equilibrium at the surface, there will indeed be a greater separation of atoms. This is achieved by the formation of a spherical surface, which also minimizes the surface area (and hence the surface energy) for a given volume of material. Droplets of liquids or bubbles of gas will be spherical because the atoms have sufficient mobility to move to equilibrium positions. The equilibrium surface energy (γ_s) is defined as

$$\gamma_s = U_{\text{surface}} - U_{\text{bulk}} \tag{4.1}$$

The units of surface energy are in energy per unit area, or specific energy (i.e., $J \cdot m^{-2}$ or, more commonly, in millijoule units of $mJ \cdot m^{-2}$). The wider spacing of atoms at the surface makes the surface behave as a stretched elastic membrane under a circumferential tension (Fig. 4.1a), because the atoms would like to return to the lower energy of the closer equilibrium distance. Thus, the surface energy can

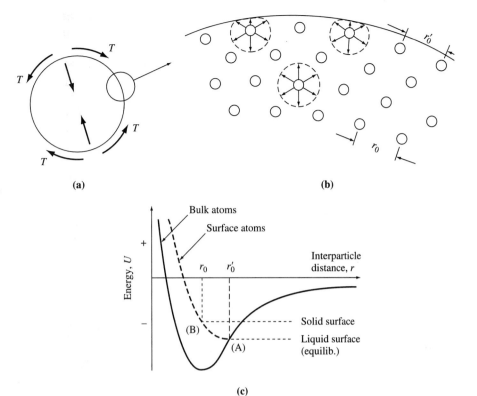

Figure 4.1
(a) Equilibrium spherical shape of fluid drop with circumferential surface stress (tension) and radial compressive stress; (b) arrangement of atoms at the surface showing increased spacing (r_0') and distorted force field seen by each atom as compared to the bulk atoms; (c) Condon-Morse curves for bulk and surface atoms. Solid surfaces have higher energies because the new equilibrium spacing r_0' is not achieved.

alternatively be viewed as *surface tension, T,* which is a tensile force per unit length and which will be shown later to be numerically equal to γ_s. The body of the material is subjected to a compressive stress which reflects the net attractive force of the surface atoms to the bulk atoms. Thus, the material within the bubble or droplet is under pressure, which is given by

$$\Delta p = \frac{2\gamma_s}{r}. \tag{4.2}$$

The equilibrium surface energy is called the *surface free energy.* However, many solid surfaces are created by cleaving the bulk phase, and because the atoms are not mobile, those at the new surfaces have an energy greater than the surface free energy because they are now at point *B* on the surface Condon-Morse curve (Fig. 4.1c). Solids, however, will also experience surface tension effects if their surfaces have formed during particle growth and will experience pressures, as indicated by Eq. 4.2. Unless *r* is very small (so that particles are of colloidal dimensions), the force will be negligible.

Surface tension was first measured experimentally by the classic soap film experiment. Consider a soap film attached to a wire frame, one side of which is moveable (Fig. 4.2). The force required to keep the moveable wire stationary is $T \cdot \ell$, where *T* is the surface tension (per unit width) and ℓ is the length of the wire. Its value can be calculated by considering energy preservation when moving the wire by an amount *dx*. The energy input is

$$2T \cdot \ell \cdot dx.$$

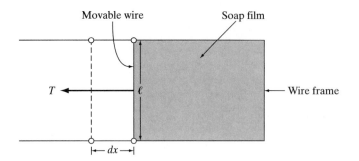

Figure 4.2
The soap film experiment.

The factor 2 recognizes that the film has two separate surfaces. As the surface area increases by this movement, the output of the work is an increase in surface energy by an amount of

$$\gamma_s \cdot 2 \cdot \ell \cdot dx.$$

On the basis of energy conservation principles,

$$2 \cdot T \cdot \ell \cdot dx = \gamma_s \cdot 2 \cdot \ell \cdot dx \tag{4.3}$$

$$T = \gamma_s. \tag{4.4}$$

The units of T are the same as those of γ_s; $N \cdot m \cdot m^{-2}$ (i.e., $J \cdot m^{-2}$ or $N \cdot m^{-1}$). Various methods have since been developed to measure the force or energy required to expand liquid surfaces.

Surface energies of solids are not easy to determine experimentally, but from consideration of breaking of short-range forces it can be shown that

$$\gamma_{s(\text{solid})} = \frac{E r_0}{\pi^2}, \tag{4.5}$$

where E is the modulus of elasticity. Using similar considerations, it can be shown that the heat of sublimation (ΔH_s) of a close-packed atomic array represents the energy required to break nine bonds (i.e., create 18 "half-bonds"), while surface atoms have three "half-bonds"; thus, the ratio of $\gamma:\Delta H_s$ should be about 1:6. Some measured and calculated values of surface energy are given in Table 4.1 for different materials. Measurements have to be made with very pure substances since one way for surface energies to be minimized is for impurity atoms that have lower surface energies to segregate at the surface.

TABLE 4.1   Measured and Calculated
Surface Energies

Material	Surface Energy (γ_s) m $\cdot$ Nm^{-1}		
	Measured	Calculated	1/6 ΔH_s
Copper	1700	—	1400
Silver	1200	—	1300
MgO	1200	1300	—
NaCl	300	310	—
CaCO$_3$	230	380	—
Water	72	—	—
Ethanol	30	—	—

It should be remembered that a surface actually represents a boundary between two phases (solid-gas, solid-liquid, etc.) and that there is thus always an *interface* between the two. We should, therefore, use the term *interfacial energy,* because that is what is measured. The interfacial energy depends on the attractions across the interface (Fig. 4.3). Thus, a solid in equilibrium with its liquid has a lower interfacial energy γ_{SL} than the energy across the boundary with its vapor γ_{SV}. While the highest possible surface energy is that of the surface in a vacuum (γ_0), for most solids $\gamma_{SV} \approx \gamma_0$. However, in the case of liquids, γ_{LV} is the measured quantity. The existence of attractions between molecules across the interface leads to three important phenomena: wetting, adsorption, and capillary effects.

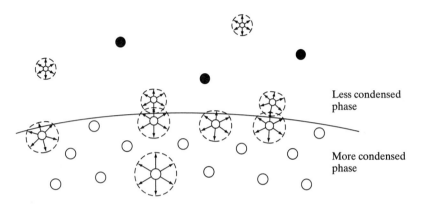

Figure 4.3
Attraction across an interface. The less condensed phase (e.g., gas) has lower force fields surrounding each atom or molecule, but this may be increased when the molecules approach atoms of the more condensed phase (e.g., solids), which have stronger force fields.

Less condensed phase

More condensed phase

4.3   WETTING

Wetting is an important aspect of the interaction between solids and liquids. When a liquid spreads readily over a surface, it is said to wet it (Fig. 4.4a). If it forms discrete drops on the surface (Fig. 4.4c), then it is nonwetting. A liquid is wetting when the attractions between its atoms and those of the surface atoms are greater than between the atoms in the liquid. If the opposite is true, then the liquid is nonwetting. The *contact angle* (Fig. 4.4) measures the degree of wetting. When a droplet is in equilibrium with a surface, then the forces are balanced, as in Eq. 4.6:

$$\gamma_{SV} = \gamma_{LS} + \gamma_{LV} \cos \theta \qquad (4.6)$$

If $\gamma_{SV} \geq \gamma_{LS} + \gamma_{LV}$, then $\theta = 0$ and wetting occurs.
If $\gamma_{LS} \geq \gamma_{LV} + \gamma_{SV}$, then $\theta > 90°$ and nonwetting occurs.
In between, when $\gamma_{SV} < \gamma_{LS} + \gamma_{LV}$ but $\gamma_{LV} < \gamma_{LS}$, $\theta < 90°$ and partial wetting occurs.

The force tending to spread the droplet is

$$\sigma_{spread} = \gamma_{SV} - \gamma_{LS} - \gamma_{LV} = \gamma_{LV}(\cos \theta - 1). \qquad (4.7)$$

Examples of each case are mercury on concrete (nonwetting), water on wood (wetting), and water on glass (partial wetting). Wetting is necessary for lubrication and flotation phenomena. The spreading of a lubricating film of liquid lowers the coefficient of friction, $\mu = \tau/\sigma$ (the ratio between the lateral force, τ, and normal

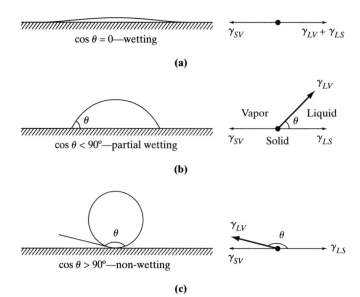

(a)

(b)

(c)

Figure 4.4
Effect of contact angle on the degree of wetting, showing equilibrium of interfacial forces.

force, σ) lateral because the value of the lateral force τ is governed by the viscosity of the liquid, which cannot maintain a large value of τ. In this case, μ is ideally <0.01, but since the lubricating film is under forces, it is in practice ~ 0.1. In contrast, μ for a solid surface without a lubricant is much higher. If a liquid which wets solid particles is foamed into a froth, then the particles will become attached to the bubbles of foam. In this way the particles achieve the equilibrium contact angle (θ), as shown in Fig. 4.5. The buoyancy of the bubbles carries the particles to the top of the liquid, where they can be drawn off. Particles which are not wetted (or are too large) fall to the bottom. This principle is used to beneficiate mineral ores. Wetting is also important in capillary action and adhesion, as discussed later.

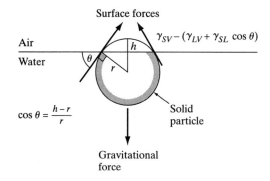

Figure 4.5
Effect of interfacial tension on flotation of particles.

4.4 ADSORPTION

Adsorption of atoms or molecules from the vapor state will reduce the surface energy by increasing the binding energy of surface atoms. Physical adsorption is the result of nonspecific van der Waals attractions between the atoms of the vapor and the atoms of the solid. The lowering of the surface energy is given by the Gibbs equation:

$$\Delta\gamma_s = \frac{nRT}{S} \Rightarrow \int d\ln P_r, \qquad (4.8)$$

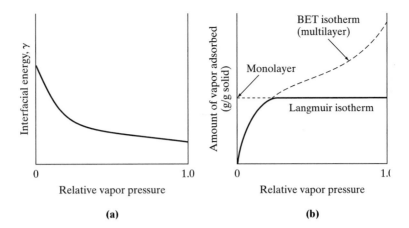

Figure 4.6
Adsorption of gases on a solid as a function of relative vapor pressure: (a) effect on interfacial energy; (b) amount adsorbed on solid.

where S is the surface area of the solid ($m^2 \cdot g^{-1}$) and P_r is the relative vapor pressure. The variation of γ_s with relative vapor pressure is shown in Fig. 4.6a; for example, adsorption of oxygen on silver lowers γ_{SV} from 1200 mJ $\cdot$ m^{-2} to 400 mJ $\cdot$ m^{-2}. Thus, surfaces must be thoroughly degassed in vacuum to remove all adsorbed impurities before adsorption characteristics are studied. Even so, impurities may not be removed if chemisorption has occurred (i.e., adsorption accompanied by the formation of strong ionic or covalent bonds). The surface of aluminum is always covered by a surface film of oxide formed by reaction with chemisorbed oxygen, and it is this film that protects aluminum from rapid corrosion. Chemisorption is an important phenomenon in corrosion processes and in the action of catalysts.

The simplest concept of physical adsorption is that the molecules of vapor form a *monolayer* on the surface. This can be described by the Langmuir equation:

$$f = \frac{aP_r}{1 + aP_r} \tag{4.9}$$

where f is the fraction of surface covered, P_r is the relative vapor pressure, P/P_0, and a is a constant. The limiting value of f is 1, corresponding to a monolayer, which can be determined experimentally (Fig. 4.6b). If the area covered by one molecule of the vapor is known, then, in principle, the surface area can be determined. The Langmuir isotherm does not allow for multilayer adsorption (which usually occurs) but is applicable to those situations in which the interactions between the first monolayer and the surface are much stronger than in the succeeding layers. In general, the adsorption isotherm continues to rise with increasing relative vapor pressure (Fig. 4.6b), indicating multilayer adsorption, and this is allowed for in the more general case described by the BET (Brunauer-Emmett-Teller) equation:

$$\frac{P_r}{W(1 - P_r)} = \frac{1}{CV_m} + \frac{(C - 1)P_r}{CV_m}, \tag{4.10}$$

where P_r is the relative pressure, P/P_0, W is the weight of vapor adsorbed, V_m is the amount of vapor required for a monolayer, and C is a constant. This equation is the one usually applied for the determination of surface areas of solids by adsorption measurements.

Adsorption of molecules onto surfaces can also occur from solution. In this case, the adsorption of a specific molecule will only be observed if it interacts much more strongly with the surface than do the more numerous solvent molecules (which, of course, will be physically adsorbed). Adsorption from solution usually involves weak chemisorption and obeys a Langmuir-type adsorption isotherm. Surfactants are a good example of this behavior.

4.5   SURFACTANTS

Surfactants are compounds which will adsorb preferentially at interfaces, thereby lowering the surface tension. The simplest surface-active molecules are the polar molecules shown in Fig. 4.7a. These molecules have a *hydrophilic* (water-loving) end and a hydrophobic (water-hating) end, so the molecules will align themselves at the boundary between two immiscible layers, with the hydrophilic end in the water (or other polar liquid) and the hydrophobic end in the nonpolar liquid (Fig. 4.7b). This alignment lowers the interfacial energy. It is now possible to disperse one liquid within the other as a mass of tiny droplets (emulsification). Normally, such emulsions are usually not stable because of the large surface area they create, but the preferential adsorption of the surfactant lowers the interfacial energy sufficiently to stabilize the emulsion. We all are familiar with oil-in-water emulsions in salad dressings; in the construction field, bitumen-in-water emulsions can be used in road paving.

These same molecules can also act at air-liquid (Fig. 4.7c) and solid-liquid or solid-air interfaces (Fig. 4.7d). Foams are bubbles of air in water, stabilized by surfactants (domestic examples include dishwashing froth; whipped cream; meringues; a construction example is entrained air in concrete). In a similar manner, an increased wetting of solids by liquids can be achieved. Lubricants adsorb on solid surfaces, thus providing liquid films that prevent attractions between the surfaces. The same principle applies for stabilized suspensions of solids in liquids (e.g., latex paints).

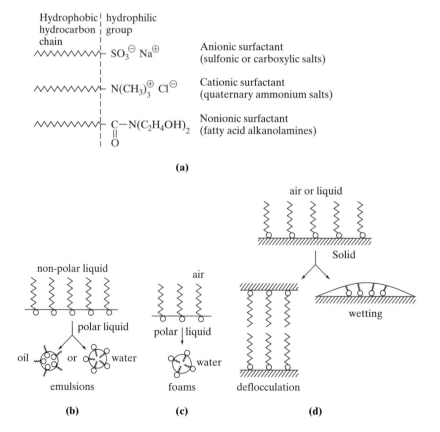

Figure 4.7
Surfactants: (a) different types of surfactants; (b) their effect at liquid-liquid interfaces to form stable emulsions; (c) formation of stable foams; (d) defloccu-lation of solids or wetting of solids in the presence of surfactants.

As the result of surface attractions between a liquid and a solid where wetting occurs, water can be drawn into small pores. Consider a glass capillary (a very fine cylindrical glass tube) standing in a reservoir of water (Fig. 4.8). The liquid is drawn up into the capillary tube by a circumferential surface force, F_{surf}:

$$F_{surf} = 2\pi r \gamma_{LV} \cdot \cos \theta. \tag{4.11}$$

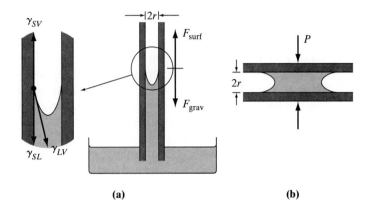

Figure 4.8
Capillarity effects in (a) vertical capillary (capillary rise); (b) horizontal capillary.

An opposing gravitational force, F_{grav}, is exerted downward on the column of liquid:

$$F_{grav} = \pi r^2 hg\rho, \tag{4.12}$$

where ρ is the density of the liquid and h is its height.

At equilibrium, $F_{surf} = F_{grav}$ and thus

$$2\pi r \gamma_{LV} \cdot \cos \theta = \pi r^2 hg\rho, \tag{4.13a}$$

or

$$h = \frac{2\gamma_{LV} \cdot \cos \theta}{rg\rho}. \tag{4.13b}$$

This is the cause of capillary action whereby a porous material may raise liquids vertically (wicking action). A good example is concrete, which when in contact with groundwater may raise the water in its pores to above ground level. (In the case of water in concrete, $\cos \theta \approx 1.0$, γ_{LV} is ~400 mJ $\cdot$ m^{-2}, and r is typically 10–100 nm. Therefore, h is about 1–10 m.)

If gravitational forces are not important (horizontal capillaries), then at equilibrium, a pressure will be exerted on the water given by

$$P = \frac{2\gamma_{LV} \cdot \cos \theta}{r} \tag{4.14}$$

The water will try to draw the walls of the capillary together to reach this equilibrium pressure. It thus exerts a tensile stress on the inside walls of the capillary. Equation 4.14 is the *Laplace equation*; for $\cos \theta = 1$ (wetting), and using two radii to describe a meniscus generally, in a noncylindrical pore it becomes

$$P = \gamma_{LV}\left(\frac{1}{r_1} + \frac{1}{r_2}\right). \tag{4.15}$$

The curvature of the air-liquid interface also leads to a lowering of the equilibrium vapor pressure of the liquid, as is described by the Kelvin equation:

$$\ln(P/P_0) = \frac{\gamma_{LV}}{RT} M\left(\frac{1}{r_1} + \frac{1}{r_2}\right) = K\left(\frac{1}{r_1} + \frac{1}{r_2}\right), \tag{4.16}$$

where p/p_0 is the ratio of vapor pressure above the meniscus to the vapor pressure of the bulk liquid (p_0) and M is the molar volume of the liquid. The Laplace equation can thus be written:

$$\Delta P = \frac{RT}{M} \ln(P/P_0) = \gamma_{LV}\left(\frac{1}{r_1} + \frac{1}{r_2}\right). \tag{4.17}$$

The Kelvin equation indicates that smaller pores will not lose water rapidly by evaporation until the partial pressure of water (relative humidity) is lowered to the equilibrium value determined by the Kelvin equation. Water will then be lost rapidly by evaporation until the pore is empty. As seen in Fig. 4.9, pores of 10 nm (0.01 μm) will not start to empty until 95% relative humidity. At the same time, the Laplace equation can be used to calculate the pressure difference that is developed. This is manifested as a tensile stress (σ_r) within the water, which in turn is exerted on the walls

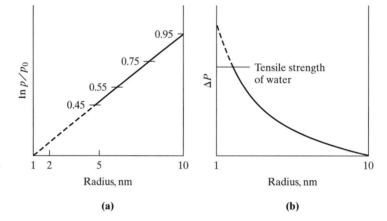

Figure 4.9 Relationships predicted by the Kelvin equation: (a) partial vapor pressure required to empty small pores; (b) stress developed as pores are emptied.

of the capillary. In a porous solid, a random array of capillaries will exert an internal isostatic stress on the solid as water is removed from the capillaries.

Consider now what happens as the relative humidity (rh) surrounding a water-saturated porous solid is lowered and then raised again. The effects are shown schematically in Fig. 4.10. For simplicity, we consider cylindrical pores of radius R. The pores remain full until $\ln(P/P_0)$ reaches the values calculated by the Kelvin equation when $r_1 = r_2 = R$. At this stage, a meniscus will form and will remain until the pores are completely empty. A force P is exerted on the solid skeleton, as calculated from the Laplace equation. When all water is removed, the stress will fall to zero again. On raising the rh, we must remember that water first condenses to form an adsorbed layer of thickness t, which is in equilibrium with the rh and can be calculated from the BET equation (Eq. 4.10). Since the layer has an effective radius of ∞, the pore radii at the point at which a meniscus can form are $r_1 = R - t$ and $r_2 = \infty$, where t is the thickness of the adsorbed layer. Thus, the meniscus does not refill by capillary condensation until a higher rh is reached, and the desorption-adsorption curves show hysteresis. In practice, there will be pores of various sizes and various shapes so that the analysis is more complicated although the principles remain the same (Fig. 4.10b). The Kelvin equation can be used with the adsorption isotherm to calculate a pore size distribution in the range of radii 1.3 nm to 15 nm.

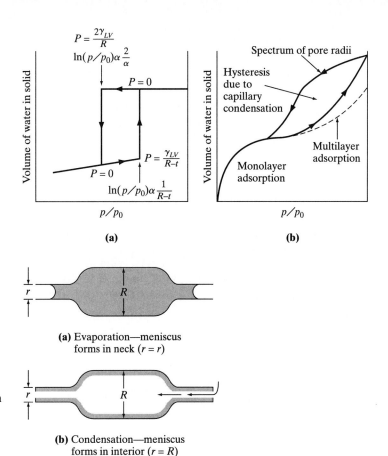

Figure 4.10
Cycling of relative vapor pressure in a porous solid: (a) single-sized pores; (b) range of pore sizes.

(a) Evaporation—meniscus forms in neck ($r = r$)

Figure 4.11
Evaporation-condensation of pore with a constricted neck ("ink-bottle" pore): (a) emptying, and (b) filling.

(b) Condensation—meniscus forms in interior ($r = R$)

All pores larger than this are filled in the rh range 0.97 to 1.0. However, it must be remembered that the values of r calculated by the Kelvin equation are only approximate values because an ideal cylindrical geometry must be assumed. Also, if pores are "ink bottle" shaped (i.e., they have restricted openings), then they behave as small pores on emptying and large pores on filling (see Fig. 4.11).

The lowering of the vapor pressure of a liquid in a capillary is an indication that the water molecules have a higher binding energy than they do in bulk water. This is the result of the additional interactions with the surface molecules of the walls, which now become an appreciable fraction of the total binding energy. Molecules not only less easily escape to the vapor state, but also are less easily ordered to form the solid state. Thus, the freezing point is lowered, too, by an amount proportional to the size of the capillaries. Water in capillaries greater than 30 nm diameter freezes very close to the normal freezing point, but in pores of 0.25 nm diameter the freezing point is $\sim -70°C$. Water in smaller pores will never freeze.

4.7   ADHESION

Surface films are most important in enhancing adhesion between two solid surfaces. Even highly polished surfaces have irregularities that are large on an atomic scale so that when put together, contacts are made only at a few points (Fig. 4.12a). High pressures at these points are likely to cause elastic deformations that outweigh the van der Waals attraction at these points. When the surfaces are wetted by an intermediate

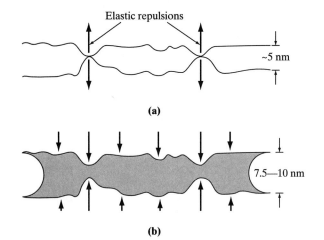

Figure 4.12
Interaction between surfaces: (a) dry surfaces in contact; (b) surfaces separated by a liquid film.

liquid film, capillary forces hold the two surfaces together at all points (Fig.4.12b). Considering the space to be a slit-shaped capillary between two plates and using Eq. 4.15 (r_2 is ∞), we get

$$P_{\text{adhesion}} = \frac{2\gamma_{\text{LV}}}{r},$$ (4.18)

where r is half the mean distance between the plates. Thus, the forces needed to overcome the pressure exerted on the film are

$$F = P \cdot A = \frac{2\gamma_{\text{LV}} A}{d} N$$ (4.19)

where A is the area of contact and d is the distance between the plates ($d \approx 2r$). Equation 4.19 emphasizes the importance of a high γ_{LV} and a low d (i.e., plates close together). In the same way, it can be shown that the water films between small particles will hold them together, requiring a force

$$P = 4\pi\gamma_{\text{LV}}(R_1 + R_2)$$ (4.20)

to separate them. Thus, damp powders become sticky and can be load bearing (e.g., sand beaches when partially saturated).

If a water film 1 μm thick holds two pieces of glass together, the force to separate them will be 72,000 N $\cdot$ mm^{-2} calculated from Eq. 4.18. However, the low shear strength of the film means that the plates can be slid apart easily. Polymers are commonly used as adhesives because the increased forces within the film (through entanglement of polymer chains as solvent evaporates or polymerization occurs) improve the shear strength of the bond.

4.8 COLLOIDS

4.8.1 Structure of Colloids

Colloids are forms of a material in such a finely divided state that the atoms at the surface become an appreciable fraction of the total number of atoms, so that the "bulk" properties of the material also reflect surface behavior. The geometry is important, as expressed by the surface to volume ratio $= B/r$, where r is the radius of average dimension and B is a constant depending on the geometry of the particle

(3 for spheres, and 6 for cubes). If r is less than 1 μm, then the surface area becomes very high and, as a result, the forces of gravity are less than the surface interactions between particles and the thermal motion of the particles (Brownian motion).

We can define a colloid as a highly dispersed phase with a mean particle dimension of less than 1 μm. Each kind of phase (solids, liquids, or gases) can exist in a colloidal form and may be dispersed in either of the other phases. Examples are given in Table 4.2. Stable dispersions of colloidal solids in a liquid are called *sols*. If sols interact to form a 3-D network, a *gel* is formed. Unlike a sol, a gel can behave as a high viscosity fluid or even a solid. Dispersions of immiscble, colloidal-sized liquid droplets in another liquid are called *emulsions*. In construction, we are concerned only with three kinds of colloidal dispersion: calcium silicate hydrate in hydrated cement paste (a solid gel), clay-water suspensions used in grouts (a high-viscosity liquid), and bitumen emulsions. A major concern is emulsion stability. Particles may aggregate and separate out from the dispersion media, either by coalescence of liquids or droplets, or by flocculation of solids. Coalescence of emulsions depends on many factors, with the most important being the interfacial surface tension (lowered by a surfactant), number and size of droplets, viscosity of the emulsion, and density differences between phases. Flocculation of the solid may be partial or complete: Hydrated calcium silicate can be regarded as a completely flocculated colloidal solid; flocculation of clays gives thixotropic properties that are useful in grouts.

TABLE 4.2   Classification of Colloids

Dispersed Phase	Surrounding Phase	Type	Examples
Gas	Liquid	Foam	Soapy water, whipped cream
	Solid	Solid Foam	Air-entrained concrete
Liquid	Gas	Aerosol (Mist or Fog)	Atmospheric smogs Aerosol sprays
	Liquid	Emulsions	Emulsified bitumens
Solid	Gas	Smokes	Silica fume, fly ash
	Liquid	Solid	Cement paste Clay/water suspensions
	Solids	Dispersed solids	ThO in nickel

4.8.2 Stability of Colloids

Although this brief discussion of the factors affecting colloid stability will be directed toward dispersions of colloidal solids in water, the principles are applicable to all colloids in general. Colloidal particles will agglomerate (flocculate) if attractive forces predominate as two particles approach each other. Four major types of interaction can occur:

1. Van der Waals attraction;
2. Electrostatic attractions between charged particles of opposite signs (not common);
3. Electrostatic repulsions between double layers of charged particles; and
4. Repulsion between adsorbed layers.

Mathematical expressions of the forces as a function of the interparticle distance (r) give a Condon-Morse type of curve for the colloidal dispersion (Fig.4.13).

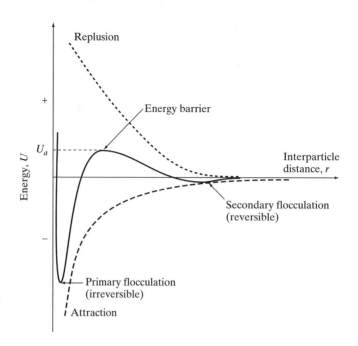

Figure 4.13
Potential energy curves for the interaction of two charged spheres.

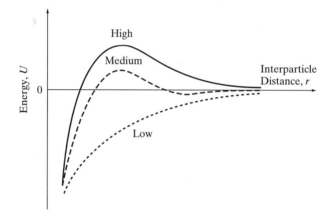

Figure 4.14
Effect of electrolyte concentration on the Condon-Morse curve and the flocculation of a colloid.

Two minima may occur: a shallow one at large r (the secondary minimum), and the deeper primary minimum at lower values of r. The secondary minimum will cause flocculation if its depth is greater than the thermal energy of the particles, but the system can be readily dispersed by shaking or stirring. If the energy barrier between the minima can be surpassed, then primary flocculation will occur, and much greater energy is needed to redisperse the colloid. Irreversible flocculation may occur. The energy barrier can be surmounted by increasing the thermal energy (temperature) of the system or by altering the electrolyte concentration (Fig. 4.14). Even though the particle may apparently be electrically neutral, there will be positive and negative charges on different parts of the surface. These may reflect the underlying structure, be caused by the presence of impurity ions, or reflect defects in atomic ordering. Good examples are clays, which, because of their layer structure, form thin platelets with residual negative charges on the surfaces of the plates and positive charges on the edges (Fig. 4.15).

Figure 4.15
Flocculation in clay-water systems: (a) enlarged view of single clay particle showing location of surface charges; (b) edge-face interaction between two clay particles; (c) partial flocculation into discrete flocs; (d) complete flocculation.

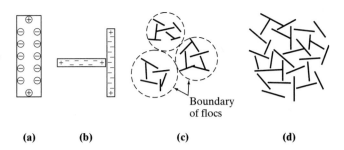

(a) (b) (c) Boundary of flocs (d)

4.9 THE DOUBLE LAYER

If ions are present in the solution (and they may come from the solid itself), then an increased number of ions of the opposite charge (counterions) will be found in the vicinity of the surface of the particle. This enhancement falls off gradually as the distance from the surface increases until a point is reached at which the surface charge "seen" is zero and a normal distribution of ions is found in the solution. This charge distribution is shown in Fig. 4.16a. If an external potential is applied to the suspension, then the particles will move along the electric field gradient. Because of the interactions between the ions and the surface, some of them (with their solvate molecules) will move with the solid. This boundary is called the *slipping plane* (or *Stern plane*), and that part of the solution moving with the solid is the *Stern layer*. The remaining, static part of the solvent with an enhanced ion distribution is the *diffuse layer*. Together, the Stern layer and diffuse layer form the *double layer*. Experimentally, only the potential of the Stern layer (the zeta potential, ψ_z) can be measured, and it is lower than the surface charge (ψ_s). The drop from ψ_s to ψ_z is sharp, while in the diffuse layer it decays exponentially to zero (Fig. 4.16a). The thickness of the Stern layer is about the diameter of the solvated ions at closest approach, while the diffuse layer is much thicker. Since the double layer forms spontaneously, it must lower the energy of the system; it can be shown that this charge is a function of ψ_z. The overlapping of double layers on adjacent particles would thus destroy part of each layer, thereby raising the energy of the system. Hence double layers repel colloidal particles.

At higher ionic concentrations, when multivalent ions are present or when specific adsorption of ions occurs (or all three), the Stern layer is compressed because the solvent sheath is tighter around the ions at the surface, and the zeta potential will be lower. Thus, the double layer is "collapsed" (Fig. 4.16b) and the particles can come closer together before double layer overlap causes repulsion (Fig. 4.17b). This may allow van der Waals or coulombic attractions to occur, causing flocculation. Figure 4.15 shows the characteristic edge-face coulombic attractions of clay particles.

If the particles are very small or the solids content is high, the suspension may gel because the particles form a complete network of particles throughout the suspensions (Fig. 4.15d). Since the attractive forces holding the particles together are weak, they can easily be broken down by the shearing forces of mixing. If the flocculated structure slowly reforms on standing, the suspension is said to be *thixotropic*. Cement paste and fresh concrete show partial flocculation (Fig. 4.15c), because although a gel is not formed, some shearing force is needed to initiate flow (Chapters 7 and 12).

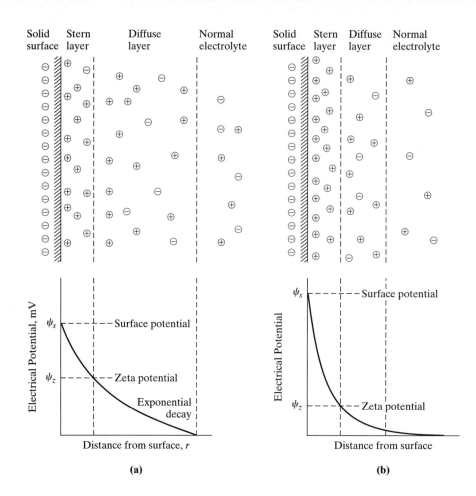

Figure 4.16
Double layer adjacent to charged solid surface: (a) diffuse (low electrolyte concentration); (b) condensed (high electrolyte concentration or surface charge).

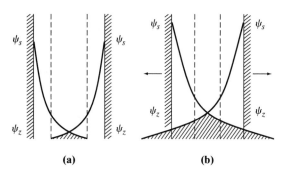

Figure 4.17
Overlap of double layers on adjacent solid surfaces: (a) condensed layers, low overlap and negligible repulsion; (b) diffuse layers, high overlap and high repulsion.

BIBLIOGRAPHY

S. LOWELL and J. E. SHIELDS, *"Powder Surface Area and Porosity"*, 2nd Ed., Chapman & Hall, 1984, 234 pp.

S. J. GREGG and K. S. W. SING, *"Adsorption, Surface Area and Porosity"*, 2nd Ed., Academic Press, 1982, 303 pp.

A. W. ADAMSON, *"Physical Chemistry of Surfaces"*, Wiley-Interscience, 1990, 777 pp.

PROBLEMS

4.1. Calculate the pressure inside a bubble of radius 100 nm in water (γ water $= 400$ mJ $\cdot$ m^{-2}). Will the bubble be stable? If a surfactant is added that reduces the surface tension to 28 dynes/cm how will this affect bubble stability?

4.2. Why doe the presence of adsorbed water on the surface of quartz (SiO_2) lower its surface energy?

4.3. Describe the difference between Langmuir and BET adsorption isotherms and the reason for this difference.

4.4. Describe the difference between an emulsion and a foam. Both are unstable dispersions; Why? How can they be stabilized?

4.5. Calculate the capillary rise above ground water level in a concrete column which has a mean continuous pore radius (i.e. all pores are connected) of 50 μm. How does this compare with a brick column? (Mean pore radius 2.5 μm. ($\gamma = 400$ mJ $\cdot$ m^{-2})

4.6. Determine the relative humidity required to remove water from a cylindrical pore of diameter 4 nm. [$\gamma = 400$ mJ $\cdot$ m^{-2}].

4.7. Describe the difference between the Stern layer and the double layer. How does the double layer influence flocculation?

Part II

BEHAVIOR OF MATERIALS UNDER STRESS

5

Response of Materials to Stress

When a solid material is subjected to external loading, it deforms instantaneously. In general, as long as the stresses are relatively small, this deformation is reversible, so when the load is removed, the material will return to its original dimensions. Such deformations are termed *elastic,* and they correspond to the region of small displacements from the equilibrium position of the atomic force-displacement curve (Fig. 1.6). For most structural materials subjected to loads within their elastic range, the relationship between stress and strain is *linear* (or nearly so), as shown in Fig. 5.1a. However, there are also important classes of materials which exhibit *nonlinear* elastic behavior, such as elastomeric materials, as shown in Fig. 5.1b.

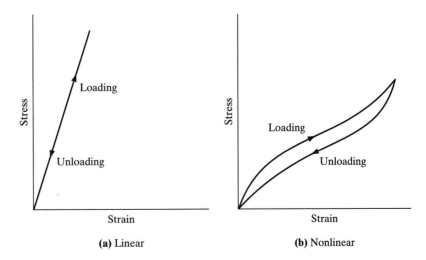

Figure 5.1
Elastic stress-strain diagrams: (a) linear, (b) nonlinear.

(a) Linear

(b) Nonlinear

Unfortunately, the response of a material to external loads cannot be predicted with sufficient accuracy merely from a knowledge of its atomic or molecular structure; empirical tests are needed to determine its mechanical properties. In this chapter, the response of solid materials to applied loads will be examined for the most common states of loading. Some of the tests required to quantify these mechanical properties will also be described.

5.1 TENSION

When a material is loaded in tension, it elongates. There are two parameters which characterize this behavior. The first is the axial stress, σ, which is defined as the total applied load, P, divided by the original cross-sectional area, A_0, of the specimen:

$$\sigma = P/A_0. \tag{5.1}$$

Stress is usually expressed in units of Pascals (1 Pa = 1 N/m²). The second parameter is the axial *strain, ε*, which is the elongation, δ, per unit length of the specimen:

$$\varepsilon = \delta/l_0, \tag{5.2}$$

where l_0 is the original length of the specimen. Note that strain is a dimensionless quantity.

In addition to the axial strain, a member will also contract laterally under tensile loading. The ratio of the transverse strains (ε_y and ε_z) to the axial strain (ε_x) is known as Poisson's ratio, ν, and is expressed as

$$\nu = -\frac{\varepsilon_y}{\varepsilon_x} = -\frac{\varepsilon_z}{\varepsilon_x}. \tag{5.3}$$

The negative sign is used because, under tensile loading, the transverse strains are contractions. For most metals, $\nu \cong 0.33$; for other engineering materials, ν ranges from about 0.16 (wood) to 0.50 (ideal elastomeric material, such as rubber). Axial loading of a material also induces a volume change. Consider a solid circular bar of original length, l, and diameter, d. When it is loaded in tension, it elongates by an amount $\delta = \varepsilon_x l$, while its diameter is reduced to $d - \nu\varepsilon_x d$. The original volume, V_0, is $V_0 = l\pi d^2/4$. The new volume, V_1, is

$$V_1 = (l + \varepsilon_x l)\frac{\pi}{4}(d - \nu\varepsilon_x d)^2 \tag{5.4}$$

$$= \frac{l\pi d^2}{4}(1 + \varepsilon_x - 2\nu\varepsilon_x + \varepsilon_x^2(\nu^2 - 2\nu) + \nu^3\varepsilon_x^3).$$

For small values of ε_x (< 0.01), we can ignore the terms containing ε_x^2 and ε_x^3. Then

$$V_1 = \frac{l\pi d^2}{4}(1 + \varepsilon_x(1 + 2\nu)). \tag{5.5}$$

Therefore, the unit volume change is

$$\frac{V_1 - V_0}{V_0} = \varepsilon_x(1 - 2\nu). \tag{5.6}$$

As long as ν is less than 1/2, this value is always positive, and so tensile loading always induces an increase in volume. For $\nu = 1/2$, there is no volume change; this is the case for metals in the plastic range and for rubber.

5.1.1 Elastic Behavior

For *linearly elastic materials* (Fig. 5.1a), the relationship between stress and strain can be described by *Hooke's law*:

$$\sigma = E\varepsilon, \tag{5.7}$$

where E is the *modulus of elasticity,* sometimes known also as *Young's modulus.* Generally, Hooke's law applies only to a relatively small range of strains. For large strains, the relationship between stress and strain is no longer linear, and Hooke's law no longer applies.

For materials which exhibit *nonlinear elastic* behavior (Fig. 5.1b), Hooke's law does not apply, since the stiffness of the material is no longer constant but varies with load. However, because the concept of a proportionality between stress and strain is such a powerful one, a number of different elastic moduli have been defined which can help to characterize the stress-strain behavior of such materials. These are defined in Fig. 5.2. The initial tangent modulus, which is the slope of the σ-ε curve at the origin, applies only to small stresses and strains. A more common measure of the stiffness is the secant modulus, which is the slope of the line joining the origin and any arbitrary point on the σ-ε curve. Clearly, its value depends on the level of applied stress chosen. The *chord modulus* is the slope of the line between any two arbitrary points on the σ-ε curve. It is often used instead of the secant modulus because it is difficult to measure the beginning of the σ-ε curve accurately. However, if one is interested in the additional strain that occurs when an additional stress is imposed on the already loaded material, the *tangent modulus* measured at the point of interest is a better measure of a material's response to small additional stresses. Elastic moduli are not generally measured on the unloading branch of the σ-ε curve.

As is the case with linearly elastic materials, it is also possible to define the strain energy per unit volume stored in a nonlinear elastic material by determining the area under the σ-ε curve. If the material exhibits *hysteresis* on unloading, as shown in Fig. 5.1b and 5.2, then not all of this energy is recoverable on unloading.

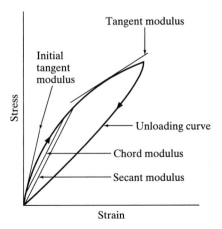

Figure 5.2
Different elastic moduli that may be defined for a nonlinearly elastic material.

Some of it may be lost as internal friction or some structural rearrangement within the material.

5.1.2 Inelastic Behavior

Elastic behavior may end in one of two ways, by *fracture* or by *yielding* of the material. In neither case will the specimen return to its original form when it is unloaded.

Fracture

The fracture of brittle materials (i.e., those that fail before any significant amount of yielding occurs) is due to the fact that flaws or imperfections in the material can raise the stress in some highly localized regions to a high enough value that the local cohesive strength is exceeded. This will cause a crack to start to grow, and its spread will bring about fracture. The mechanics of the fracture of brittle materials will be discussed in detail in Chapter 6. In any solid material, flaws and imperfections—either those inherent in the structure of the material or those induced by the processing of the material—tend to be randomly distributed in both orientation and size. Therefore, fracture strengths show an appreciable amount of scatter, so that for design purposes it is necessary to look at the *statistical distribution* of fracture strengths rather than simply average values. Indeed, this concept is already built into most of the building code specifications for both concrete and timber. The random distribution of flaws also leads to a considerable *size effect* in terms of fracture strengths, which means that the larger the test specimen, the more likely it is to have flaws of greater severity; hence, the more likely it is to exhibit a lower fracture strength. This effect must be taken into account when testing brittle materials such as wood or concrete, since the effects on strength can be large.

Yielding

The second way in which elastic behavior can be terminated is by *yielding* of the material. Yielding involves a permanent deformation of the material, which is often referred to as *plastic deformation*. This type of behavior is associated, in particular, with crystalline or polycrystalline materials, most notably metals. Yielding implies that there has been a permanent structural rearrangement within the material (i.e., that under tensile loading some atoms in the crystal structure have moved to new equilibrium positions, in which they have formed new bonds, with no tendency to return to their original positions). Hence, deformations beyond the elastic range are at least partly irrecoverable. In many crystalline materials, like metals, the end of the elastic range and beginning of some plastic behavior occurs at the point when the σ-ε curve deviates from linearity.

The idealized stress-strain diagram for such a material is shown schematically in Fig. 5.3. If the specimen is unloaded at some time after it has yielded, the unloading curve will be parallel to the original loading curve, since the same type of interatomic bonds as existed in the elastic region are still there. While there is some *elastic recovery,* a portion of the total deformation is irrecoverable; this is known as *permanent set.* If the specimen is then reloaded, it will follow the unloading line until the previously applied maximum load is reached, at which point further yielding may take place. In practice, however, loading and unloading are not completely reversible, and an unloading-reloading cycle once the specimen has yielded will show a slight hysteresis loop. A yielding type of behavior induces a mechanism which enables the material to support larger strains prior to failure. This is often referred

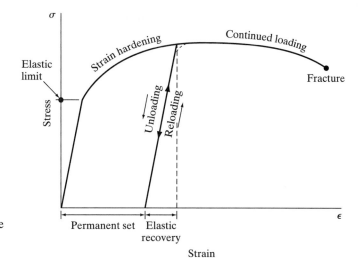

Figure 5.3
Typical σ-ϵ diagram for a
ductile material. The un-
loading branch of the curve
is parallel to the loading
branch.

to as *ductility* or *ductile behavior*, in contrast to *brittle behavior*, in which fracture occurs at relatively low strains. The quantification of ductile and brittle behavior will
be dealt with in Chapter 6.

Yield Mechanism

Yielding is a *shear* phenomenon. For a material subjected to a uniaxial tensile
stress, σ_t, shear stresses exist on all planes inclined to the axis of loading, as shown
in Fig. 5.4. If we consider a plane whose normal makes an angle θ with the tensile
axis, then the resolved shear stress, τ, on that plane may, in general, be given by

$$\tau = \sigma_t \cos\theta \cos\phi, \qquad (5.7)$$

where ϕ denotes the angle (on the inclined plane) between the tensile force and the
direction of interest. For a given tensile stress, the maximum value of τ occurs when
both ϕ and θ are equal to 45°, giving $\tau_{max} = \sigma_t/2$.

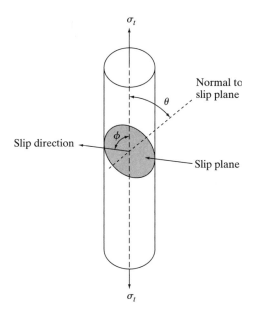

Figure 5.4
Geometry of inclined plane
in a specimen under uniaxial tension, for determination of the resolved shear
stress.

In single crystals, yielding occurs when the shear stress along some particular crystallographic plane (i.e., the *shear plane*) reaches a critical value, the *critical resolved shear stress*, τ_{cr}. Equation 5.7 is also known as Schmid's law, which embodies the empirical observation that, for single crystals, τ_{cr} is a material constant at a given temperature. The most common yield mechanism for crystalline materials is *slip*, and in metals it is associated with movement of dislocations (Chapter 2). In *polycrystalline* materials, in which there is a random orientation of individual crystals (or *grains*), yielding is a much more complicated process, arising from the presence of *grain boundaries* and from constraints on the deformation of an individual grain due to the presence of neighboring grains. To avoid failure at the grain boundaries, the deformation of each grain must be compatible with the deformations of its neighboring grains. Indeed, unless the regions around the grain boundaries can undergo rather complex shape changes, voids may be opened up at the grain boundaries.

Since each grain in a polycrystalline material has a different crystallographic orientation, each will respond differently to an applied stress. In general, in polycrystals, slip will begin first in grains that are most favorably oriented with respect to the maximum shear stress and then will progress to successively less favorably oriented crystals. During deformation, each individual grain rotates so that its active slip direction becomes more nearly parallel to the tensile axis and the grains elongate in this direction. In addition, there will be movement along the grain boundaries. Since yielding does not occur throughout the material at the same time, the yield point on the σ-ε curve is less sharply defined. The yield strength of a polycrystalline material is considerably higher than that of a single crystal of the same material, since many different slip systems must become active before generalized yielding can occur. The grain boundaries act as barriers to the movement of dislocations, which further delays the onset of plastic deformation. Thus, polycrystallinity and a decrease in grain size increase the strength and hardness of materials.

In *amorphous* materials, a variety of mechanisms are responsible for yielding. In the case of thermoplastic polymers, yielding is due to the slip of long-chain molecules past each other. However, since this process is highly time dependent, the response of these materials depends on the rate of loading.

Strain Hardening

Almost all of the mechanical properties of a material, except *E,* are affected by plastic deformation. One consequence of yielding is that, as plastic deformation proceeds, an increase in stress is required to produce additional strain. This is called *strain hardening* or *work hardening*. It may be seen by reference to Fig. 5.3; on reloading, yielding will not take place until a higher stress than the original elastic limit is reached. Strain hardening is related to the increase in dislocation density that occurs as slip progresses; it is due to the interactions of the stress fields associated with the dislocations. In polycrystalline materials, both crystal boundaries and the presence of foreign atoms act as additional obstacles to dislocation movement. Strain hardening is not common to polymers, although it does occur when the material can be oriented into a fiber-like structure.

Necking in Tension

Under tensile loading, materials which undergo yielding eventually begin to *neck*— that is, they exhibit a *plastic instability,* where the cross-sectional area in one part of the specimen is significantly reduced during the course of a test due to the fact that yielding becomes localized in this region. If the material continues to be strained, failure will occur there. The necking of a mild steel specimen is shown in Fig. 5.5, at

Figure 5.5
Necking in a mild steel tensile specimen, at a conventional strain of about 15%.

a conventional strain of about 15%. What happens physically is that, as the material begins to yield, the cross-sectional area in the necked region decreases slightly. Initially, this is more than compensated for by the strain hardening that is occurring simultaneously, and so the conventional σ-ε curve continues to rise. Eventually, however, the rate of strain hardening decreases, and the strain hardening can no longer compensate for the reduction in cross-sectional area. This corresponds to the point of maximum stress in the conventional σ-ε curve. Beyond this point, the specimen becomes unstable with additional strain, and failure occurs.

5.1.3 Definitions of Stress and Strain

During *elastic deformation,* the dimensional changes in the material are negligible; so the definitions of stress and strain given by Eqs. 5.1 and 5.2, based on the *original* cross-sectional area and length, are perfectly adequate. Such stresses and strains are referred to as *engineering,* or *conventional,* stress and strain. However, during plastic deformation, both the reduction in cross-sectional area and the elongation of the specimen can be appreciable. To represent better the behavior of materials during plastic deformation, it is now common to use different definitions of stress and strain, which are generally referred to as "true stress" and "true strain" and which are based on the *instantaneous* dimensions of the specimen during the course of a test. *True stress* is defined as

$$\sigma' = P/A', \tag{5.8}$$

where A' is the actual cross-sectional area (taking into account the Poisson effect) corresponding to the load P.

The basis of the definition of *true strain* is that any increment of strain is the incremental change in length, dl, divided by the instantaneous length, l (rather than the original length, l_0). The true strain is then the sum of all of these instantaneous strains:

$$\varepsilon' = \int_{l_0}^{l} \frac{dl}{l} = ln\left(\frac{l}{l_0}\right). \tag{5.9}$$

It can be shown that

$$\varepsilon' = ln(1 + \varepsilon). \tag{5.10}$$

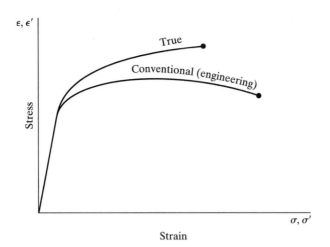

ϵ, ϵ'

Stress

True

Conventional (engineering)

σ, σ'

Strain

Figure 5.6
Comparison of true and
conventional stress-strain
curves in tension.

Once *necking* has occurred, it is no longer meaningful to consider the total length of the specimen, since necking is a localized phenomenon. However, measurement of the minimum area in the necked region of the specimen provides a basis for determining the true strain, since $\varepsilon' = ln(A_0/A)$, where A_0 is the original cross-sectional area. A schematic representation of the differences between conventional and true stress-strain curves is given in Fig. 5.6.

5.1.4 Experimental Determination of Tensile Properties

Specimen Geometry

Specimens for tensile tests on metallic or other reasonably ductile materials may be either cylindrical or prismatic in cross section. The dimensions will vary according to the type of material and the procedure being followed. Typical direct tension specimens are shown in Fig. 5.7; in all cases, the ends of the specimens are enlarged relative to the central, or reduced, section. This geometry reduces the stress concentrations in the central portion caused by gripping the specimen in the test system. The gauge length should be no closer than one specimen diameter (or width) to the point of the change in cross section. If *eccentricities* in loading are introduced by poor specimen fabrication or improper gripping techniques, then a *bending moment* will also be introduced in the specimen, in addition to the axial load. The maximum tensile stress will then be increased to

$$\sigma_{max} = \frac{P}{A} + \left(\frac{Mc}{I}\right) \tag{5.11}$$

where M is the bending moment, c is the distance from the neutral axis to the outer fibers of the specimen, and I is the moment of inertia. For a *cylindrical* specimen, it can easily be shown that an eccentricity of 1% will lead to an 8% error in the maximum stress.

For relatively ductile materials such as steel, using specimens with enlarged ends is usually sufficient to overcome the problem of the stress concentrations induced by the gripping device. However, for brittle materials such as glass or concrete, the stress concentrations in direct tension tests are of such a magnitude that failure at the grips can be extremely difficult to avoid. Therefore, a number of indirect tensile tests have been developed, in which a uniform, uniaxial tensile stress is induced by the application of other types of loading.

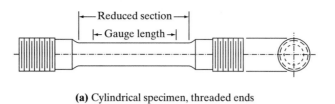

(a) Cylindrical specimen, threaded ends

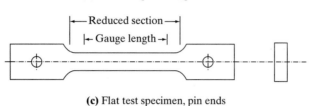

(b) Flat test specimen, plain ends

Figure 5.7
Typical tensile specimen geometries: (a) cylindrical specimen, threaded ends; (b) flat test specimen, plain ends; (c) flat test specimen, pin ends.

(c) Flat test specimen, pin ends

The most common such test used in civil engineering applications is the splitting tension test, which is more properly referred to as diametrical compression of a solid cylinder (also sometimes known as the Brazil test). This test utilizes the secondary tensile stresses which are developed when a solid cylinder is subjected to diametrical loading, as shown in Fig. 5.8. The maximum tensile stress, which acts across a substantial part of the loaded diameter, is given by

$$\sigma_{\text{max tensile}} = \frac{2P}{\pi D L}, \tag{5.12}$$

where P is the diametrical load, D is the specimen diameter, and L is the length of the cylinder. As can be seen in Fig. 5.8, about 80% of the vertical diameter is

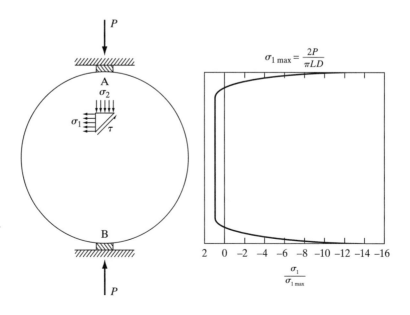

Figure 5.8
Indirect tension test on brittle materials: stress distribution across loaded diameter for cylinder in diametrical compression (splitting tension test, Brazil test).

essentially in pure tension. To avoid stress concentrations along the line of loading, a narrow, relatively soft pad is used to help distribute the load. This test is most commonly used for concrete.

Test Parameters

As with all tests of mechanical properties, the measured tensile properties of a material will depend to some degree on how the test is carried out. It is beyond the scope of this book to examine test procedures in detail. However, four experimental variables should be noted:

1. *Surface imperfections:* Surface imperfections, due to improper specimen preparation or handling or due to corrosion, may cause stress concentrations. This effect is most serious for brittle materials.
2. *Rate of loading:* In general, the more rapidly the specimen is loaded, the higher the apparent yield point and maximum stress, and the higher the elastic modulus (see Chapter 6). The material will also behave in a more brittle manner. Most test methods specify both a maximum and a minimum loading rate.
3. *Temperature:* At higher temperatures, the ductility increases; the yield strength, fracture strength, and modulus of elasticity are decreased.
4. *Specimen Size:* As stated earlier, the larger the specimen, the more likely it is to contain severe flaws of one kind or another. Therefore, larger specimens tend to give lower ultimate strengths.

5.2 COMPRESSION

Ideally, the compression test consists of applying an axial compressive force to the ends of a specimen; a uniform compressive stress, *P/A,* results. Most of the considerations involved in tensile testing, such as specimen size, rate of loading, and so on, apply equally to compression testing. Generally, in the elastic range, the σ-ε curve for compression is simply a linear extension of the σ-ε curve for tension, and the values of E and ν are the same. However, it is difficult to bring about compressive failure in ductile materials, since they tend simply to deform rather than break; such materials are better characterized by tensile behavior. Compressive behavior is, therefore, used primarily to characterize brittle materials. The axial stress required to cause failure in a brittle material is much greater in compression than in tension. Cracks or flaws that initiate failure in tension tend to close up in compression and no longer act as stress raisers. Therefore, higher applied loads are needed to cause failure; the failure mechanism is often due to secondary tensile stresses acting perpendicular to the applied compressive stress.

As a material is loaded in compression, the cross-section area increases because of the Poisson effect; therefore, the engineering σ-ε curve in compression lies above the true σ-ε curve.

5.2.1 Experimental Determination of Compressive Properties

Compressive tests are usually carried out on cylindrical or cube specimens which are loaded on two opposite faces. Although simple to execute, there are two main problems to be overcome; these are discussed next.

Difficulty in Applying a Truly Axial Load

To apply pure uniaxial compression, the ends of the specimens must be both *plane* and *parallel*. If the ends cannot be machined or ground to be smooth enough, then an *end cap* should be provided for the specimen, using an appropriate capping jig and a quick-setting capping material, such as high-strength gypsum or sulphur. Very soft end caps, such as rubber pads or soft fiberboard, should be avoided since they will flow laterally (outward) under load, superimposing tensile stresses at the ends of the specimen, which in extreme cases may cause a tensile splitting failure perpendicular to the direction of loading. For most specimens, a small amount of nonparallelism of the ends is inevitable. Therefore, the testing machine should be fitted with a spherically seated bearing block so that the load can be evenly distributed across the ends of the specimen.

End Restraint

The most difficult problem in trying to apply a pure uniaxial compressive stress is to overcome end restraint. End restraint is caused by the lateral expansion of a specimen that accompanies axial compression. As a specimen becomes shorter under the load from the bearing platens of the machine on its ends, its diameter increases due to the Poisson effect ($\varepsilon_x = -\nu\varepsilon_y$). However, expansion of the ends of the specimen is hindered by the frictional forces that develop between the specimen ends and the platens. This tends to hold the ends of the specimen near their original dimensions, while the center portion expands laterally, resulting in a barrel-shaped specimen, as shown in Fig. 5.9.

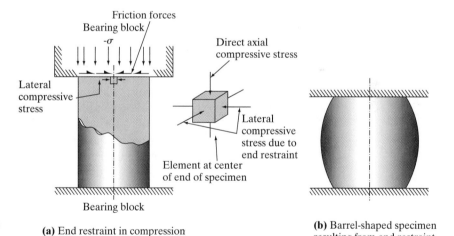

Figure 5.9
The effects of end restraint in the compression test.

(a) End restraint in compression

(b) Barrel-shaped specimen resulting from end restraint

This end restraint produces transverse compression near the specimen ends. The transverse compression is greatest right at the ends and then diminishes toward the center of the specimen; pure uniaxial compression occurs only beyond a distance of about one specimen diameter from the ends. The net effect is to produce an apparently higher compressive strength; this is particularly noticeable in shorter specimens. For instance, for concrete, reducing the l/d ratio from the standard 2.0 to 1.0 increases the apparent compressive strength by about 20%. There is no way to eliminate end effects completely, though special platens have been developed for particular materials which reduce these effects. If the specimen is too long, on the other hand, buckling may occur. For instance, when wood is tested in compression parallel to the grain direction, failure may occur by local buckling of the long individual fibers that comprise the material.

5.3 BENDING

In bending tests carried out primarily on nonmetallic materials, members are loaded so that one side is elongated while the other side is compressed. This is, therefore, a *nonuniform* stress test; on every cross section of the member, there are normal stresses that vary from tension on one side to compression on the other. In the simple case of pure bending, only these normal stresses are present on a cross section. Making the usual assumptions of engineering beam theory, it can be shown that the stress distribution on any cross section is given by

$$\sigma = \frac{Mc}{I},\qquad(5.13)$$

where M is the bending moment, y is the distance above or below the neutral axis, and I is the moment of inertia. The strength of a flexural specimen, referred to as *flexural strength* or *modulus of rupture,* is given by Eq. 5.12, with y set equal to the distance from the neutral axis to the extreme outer fibers.

5.3.1 Behavior in Pure Bending

In pure bending, the graph of bending moment versus maximum strain (i.e., the strain at the outer fibers), M versus ε_{max}, looks like a conventional σ-ε diagram in tension (Fig. 5.10). The Poisson effect also occurs in bending, just as it does in tension or compression. This means that in the tension side of the member there is a lateral contraction while on the compression side there is lateral expansion. Since the normal strain, ε_x, varies linearly across the cross section, so too does the lateral strain, ε_z ($\varepsilon_z = -\nu\varepsilon_x$). This leads to a transverse curvature of the beam, "anticlastic" curvature (inset in Fig. 5.10), which is opposite in sense to the bending curvature (this effect can be easily observed by bending a soft rubber eraser).

If the σ-ε diagram for a material is nonlinear or if the material is stressed in the plastic range (i.e., beyond the proportional limit), then Eq. 5.13 no longer applies, and the relationship between the applied moment and the stress in the material must be determined from the material properties and for the particular stress distribution which is developed.

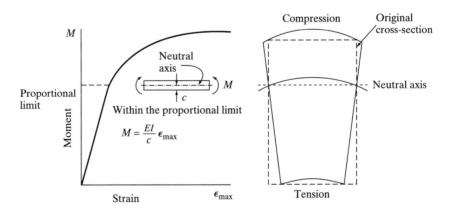

Figure 5.10
Bending moment, *M,* versus extreme fiber strain, ε_{max}. The inset shows the phenomenon of anticlastic curvature.

Chap. 5   Response of Materials to Stress

5.3.2 Failure in Pure Bending

Ductile Materials

The mechanism for *yielding* in bending, as it is in tension, is slip along planes in the direction of the maximum shear stress (i.e., inclined 45° from the longitudinal axis of the beam). Yielding first occurs locally, at the extreme outer fibers, when they reach the stress at which yielding would occur in simple tension (or compression). As the loading is further increased, yielding slowly progresses inward, toward the neutral axis, until the entire cross section has yielded. Thus, during a bending test, the nature of the stress distribution across the cross section changes gradually, as shown in Fig. 5.11. For a rectangular cross section, the *ultimate moment, M_u*—that is, the moment when the beam becomes fully plastic, as in Fig. 5.11d—is 1.5 M_y, (Fig. 5.11b). Because the yielding takes place gradually over the entire cross section, the departure from linearity of the M versus ε_{max} curve (Fig. 5.10) is also gradual, which makes it difficult to determine accurately the proportional limit in a bending test.

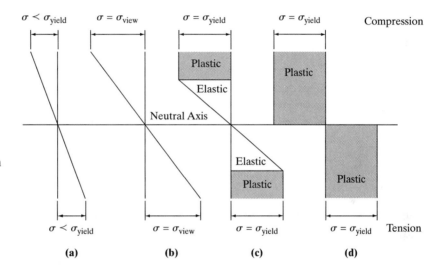

Figure 5.11
Stress distribution across the cross section of a beam in pure bending: (a) stress below yield stress in the outer fibers; (b) stress ≈ yield stress in outer fibers; (c) moment somewhat beyond the yield stress; (d) fully plastic section.

Brittle Materials

In brittle materials, the type of yielding shown in Fig. 5.11 does not occur. Instead, the stress distribution at failure tends to be nonlinear, as shown in Fig. 5.12. For brittle materials in bending, failure always is initiated as a tensile failure, since the compressive strength of brittle materials is much higher than the tensile strength. However, because of the nonlinearity of the stress distribution, the flexural strength (modulus of rupture) computed using the assumptions of engineering beam theory is considerably higher than the uniaxial tensile strength of the material. For instance, for concrete, the modulus of rupture is about 50% greater than its direct tensile strength.

5.3.3 Types of Bending Tests

There are two common bending test arrangements:

    1. Three-point (or center-point) bending (Fig.5.13a)
    2. Four-point bending (Fig. 5.13b).

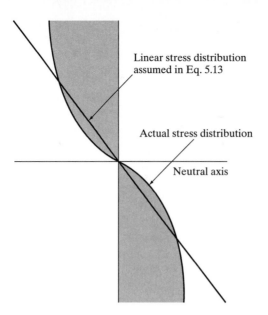

Linear stress distribution
assumed in Eq. 5.13

Actual stress distribution

Neutral axis

Figure 5.12
Stress distribution across
the cross section of a brittle
material near failure in
bending.

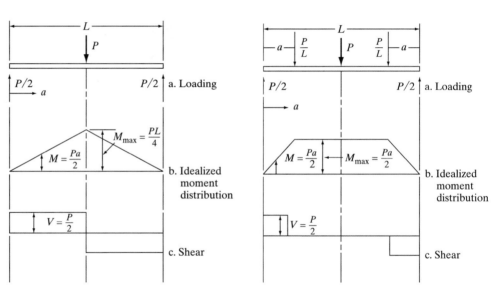

Figure 5.13
(a) Idealized loading mo-
ment and shear diagrams
for three-point bend test;
(b) idealized loading mo-
ment and shear diagrams
for four-point bend test.

The choice of which test to use depends on the material, the specimen dimensions, and the particular test specification being followed. The four-point bending test is generally preferred, since it provides a constant moment over a substantial length of the specimen. The shear stress in this portion of the beam is zero, which closely approximates the case of pure bending, on which engineering beam theory is based. For this reason, failures occurring outside of the inner loading points should be discounted. Most commonly, the loads are applied at the third points of the beam (third point loading), though any other spacings may be used. On the other hand, the bending moment in a three-point bending test varies along the length of the beam and has a maximum directly under the central load point. Now there are also shear stresses present everywhere in the beam, so that if failure does not occur right under the load point, the value of the maximum moment is uncertain. Although it is also necessary to assume that the effects of shear can simply be ignored, this test is easier to carry out, particularly on small specimens.

The three-point bend test generally gives higher flexural strength (modulus of rupture) values than does the four-point test. This is due primarily to the size effect, since in three-point bending only a very small volume of the beam is subjected to the maximum stress. In four-point bending, a much larger volume of material is subjected to the maximum stress. (Indeed, even for four-point bending, it has been found that as the inner load points are moved farther apart, the apparent bending strength decreases, again because of the size effect.) For instance, for wood, the difference in modulus of rupture between three-point loading and four-point loading is about 25%.

5.3.4 Limitations in Bending Tests

Specimen Size

Although specimen dimensions for bending tests vary greatly, depending on the material and the type of information being sought, there are limitations on specimen geometry which must be considered. In the foregoing analysis, shear effects have been ignored. However, if the ratio of length to depth (l/d) of the specimen is too low, then shear effects must be considered. For a rectangular beam under center-point loading, it can easily be shown that the ratio of the maximum shear stress (which occurs at the neutral axis) to the maximum bending stress (which occurs at the outer fibers) is

$$\frac{\tau_{max}}{\sigma_{max}} = \frac{1}{2}\frac{d}{l}. \tag{5.14}$$

Thus, to provide a beam in which the maximum shear stress is, say, less than 10% of the maximum bending stress, an l/d ratio of at least 5 should be used. Ideally, for shear effects to be ignored completely, the l/d ratio should be at least 10, though this is not always possible in practice. The major effect of the shear stresses is to increase the beam deflection (by superimposing shear deflection upon the bending moment deflection). This will make short beams appear to have a lower modulus of elasticity (E) than longer beams of the same material.

Sources of Error in Bending Tests

In addition to the aforementioned difficulties in interpreting bending test results, there are a number of sources of experimental error which must be considered. Loads on flexural specimens are usually applied through knife edges or small diameter rollers, and this gives rise to high localized stresses, both vertical and longitudinal, at the loading points. This may cause local crushing or, in extreme cases, failure by local shear. Failures which occur right at the load points in four-point bending tests should, therefore, be examined carefully. In three-point tests, failure occurs under the load point, and the effect of local stress concentrations is simply ignored. It has been shown that, for a rectangular beam with an l/d ratio of 10, the actual stress at midspan is about 4% less than that calculated from Eq. 5.13 due to the local longitudinal stresses caused by the loading device. In addition, because of the high localized stresses, the loading points tend to "dig in" slightly. This can induce frictional forces, as the loading device tends to restrain the changes in length of the beam surfaces.

If the load is not applied uniformly across the width of the specimen, it will tend to twist about its longitudinal axis. The resulting torsional stresses will then be superimposed on the bending stresses. If the beam is not symmetrically loaded, then an error in the computed moment will also occur.

A torque, or twisting moment, applied around the longitudinal axis of a specimen with a circular cross section results in *pure shear strain* acting on the cross section. No direct tensile or compressive strains occur in either the longitudinal or radial directions, and hence the overall specimen dimensions remain unchanged. The distortion of a body in torsion is shown in Fig. 5.14a; a schematic diagram of the distortion of the crystal structure in pure shear is shown in Figs. 5.14b and 5.14c. As long as the distortions are small enough that none of the atoms slip to new equilibrium positions, the action will be completely elastic.

In the following discussion, only specimens with circular cross sections are considered. While it is possible to carry out torsion tests on specimens with prismatic sections (and, of course, such shapes are frequently used in practical engineering projects), the deformations and stresses of such shapes under torsion are very complicated.

5.4.1 Stress and Strain Relationships in Torsion

To derive the elastic stress-strain relationships in torsion, we must make the following assumptions: (1) A plane section perpendicular to the axis of a circular shaft remains plane after torque is applied; (2) shear strains vary linearly from the longi-

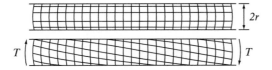

(a) Circular Shaft before and after torque T is applied

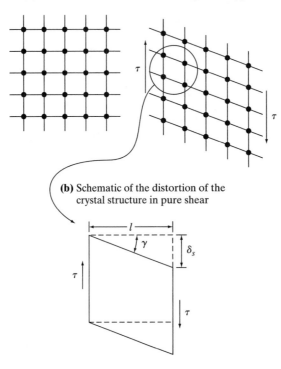

(b) Schematic of the distortion of the crystal structure in pure shear

(c) Shear distortion

Figure. 5.14
Torsional deformations:
(a) circular shaft before and after torque is applied;
(b) schematic of the distortion of the crystal structure in pure shear; (c) shear distortion.

Chap. 5   Response of Materials to Stress

tudinal axis to the outside of the shaft; and (3) shearing stress is proportional to shearing strain.

The *shearing strain,* γ, may be defined in terms of the tangent of the angle of rotation of originally perpendicular lines on the surface of the shaft. Referring to Fig. 5.14c,

$$\gamma = \frac{\delta_s}{l}. \tag{5.15}$$

From the theory of elasticity, the *maximum shearing stress,* τ, at the outside of the shaft is related to the twisting moment, T, and the specimen dimensions by

$$\tau = \frac{Tr}{J} \tag{5.16}$$

where r is the radius of the shaft and $J(= \pi r^4/2)$ is the *polar moment of inertia.* For a linearly elastic material, we get the Hooke's law relationship in torsion:

$$\tau = G\gamma, \tag{5.17}$$

where G is referred to as the *modulus of elasticity in shear* (or *shear modulus*). For orthotropic materials, it is related to the elastic modulus by

$$G = -\frac{E}{2(1 + \nu)}. \tag{5.18}$$

For metals, in which $\nu \cong 1/3$, $G \cong 3/8 E$.

The relative rotation, or *angle of twist* (in radians), between two ends of a solid shaft of length L under torsion is given by

$$\phi = \frac{TL}{GJ}. \tag{5.19}$$

The term GJ is referred to as the *torsional stiffness* of the shaft. In addition to the shearing stresses, a circular shaft in torsion also develops normal stresses. The maximum normal stresses, both tensile and compressive, are numerically equal to the maximum shear stress but act on planes inclined 45° to the axis of the shaft (and hence to the direction of the maximum shear stresses).

5.4.2 Failure in Torsion

Ductile Materials

As a ductile material is loaded in torsion, yielding will begin at the outer surface at some critical value of the shear stress, τ_{yp} (see Fig. 5.15a), which is approximately equal to one-half the yield stress in simple tension. As the twisting continues, the yield surface moves in toward the axis of the shaft (Fig. 5.15b), until the entire cross section has yielded (Fig. 5.15c). It can be shown that the twisting moment when the entire cross section has yielded is approximately equal to 4/3 of the twisting moment at the onset of yielding.

As is the case in bending, the departure from linearity of the τ versus γ curve is gradual and it may be difficult to define the proportional limit accurately. *Failure* eventually begins with the formation of an irregular circumferential shear crack at the outer surface. As twisting continues, the fracture surfaces rub against each other and the two pieces of the bar are slowly forced apart, producing the mode of failure shown in Fig. 5.16a.

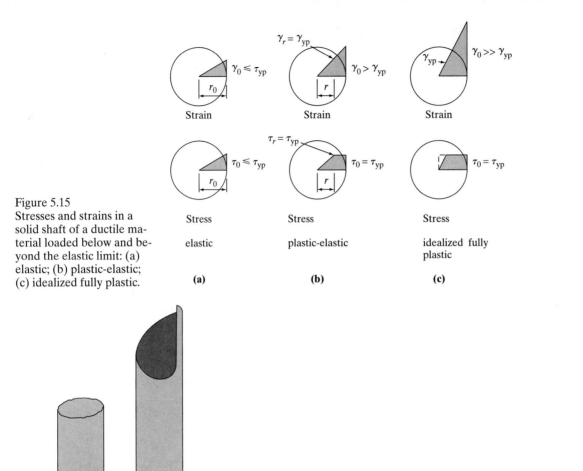

Figure 5.15
Stresses and strains in a solid shaft of a ductile material loaded below and beyond the elastic limit: (a) elastic; (b) plastic-elastic; (c) idealized fully plastic.

Strain | Strain | Strain
elastic | plastic-elastic | idealized fully plastic

(a) | **(b)** | **(c)**

Figure 5.16
Modes of failure in torsion: (a) solid bar of ductile material, plane fracture; (b) solid bar of brittle material, helicoidal fracture; (c) buckling of a thin-walled hollow tube.

Since yielding in pure shear takes place when the maximum tensile stress (acting at 45° to the maximum shear stress) is only about half that required to cause yielding in simple tension, materials yield more easily in pure shear, and yielding may continue for a much longer period before it is interrupted by fracture, which is governed by the tensile stress. Thus, materials appear to be much more ductile in torsion than in tension. For instance, in a torsion test carried out on a mild steel rod about 250 mm long and 25 mm in diameter, it is possible to obtain a relative rotation of one end with respect to the other of about 3600° (i.e., 10 complete rotations) before failure occurs, though yielding begins at a relative rotation of only about 50°.

Brittle Materials

Brittle fracture in torsion is governed by the maximum tensile stress and therefore occurs along planes inclined at 45° to the axis of the shaft. This results in a helicoidal fracture, as shown in Fig. 5.16b. This can easily be demonstrated by twisting a piece of chalk between the fingers. The tensile stress at failure in torsion is approximately equal to that in direct tension; this indicates that the compressive stresses acting perpendicular to the tensile stresses have little effect. These effects are considered in the Mohr-Coulomb criteria for failure in Chapter 6.

5.4.3 Test Methods in Torsion

Special testing machines are generally used for torsion tests. These machines are capable of applying a torque at one end of the specimen and of measuring the angular twist over the gauge length of the specimen. Specimens may be of any convenient size, but a circular cross section is preferable to simplify the analysis. The specimen ends are generally enlarged relative to the gauge section of the specimen, for the same reason that this configuration is used in tension tests. If tests are being carried out on hollow tubes, the ends must be plugged to avoid crushing in the jaws of the machine.

One major difficulty in interpreting torsion tests is the fact that the stresses and strains vary radially across the cross section of a solid specimen, from a maximum at the outside to zero at the center. To determine the proportional limit accurately, it is common to use thin-walled, hollow cylindrical specimens. If the ratio of specimen diameter to wall thickness is chosen appropriately (typically in the range of 8 to 10), then there is almost a constant shearing stress across the wall thickness, and this simplifies the determination of the yield shear stress. On the other hand, it may be more difficult to determine the ultimate shear strength using thin-walled tubes, since they may fail by torsional buckling, as shown in Fig. 5.16c.

5.4.4 Sources of Error in Torsion Tests

There are a number of possible sources of error in torsion testing. These include the following: (1) The specimen radius must be determined very accurately, since J is a function of r^4 (2) bending may occur, due to either misalignment of the torquing heads or to warped specimens; and (3) if the torquing heads are not properly aligned, some tension or compression may be superimposed on the specimens during a test.

5.5 DIRECT SHEAR

It is most common to determine the shear properties of materials using the torsion tests described previously. Nevertheless, there are some circumstances in which *direct shear tests* may be desirable, since they may more closely approximate the actual service conditions. Therefore, a number of direct shear tests have been developed, for both metallic and nonmetallic materials.

There are a number of problems associated with the use of direct shear tests:

1. It is not possible to devise a test in which bending stresses are completely absent.

2. The shear stresses are not uniformly distributed across the cross section; they must be zero at the top and bottom, with some indeterminate (though probably fairly uniform) distribution across the rest of the cross section.

3. It is not possible to make sensible strain measurements, and so these tests cannot be used to determine such properties as the proportional limit or the shear modulus.

Given the differences in experimental techniques and in analysis, it should not be surprising that there is not very good agreement between shear strengths obtained from different direct shear tests or between direct shear and torsion tests.

5.6   MULTIAXIAL LOADING

In the foregoing sections we have considered only relatively simple forms of loading. However, in practice, materials are often loaded in ways which produce *complex stress distributions* (for example, plates in bending or notched bars in tension). Unfortunately, the effects of complex stresses are not always intuitively obvious, and materials undergoing such stresses may behave in rather unexpected ways.

It is often convenient to discuss multiaxial stress states in terms of the *principal stresses*. It can be shown that in every state of stress there are three *principal directions,* at right angles to each other, along which the principal stresses act. Each principal stress represents the maximum (or minimum) normal stress for one set of plane stresses. *No shearing stresses act on the principal planes.* The three principal stresses are usually designated as $\sigma_1 \geq \sigma_2 \geq \sigma_3$, where tension is considered positive and compression negative. Then the maximum shearing stress acting in a body is given by

$$\tau_{\text{max}} = \frac{\sigma_1 - \sigma_3}{2}.$$ (5.20)

We may similarly define three *principal strains* associated with the principal stresses. They may be expressed as

$$\varepsilon_1 = \frac{1}{E}[\sigma_1 - \nu(\sigma_2 + \sigma_3)]$$ (5.21)

$$\varepsilon_2 = \frac{1}{E}[\sigma_2 - \nu(\sigma_1 + \sigma_3)]$$

$$\varepsilon_3 = \frac{1}{E}[\sigma_3 - \nu(\sigma_1 + \sigma_2)].$$

With these definitions in mind, we can now discuss the effect on the strength of materials of *multiaxial* stresses (i.e., stresses in several directions), which may produce complex stress distributions. Highly specialized equipment and techniques are required to carry out multiaxial loading tests, which are beyond the scope of this book. Biaxial stresses often occur in thin plates, with triaxial stresses being introduced if the plates become thick. Here, we will only discuss the simplest case of multiaxial loading (i.e., hydrostatic stresses).

Hydrostatic Stresses

Hydrostatic compression is the state of stress that exists in a body submerged in a liquid under pressure p. The pressure is the same in all directions and is always normal to any surface on which it acts. No shearing stresses are possible, since the static shear resistance of the liquid is zero. Thus, any three orthogonal directions are principal directions, and $\sigma_1 = \sigma_2 = \sigma_3 = -p$. In an ideal elastic material, the mechanism of hydrostatic compression is simple. The atoms and molecules are merely pushed closer together; their spatial configuration remains geometrically similar at all times. The repulsive forces are increased proportionately in all interatomic bonds, but there is no tendency for atoms to change bonds. Thus, the action remains purely elastic; no inelastic behavior can occur. Because there are no shearing stresses, and hence no possibility of slip, materials appear to behave in a perfectly brittle manner.

Elastic materials also follow a form of Hooke's law in hydrostatic compression, at least at moderate pressures. Hooke's law for hydrostatic compression can be written as

$$\sigma = K\frac{\Delta V}{V}, \tag{5.22}$$

where $K = $ *volume modulus of elasticity,* or *bulk modulus,* and $\Delta V/V = $ unit volume change.

If we consider the change in volume of a rectangular prism under a hydrostatic stress σ_{hydro} (see the derivation of Eq. 5.4), then

$$\frac{\Delta V}{V} = \frac{3\sigma_{\text{hydro}}}{E}(1 - 2\nu). \tag{5.23}$$

Therefore, K can be related to the elastic constants E and ν by

$$K = \frac{E}{3(1 - 2\nu)}. \tag{5.24}$$

Note that for $\nu = 1/3$, $K = E$, and this is at least approximately true for metals.

As the pressure is increased and the atoms are pushed closer together, the repulsive forces in the interatomic bonds increase rapidly (Fig. 1.6). At higher pressures, the nonlinearity of the interatomic bond versus spacing curve produces an upward curvature in the pressure versus volumetric strain curve (Fig. 5.17). However, the deformation remains purely elastic. (Note that for some real materials, the pressure versus $\Delta V/V$ curve may turn down, due to the collapse of a rather open structure under high pressures; this may occur in glasses where there are open spaces in the network or in porous materials.)

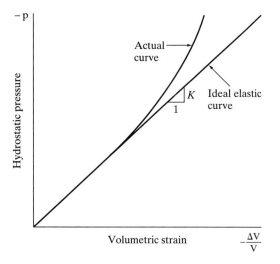

Figure 5.17
Pressure versus volumetric strain in hydrostatic compression.

Hydrostatic tension rarely occurs (except perhaps in thick-notched plates in bending), but it is of interest as the limiting case opposite to hydrostatic compression. Again, since there are still no shearing stresses, the action must remain elastic, and yielding cannot occur. The atoms are pulled apart in all directions, with the body remaining geometrically similar to its original state. However, in hydrostatic tension, there is a limit to the amount of strain the bonds can withstand, and failure by cleavage can take place, though this will be a completely brittle failure. Fracture in hydrostatic tension is governed by the same mechanisms as for the fracture of brittle materials in simple tension (Sec. 5.1.2).

5.6.1 Transverse Stresses

The more common problems of triaxial loading lie somewhere between the two extremes of hydrostatic compression and tension: What happens to a body loaded in uniaxial tension or compression when transverse (all-around) stresses are also imposed, as shown in Fig. 5.18?

When transverse compressive stresses are added to a simple uniaxial compression, there is an increase in the apparent strength of the material and a decrease in ductility. The transverse compression combines with an equal part of the

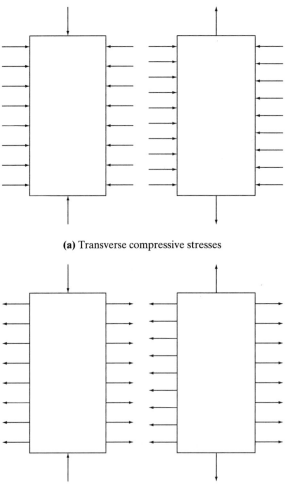

(a) Transverse compressive stresses

(b) Transverse tensile stresses

Figure 5.18
Transverse stresses: (a) transverse compressive stresses; (b) transverse tensile stresses.

Chap. 5   Response of Materials to Stress

axial compression to form a hydrostatic component, which does not contribute to fracture or to yielding. The result is that the axial compressive stress can be raised to higher values before fracture or yielding take place, thereby increasing the apparent strength and decreasing the amount of yielding for a given axial stress. On the other hand, a transverse compressive stress superimposed on a uniaxial tensile stress will tend to increase the maximum shear stress $(\sigma_1 - \sigma_3)/2$ rather than decrease it as in the preceding case, leading to a lower apparent strength and more ductile behavior.

Similarly, when a transverse tensile stress is imposed on simple uniaxial tension, a hydrostatic tension component results, which reduces the ductility and increases the apparent yield strength of ductile materials. For brittle materials, there is little or no increase in strength, since the failure mechanism depends not on yielding but on the maximum tensile stresses acting at imperfections or other stress raisers. A transverse tensile stress acting on a body in uniaxial compression again tends to increase the shear stresses, reducing the strength but leading to more ductile behavior.

5.7 HARDNESS

Hardness is a difficult term to define because it means different things to different people. In common engineering usage, hardness refers to the ability of a material to resist permanent deformation of its surface, in the form of scratching, indentation, abrasion, or cutting. However, hardness is not a fundamental material property. Rather, it depends almost entirely on the type of hardness test employed, since each hardness test measures a different combination of material properties. Therefore, while we know intuitively that materials of different hardnesses are not alike, we must realize that materials that have the same hardness may also not be alike. For instance, quartz is considered to have about the same hardness as tool steel, but clearly their other physical and mechanical properties are very different.

Because hardness measurements can sometimes be used as a strength criterion, and because they are both simple and nondestructive, a great many hardness tests, all of them completely empirical, have been devised over the years.

5.7.1 Scratch Hardness

Probably the oldest hardness test is the *scratch hardness test,* devised in 1822 by the mineralogist Friedrich Mohs. This type of test is used to determine the relative hardness of two materials; the harder solid is capable of scratching the softer one, but the reverse is not true. Mohs selected a series of 10 minerals, which he ranked in order of increasing hardness, from 1 to 10. A mineral which will scratch another is given a higher *Mohs hardness number*. This test is still in common use by mineralogists, primarily as a means of helping to identify and classify minerals, though other materials may also be fitted into this scale. The Mohs hardness scale is given in Table 5.1.

5.7.2 Indentation Hardness

Indentation hardness refers to the resistance of a material to permanent indentation under a localized pressure. Clearly, the results of such tests will depend on the size and geometry (ball, cone, pyramid, etc.) of the indenter and on the magnitude of the loading. Nevertheless, tests of this type make up the most important class of hardness tests. Though they were developed primarily for metals, some indentation hardness tests are now also being applied to brittle materials, such as ceramics. A number

TABLE 5.1 Mohs Hardness Scale

Mohs Number	Mineral
1	Talc
2	Gypsum
3	Calcite
4	Fluorite
5	Apatite
6	Orthoclase (feldspar)
7	Quartz
8	Topaz
9	Corundum (or sapphire)
10	Diamond

of indentation hardness tests have been developed over the years; only the most common ones will be discussed here.

Brinell Test

The Brinell hardness test is the oldest of the indentation hardness tests still in use. It was developed in 1900 by the Swedish engineer S. A. Brinell, who was looking for a rapid, nondestructive test on metals that would provide information that could be related to that obtained from a standard tension test. For this purpose, he chose an indentation hardness test, using a hard steel ball as the indenter. He found, empirically, that for an appropriate combination of ball diameter and indenting load, the average pressure (in kg/mm^2) calculated on the spherical surface of the indentation (the Brinell hardness number, BHN) was about three times the ultimate tensile strength for a variety of steels. This became the basis for the Brinell hardness test.

The Brinell hardness test is shown schematically in Fig. 5.19. The spherical surface of the indentation, *A,* is given by

$$A = \frac{\pi D}{2}[D - \sqrt{(D^2 - d^2)}\,]\tag{5.25a}$$

or

$$A = \frac{\pi D^2}{2}\left[1 - \cos\frac{\phi}{2}\right],\tag{5.25b}$$

where the symbols are defined in Fig. 5.19. Therefore, the Brinell hardness number is

$$\text{BHN} = \frac{P}{\dfrac{\pi D}{2}[D - \sqrt{(D^2 - d^2)}]}\tag{5.26a}$$

or

$$\text{BHN} = \frac{P}{\dfrac{\pi D^2}{2}\left[1 - \cos\dfrac{\phi}{2}\right]}.\tag{5.26b}$$

In the standard Brinell test, a 10-mm steel ball is used with a load of 3000 kg applied for 10 to 15 s. For softer metals, loads of 1500 kg or 500 kg may be used, chosen so that the magnitude of the diameter of the impression lies in the range of 25% to 60% of the ball diameter. To maintain geometrically similar indentations (i.e., so

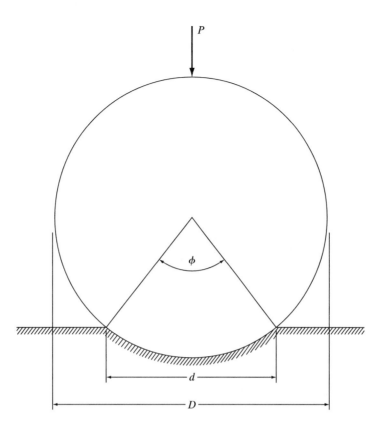

Figure 5.19
The Brinell hardness test.

that the indentation angle ϕ remains constant), the load must be varied as the diameter of the ball changes. It can easily be shown that, for equal mean pressures and equal indentation angles, the ratio P/D^2 must be a constant. For the three standard loads, using a 10-mm diameter ball,

$$P/D^2 = 30 \text{ for 3000 kg}$$
$$P/D^2 = 15 \text{ for 1500 kg}$$
$$P/D^2 = 5 \text{ for 500 kg.}$$

Thus, for a 5-mm ball, a 750-kg load will approximate the standard 3000-kg load on a 10-mm ball; for a 2-mm ball, a load of 20 kg will approximate a 500-kg load on a 10-mm ball.

The localized state of stress, and the localized deformations which occur in a ball test, are both complex. As the hard ball is pressed against the surface, the initial behavior is elastic. Very quickly, however, some plastic deformation begins under the ball; as the load is increased, the indentation becomes deeper, and the plastic region increases in volume. The plastic region extends approximately three times the radius of the indentation itself. Thus, as shown in Fig. 5.20, there is a plastic region which is surrounded, and restrained, by the remaining elastic material. It should be remembered that, during plastic flow, Poisson's ratio becomes 1/2, and thus the material that has become plastic maintains a constant volume. Therefore, there is some buildup of material around the ball, to accommodate the material displaced by the ball. The exact shape of this built-up edge depends largely on the frictional forces between the ball and the material; for some softer materials, the edges of the impression may be pulled down somewhat.

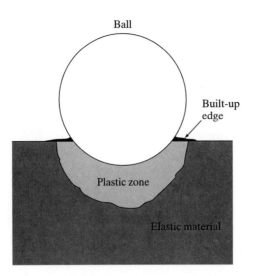

Figure 5.20
Formation of a plastic zone under the ball during a Brinell hardness test.

If the appropriate equations of elasticity and plasticity are applied to the indentation hardness problem, it can be shown that the Brinell hardness number (i.e., the force per unit area) is about three times the tensile yield strength of the material. The details of these calculations, however, are beyond the scope of this book.

There are a number of factors which may influence the Brinell test (as well as the other indentation hardness tests, which will be described later). These include the following:

1. The relative hardnesses of the ball indenter and the test material, since the hardened steel ball itself undergoes elastic deformation during a test.
2. The condition of the surface of the test material. Particularly for metals, the fabrication processes involved in producing the test specimen in question may develop a surface layer which is harder than the interior of the specimen; the hardness number will then depend, in part, on the depth of penetration.
3. Distance from the edge of the specimen. Since the plastic zone is about three times the radius of the indentation, tests should not be carried out closer to the edge (or to another indentation) than that.
4. The Brinell test in particular does not give geometrically similar indentations for different diameters of the indentation.
5. The Brinell test may not be carried out on specimens that are too thin.

Rockwell Hardness Test

The Rockwell hardness test, developed in 1919, is probably the most common hardness test because it is simple and can be done very quickly. In this test, the depth of penetration of the indenter is measured rather than the diameter of the indentation. The principles of the regular Rockwell test are shown schematically in Fig. 5.21. First, a 10-kg *minor load* is applied, to seat the indenter firmly on the specimen, and then the *major load* is applied. When the major is removed, the difference in depth of penetration between the minor and major loads is automatically indicated on a dial gauge. Typically, a test takes from 5 to 10 s. However, the Rockwell hardness number, unlike the Brinell hardness number, does not have any dimensions associated with it.

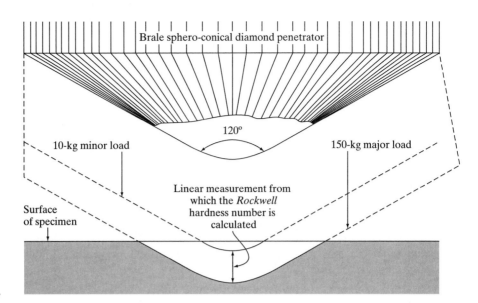

Figure 5.21
Principle of Rockwell test.

Various indenters may be used in the regular Rockwell test. In addition, the major load may be either 60, 100, or 150 kg, though the minor load is always 10 kg. Thus, both the type of indenter and the major load must be specified. The two most common indenters are a 1.59-mm (1/16-inch) ball, and a 120° diamond cone (or *brale*). This availability of different indenters and different major loads allows the Rockwell test to be used on a wide variety of materials.

One major advantage of the Rockwell test compared to the Brinell test is that the size of the indentation is much smaller (Fig. 5.22), and hence the Rockwell test may be considered to be nondestructive in most applications. In addition, for extremely thin specimens or for measurements of the hardness of the surface of a surface-hardened material, a *superficial* Rockwell hardness machine may be used. It differs from the normal Rockwell machine in that the minor load is only 3 kg (instead of 10 kg), and lower major loads are used, yielding an even smaller indentation than the normal Rockwell test (Fig. 5.22).

One disadvantage of the Rockwell test is that, because of the smaller indentation, more care must be taken to ensure a smooth and clean surface, compared to the Brinell test. In addition, since the measurement of penetration depth includes both elastic and plastic deformations, the type of support and the type of backing material between the specimen and the anvil may affect the results.

The mechanics of the Rockwell test are similar to those of the Brinell test. That is, the material beneath the indenter becomes plastic and has a tendency for flowing

Figure 5.22
Comparative impressions in steel using Brinell, common Rockwell, and Rockwell superficial testers: (*A*) Superficial Rockwell, 30-kg load; (*B*) Common Rockwell, 150-kg load; (*C*) Brinell 10-mm ball, 3000-kg load.

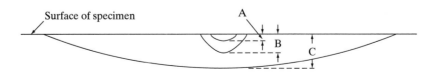

outward, but it is restrained by the surrounding elastic material. For a conical indenter, it can be shown that the ratio of the deformed volume of internal, v_d, to the volume of the indentation, v_i, may be expressed approximately as

$$\frac{v_d}{v_i} = \frac{3.2}{\bar{\varepsilon}},$$

(5.27)

where $\bar{\varepsilon}$ is the average shear strain.

5.7.3 Microhardness Tests

The Brinell and Rockwell tests described previously use high indentation loads and therefore make relatively large indentations in the material. However, there are some applications in which it is necessary to produce extremely small indentations, such as the following:

1. Measurements of different phases of a composite material on the microstructural scale
2. Tests on very thin specimens or surface layers
3. Tests on brittle materials such as minerals or ceramics, which might fracture completely in a normal hardness test.

The reproducibility of microhardness tests is, however, less than that of normal hardness tests because of the difficulties in uniform surface preparation of the specimens and because only a very small volume of material is sampled. In general, the term *microhardness* refers to tests in which the applied load is less than 1000 g; the corresponding size of the indentation may be up to about 50 μm. Because of the very small indentations, microhardness tests require more complicated machines. Usually, the apparatus is combined with a microscope, which is used both to locate and to measure the size of the indentations.

5.7.4 Vickers Diamond Pyramid

The Vickers diamond pyramid was first proposed in 1922, by Smith and Sandland. The indenter, shown in Fig. 5.23, is in the form of a pyramid with a square base, with an angle of 136° between the faces. The diamond pyramid hardness value, DPH, is given by the load (in kg) divided by the contact area of the impression, which becomes

$$\text{DPH} = \frac{1.845P}{D^2},$$

(5.28)

where D is the mean length of the diagonal of the indentation (in mm).

The Vickers and Brinell hardness numbers are very similar at low and medium hardnesses. For very hard materials, there is a discrepancy between the two, since the steel ball used in the Brinell test will deform more than the diamond indenter of the Vickers test.

Knoop Indenter

The Knoop diamond indenter was developed at the National Bureau of Standards (1939) to provide an even more sensitive microhardness test than the Vickers test. The Knoop indenter (Fig. 5.24) is a diamond pyramid ground so that the ratio

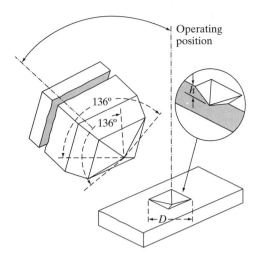

Figure 5.23
Diamond pyramid indenter.

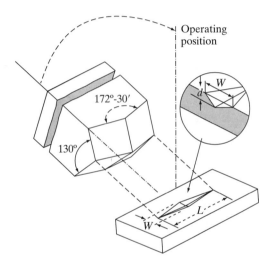

Figure 5.24
Knoop indenter.

between the long and the short diagonals is 7.11 : 1 The Knoop hardness number, KHN, is based on the load divided by the *projected area* of the indentation:

$$\text{KHN} = \frac{14.2P}{L^2}, \tag{5.29}$$

where L is the length of the long diagonal.

The Knoop test is more sensitive than the Vickers test because, for a given depth of impression, a longer diagonal is obtained. For instance, for a given load, the Vickers indenter will penetrate about twice as far into a specimen, but the length of the diagonal will be only about one-third that of the long diagonal in the Knoop test.

PROBLEMS

1. Describe the mechanism of *slip* in ductile materials.
2. Why do both yielding and brittle fracture occur at stresses far below the theoretical tensile (cohesive) strength of a material?

3. A steel rod is 500 mm long, and 25 mm in diameter. If it is subjected to a tensile load of 6×10^4 N, determine its total change in (a) length, (b) diameter, and (c) volume.
4. Repeat question 3 for the case of pure compression.
5. Explain the phenomenon of *barrelling* which generally occurs during compression tests of cylindrical specimens. Is there some ratio of height to diameter below which one cannot carry out a sensible compression test? Explain.
6. Would you expect the elastic modulus computed from a bending test to be exactly the same as that found from a tension test? Discuss.
7. Why would you not want to test rectangular specimens in torsion to determine the shear stress vs. shear strain curve?
8. What are the experimental problems in trying to carry out tests in direct shear?
9. For what sorts of materials are hardness tests (a) most appropriate, and (b) least appropriate?
10. Explain why the apparent strength of a material appears to depend upon its size.

BIBLIOGRAPHY

BORESI, A. P., SCHMIDT, R. J. and SIDEBOTTOM, O. M., *Advanced Mechanics of Materials,* 5th ed., John Wiley & Sons, Inc., New York, 1993.

HIBBELER, R. C., *Mechanics of Materials,* 2nd ed., Macmillan College Publishing Company, New York, 1994.

POPOV, EGOR P., *Engineering Mechanics of Solids,* Prentice Hall, Englewood Cliffs, New Jersey, 1990.

6

Yielding and Fracture

A structural material can be said to have failed when it can no longer perform its design function, either through complete fracture (*brittle* material) or by excessive deformation (*ductile* material). If we know that a structural member will fail by excessive deformation before it actually breaks, then we are primarily concerned with its *yield* behavior. On the other hand, if the member is expected to break after only very little yielding, then its *fracture* behavior becomes most important. In this chapter, we will consider methods of predicting the circumstances in which failure will occur for both types of materials, under a generalized state of loading. Some materials may behave in a ductile fashion under some conditions, but in a brittle fashion under other conditions. Therefore, the transition from ductile to brittle behavior will also be discussed.

6.1 FAILURE THEORIES

The mechanical properties of structural materials are usually determined from tests which subject the specimens to simple stress conditions, generally simple tension or compression, and occasionally shear. However, the strength of materials under more complicated stress conditions has been investigated only in a few cases. To determine suitable allowable stresses for the complicated loading conditions which arise in practical design, various strength theories have been developed. The purpose of these theories is to predict when failure will occur under multiaxial stresses, assuming that the material behavior in simple tension or compression is known. In the discussion which follows, *failure* will mean either *yielding* or *fracture;* these theories have been applied to both ductile and brittle materials, though they are probably most applicable for ductile materials.

As was stated in Chapter 5, the most general state of stress which can exist in a body is completely determined by specifying the three principal stresses. In the discussion which follows, tension is considered as positive, compression as negative, and the principal stress axes will be so chosen that $\sigma_1 \geq \sigma_2 \geq \sigma_3$; σ_{ft} and σ_{fc} will refer to the failure (yield or fracture) stresses in simple tension or simple compression, respectively. When they are assumed to be of equal absolute magnitude, they will be referred to simply as σ'. The graphical representations of the failure theories will, for simplicity, all be shown for the biaxial stress state (i.e., with $\sigma_3 = 0$).

Over the years, many failure theories have been proposed. The two oldest of these assumed that failure would occur either when the maximum principal stress reached the failure stress in simple tension (*Rankine* theory) or when the maximum principal strain reached the failure strain in simple tension (*St. Venant* theory). However, these theories, and various others, often do not agree well with experimental data and so are rarely used today. Here, we will consider the three most important and widely used failure theories:

1. Maximum shear stress theory
2. Maximum distortional strain energy theory
3. Mohr strength theory.

6.1.1 Maximum Shear Stress Theory

This theory was first proposed by Tresca in about 1865 and is still widely used. It results from the observation that in ductile materials, slip occurs during yielding, and this suggests that the maximum shearing stress plays the key role. If we assume that $\sigma_{fc} = \sigma_{ft}$, then the maximum shearing stress theory states that *yielding* begins when the maximum shear stress in the material, τ_{max}, becomes equal to the maximum shear stress at the yield point in a simple tension, test, τ_f. That is, $\tau_{max} = \tau_f$. Since $\tau_{max} = (\sigma_1 - \sigma_3)/2$ and $\tau_f = \sigma_f/2$, this can be written as

$$\sigma_1 - \sigma_3 = \sigma_f. \tag{6.1}$$

This expression is shown graphically in Fig. 6.1. The Tresca yield criterion gives good agreement with experimental results for ductile materials; because of its simplicity, it is the most often used yield theory. The main objection to this theory is that it ignores the possible effect of the intermediate principal stress, σ_2. However, only one other theory, the maximum distortional strain energy theory, predicts yielding better than does the Tresca theory, and the differences between the two theories are rarely more than 15%.

6.1.2 Maximum Distortional Strain Energy Theory

This theory is also referred to as the *octahedral shear stress theory*, or the *Huber-Hencky-von Mises theory*. (It was first proposed by Huber in 1904, developed independently by von Mises on purely mathematical grounds in 1913, and later extended by Hencky.) This theory provides the best agreement between experiment and theory and, along the Tresca theory, is very widely used today. In this theory, the total stress acting on the body is divided into two parts:

1. Hydrostatic principal stresses, which produce only a volume change, while the body remains geometrically similar

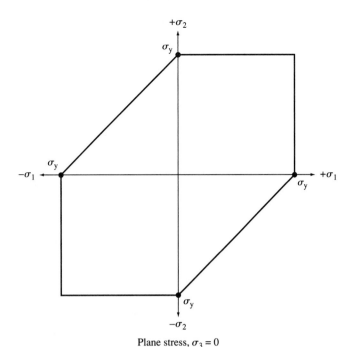

Figure 6.1
Maximum shear stress
theory.

Plane stress, $\sigma_3 = 0$

2. The remaining principal stresses, which produce distortion of the body from its original geometry.

As we have seen in the previous chapter, hydrostatic stresses do not contribute toward yielding, in either tension or compression. Therefore, these stresses are ignored in this theory. Only the strain energy associated with the distortional principal stresses are used to predict failure. Thus, according to this theory, *yielding* will occur when the energy of distortion under a general state of stress becomes equal to the energy of distortion at yield in a simple tension test. Mathematically, this can be expressed as

$$(\sigma_1 - \sigma_2)^2 + (\sigma_2 - \sigma_3)^2 + (\sigma_1 - \sigma_3)^2 = 2(\sigma_{ft})^2 \qquad (6.2)$$

Graphically, this is shown in Fig. 6.2.

6.1.3 Comparison of the Failure Theories

It is of interest to compare these two failure theories for several different stress conditions, as is shown in Table 6.1. From this table, it may be seen that for tests near case (c) (hydrostatic stresses), the maximum shear stress and the maximum distortional strain energy theory predict no yielding, which is correct. They are in good agreement both with experimental results and with each other for the other cases and so remain the most popular theories today.

6.1.4 Mohr's Strength Theory

The two aforementioned failure theories were really developed to establish criteria for the failure of *ductile* materials, though they have sometimes been applied to brittle materials as well. The Mohr strength theory, first published in 1900, deals explicitly with both yielding and brittle fracture and with materials whose strengths in

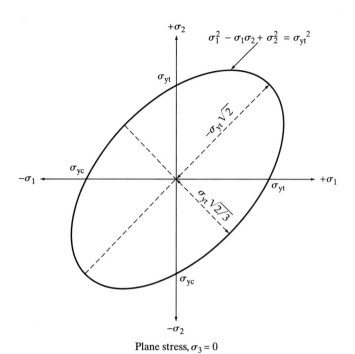

$$\sigma_1^2 - \sigma_1\sigma_2 + \sigma_2^2 = \sigma_{yt}^2$$

Plane stress, $\sigma_3 = 0$

Figure 6.2
Maximum distortional
strain energy theory.

Table 6.1   Comparisons of Failure Theories for Different
Stress Conditions

| | σ_{ft} at yield ($\nu = 0.3$) | | |
Theory	Case (a)	Case (b)	Case (c)
1. Maximum shear stress	$0.50\,\sigma_{ft}$	$1.00\,\sigma_{ft}$	—
2. Maximum distortional strain energy	$0.58\,\sigma_{ft}$	$1.00\,\sigma_{ft}$	—

Assume $\nu = 0.3$.
(a) Plane stress, pure shear: $\sigma_1 = \sigma_3$, $\sigma_2 = 0$.
(b) Plane stress, equal stresses in other two directions: i.e., $\sigma_1 = \sigma_2$, $\sigma_3 = 0$.
(c) Hydrostatic stress state: $\sigma_1 = \sigma_2 = \sigma_2$.

uniaxial tension and compression may be different. In the maximum shear stress theory, failure was defined in terms of the maximum shear stress occurring in the material. The Mohr theory is an extension of this; it defines failure in terms of a limiting combination of normal and shear stresses acting on a body.

It is easiest to describe the Mohr theory by making use of *Mohr's circle,* which is a graphical representation of the state of stress on an element in a body. The student should refer to any basic text on mechanics of materials for a detailed account of Mohr's circle. Briefly, consider an element in plane stress, Fig. 6.7a, with σ_x, σ_y, and τ_{xy} as the initially known normal and shear stresses acting on the element. On any plane perpendicular to the xy plane, as defined by the angle θ, the normal stress, $\sigma_{x'}$, and the shearing stress $\tau_{x'y'}$, can be given by the equations

$$\sigma_{x'} = \frac{\sigma_x + \sigma_y}{2} + \frac{\sigma_x - \sigma_y}{2}\cos 2\theta \tag{6.3}$$

$$\tau_{x'y'} = \frac{\sigma_x - \sigma_y}{2}\sin 2\theta. \tag{6.4}$$

These results can be represented on the typical Mohr's circle shown in Fig. 6.3b, with σ and τ as the coordinate axes. Any point on the circle corresponds to a set of

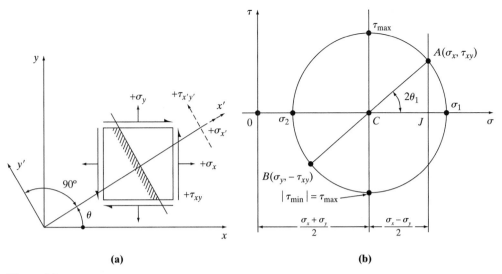

Figure 6.3
Typical Mohr's circle for plane stress.

$(\sigma_{x'}, \tau_{x'y'})$ for a particular value of θ. Thus, the Mohr's circle can be used as a graphical means of finding the normal and shear stresses on any plane defined by the angle θ. The important points to note from the Mohr's circle are as follows:

1. The maximum normal stress is σ_1, and the minimum is σ_2. These are referred to as the *principal stresses;* no shearing stresses exist with either of the principal stresses.
2. The maximum shearing stress, τ_{max}, is equal to the radius of the circle, $(\sigma_1 - \sigma_2)/2$.
3. If $\sigma_1 = \sigma_2$, Mohr's circle degenerates to a point, and no shearing stresses at all can develop.
4. If $\sigma_x + \sigma_y = 0$, the center of Mohr's circle is at the origin of the $\sigma\text{-}\tau$ coordinates, and a state of *pure shear* exists.

Consider the Mohr's circles shown in Fig. 6.4. These three circles represent the stresses on three sets of sections, each corresponding to one of the biaxial

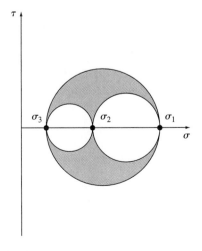

Figure 6.4
Mohr's stress circle for a triaxial state of stress.

principal states of stress (i.e., to one of the three pairs of principal stresses). It can be shown that the normal and shearing stresses acting on *any* plane can be represented by a point lying within the shaded area. Mohr assumed that, on any plane, failure is governed by the maximum shearing stress. Therefore, the underlying assumption in the Mohr theory is that the largest (or outer) circle alone, defined by σ_1 and σ_3, is sufficient to determine the failure condition; the value of the intermediate principal stress may be disregarded (as in the maximum shear stress theory).

Now to determine the stresses at failure (either yielding or fracture), consider the set of Mohr's circles shown in Fig. 6.5. Circle 0-$\sigma_{tension}$ represents the condition for failure in pure tension; circle 0-$\sigma_{compression}$ represents the condition for failure in pure compression, and circle 0-τ_{shear} represents the condition for failure in pure shear. If several circles of these types are obtained by carrying out suitable tests on a given material, envelopes tangent to these circles can then be drawn, as shown in Fig. 6.5. The failure region lies outside of these envelopes. In other words, failure is assumed to occur when the largest Mohr's circle representing the state of stress at a given point becomes tangent to (or exceeds) the failure envelope defined previously.

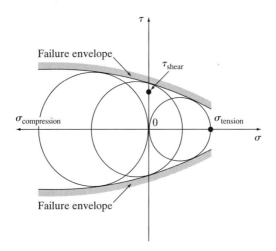

Figure 6.5
Mohr's failure envelope.

If we compare the Mohr theory to the maximum shear stress theory (Fig. 6.6), the two theories give the same failure criterion if the failure stresses in pure tension and compression are the same. For ductile materials with unequal tensile and compressive failure stresses, the Mohr failure criterion is shown in Fig. 6.6b. For brittle materials, particularly when the tensile and compressive strengths of the material are very different (as is the case for portland cement concrete), a modified form of Mohr's theory seems to work best. This modification requires an extension of the failure boundaries to a value of $-\sigma_t$ into the second and fourth quadrants, as shown in Fig. 6.6c.

6.2 FRACTURE MECHANICS

Brittle fractures have long been known to occur in structures and structural components. The best known of these are probably the failures that occurred in the Liberty merchant ships and T-2 tankers built during World War II. The failure of

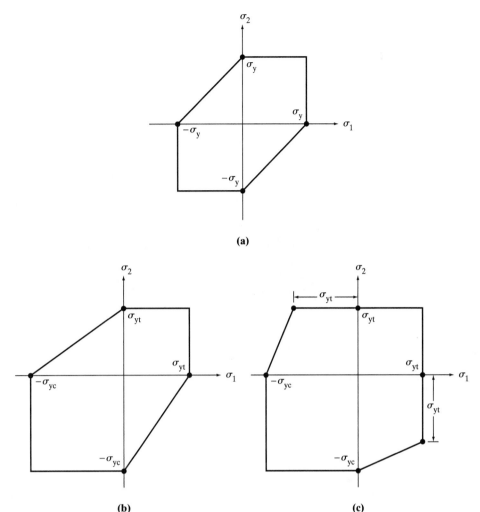

Figure 6.6
Maximum shear stress theory compared to Mohr's theory for a biaxial state of stress: (a) maximum shear stress theory (ductile); (b) Mohr's theory (ductile); and (c) modified Mohr's theory (brittle).

one such ship is shown in Fig. 6.7. This failure occurred while the T-2 tanker *Schenectady* was docked, in mild weather and with the sea calm. The maximum bending moments in the ship at the time of failure were later calculated to be only about one-half of the design moments. Clearly, then, conventional design procedures based only on some maximum stress criterion are not adequate under all circumstances.

The field of *fracture mechanics* provides a very different approach to failure prediction. Fracture mechanics is primarily concerned with the stress and displacement fields in the region of a crack tip in materials under stress, particularly at the onset of unstable crack growth (or fracture). The failure theories described in Sec. 6.1 defined failure in terms of the applied stresses and the tensile and compressive strengths of the material. Fracture mechanics, on the other hand, deals with the interrelationships between the *applied stress,* the *crack* (or flaw) *size* in the material, and a material parameter, to be discussed later, called the *fracture toughness*. The concepts of fracture mechanics are particularly applicable to brittle materials, in which inelastic behavior is at a minimum, but under certain conditions they may be applied to other materials as well. They are also useful to account for the influence of the rate of loading on strength.

Figure 6.7
Photograph of I.O.S. 3301 barge failure. (From S. T. Rolfe and J. M. Barsom, *Fracture and Fatigue Control in Structures*, (Prentice Hall, 1977.)

6.2.1 Griffith Theory

In Chapter 1, it was shown that the theoretical fracture (cohesive) strength of a material could be given by the equation

$$\sigma_{ft} = \left(\frac{E \cdot \gamma_s}{r_0}\right)^{1/2},$$ (6.5)

where E is the modulus of elasticity, γ_s is the surface energy, and r_0 is the equilibrium atomic spacing. For typical values of γ_s and r_0, a reasonable estimate of the theoretical cohesive strength of solids would be of the order of $E/10$.

Based on thermodynamic considerations, Griffith arrived at a similar solution of the theoretical cohesive strength. Considering an elastic body containing a crack and subjected to external loads, he calculated the condition at which the total free energy of the system was minimized. The total energy in the system is

$$U = (-W_L + U_E) + U_s,$$ (6.6)

where $-W_L$ = work due to the applied loads, U_E = strain energy stored in the system, and U_s = free surface energy in creating a new crack surface. A crack would propagate when $dU/dc < 0$, where c is the size of flaw (crack) in the material. Using this theory, we can derive the Griffith equation, which gives the theoretical fracture strength:

$$\sigma_{ft} = \left(\frac{2E\gamma_s}{\pi C}\right)^{1/2},$$ (6.7)

where C is one-half the crack length. When $C = r_0$ (which is the condition existing for the calculation of the theoretical cohesive strength), Eq. 6.7 becomes very similar to Eq. 6.5, even though it was derived in a different fashion.

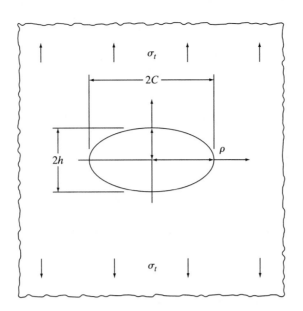

Figure 6.8
Elliptical hole in an infinite
plate in tension.

The question then arises: Why is there this enormous discrepancy, perhaps two or three orders of magnitude, between theoretical values of strength (as predicted by Eqs. 6.5 and 6.7) and those actually measured? The answer to this question was first suggested by Griffith in 1920, who concluded that any real material must contain flaws, microcracks, or other defects that would have the effect of concentrating the stress sufficiently to reach the theoretical fracture (cohesive) stress in highly localized regions of the specimen. Cracks would thus grow under an applied stress until failure occurred. It is easy enough to show that cracks can indeed concentrate the stress sufficiently to achieve these very high stresses locally. If we consider a plate in uniform tension, under a stress σ_t, containing an elliptical hole (which in the limit might represent a flaw, in the form of a crack or notch), as shown in Fig. 6.8, then the stress at the crack tip can be written as

$$\sigma_{max} = \sigma_t\left(1 + 2\left(\frac{C}{\rho}\right)^{1/2}\right),$$ (6.8)

where ρ is the radius of the crack tip. Based on Eq. 6.8, and assuming that $(C/\rho) \geq 1$, one may define the stress concentration factor, K_t:

$$K_t = \frac{\sigma_{max}}{\sigma_t} = 2\left(\frac{C}{\rho}\right)^{1/2},$$ (6.9)

where K_t is the ratio of the maximum stress at the root of the crack tip calculated from the elastic theory to the nominal stress at the same point in the absence of a crack. Clearly, for $C \gg \rho$—that is, for a very "sharp" crack—this factor can become very large. The form of the stress distribution ahead of the crack is shown in Fig. 6.9.

6.2.2 Stress Intensity Factor

Although we can distinguish three modes of crack displacement, as shown in Fig. 6.10, the crack-opening mode (Mode I) is the most important one to consider for brittle materials. If we consider the stress field created by a uniform tensile loading

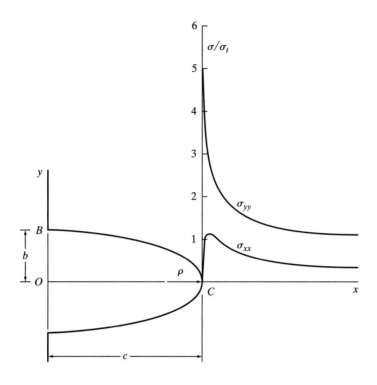

Figure 6.9
Stress concentration at elliptical hole; $c = 3b$. Note that concentration is localized within $\approx c$ of the tip, with high stress gradients within $\approx \rho$ of the tip.

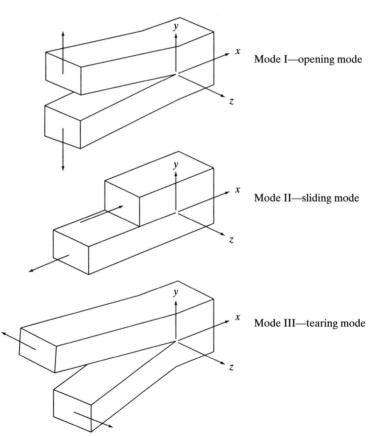

Mode I—opening mode

Mode II—sliding mode

Mode III—tearing mode

Figure 6.10
The three basic modes of crack surface displacements.

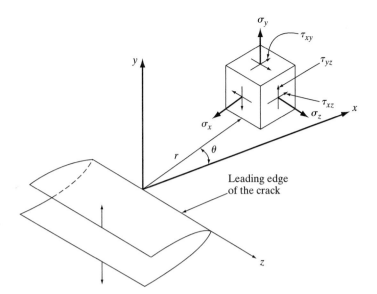

Figure 6.11
Coordinate system and stress displacement ahead of a crack tip (mode I displacement).

stress near a sharp crack, as shown in Fig. 6.11, the stresses can be written as

$$
\begin{Bmatrix} \sigma_x \\ \sigma_y \\ \tau_{xy} \end{Bmatrix} = \frac{K_I}{(2\pi r)^{1/2}} \begin{Bmatrix} \cos(\theta/2)[1 - \sin(\theta/2)\sin(3\theta/2] \\ \cos(\theta/2)[1 + \sin(\theta/2)\sin(3\theta/2] \\ \sin(\theta/2)\cos(\theta/2)\cos(3\theta/2] \end{Bmatrix} \tag{6.10}
$$

$$
\tau_{xy} = \nu(\sigma_x + \sigma_y) \qquad \tau_{xz} = \tau_{yz} = 0,
$$

where ν is Poisson's ratio. The parameter K_I, the stress intensity factor, has the form $K_I = \sigma(\pi C)^{1/2}$, for an infinite solid, where σ is the uniform loading stress and C is the half-crack width. For specimens with finite dimensions, this becomes $K_I = Y\sigma(\pi C)^{1/2}$, where Y is a modification factor that takes into account the geometry of the specimen. K_I has the dimension of stress $\times$ (length)$^{1/2}$, has units of $N \cdot m^{-3/2}$ (or lb in$^{-3/2}$), and may be considered to be a single-parameter description of the stress and displacement fields in the region of a crack tip. Its calculation assumes a linearly elastic material which is both homogeneous and isotropic. Although these assumptions are really incorrect for most materials, we generally assume that the approximations involved in applying linear elastic fracture mechanics to them are reasonable. The underlying assumption is that when the stress intensity factor reaches some critical value, unstable fracture occurs. This critical stress intensity factor is designated K_{IC} and is sometimes referred to as *fracture toughness*. It should be a fundamental material property, independent of how it is measured.

An alternative way of considering fracture involves not K_I, but the strain energy release rate, G_I. Unstable crack extension occur when G_I reaches the critical strain energy release rate, G_{IC}. It can be shown that

$$
G_{IC} = \frac{K_{IC}^2}{E} \qquad \text{Plane stress} \tag{6.11}
$$

$$
G_{IC} = \frac{K_{IC}^2}{E(1 - \nu^2)} \qquad \text{Plane strain}
$$

so that these are equivalent ways of expressing the fracture criterion. It can also be shown that $G_{IC} = 2\gamma_s$.

6.2.3 Compressive Failure

So far, we have only considered tensile stresses, but materials are also used in compression. However, the Griffith analysis can be extended to include uniaxial compression as well as biaxial stress states. Griffith showed that under biaxial compressive stresses, the presence of small cracks leads to tensile stresses at some points along the edge of the flaw, as long as the stress components are not equal. Using the normal convention that $\sigma_1 > \sigma_2 > \sigma_3$ with tension positive, the Griffith criterion becomes

$$\sigma_1 = \sigma_{ft} \qquad \text{if } 3\sigma_1 + \sigma_3 > 0 \tag{6.12}$$

and

$$(\sigma_1 - \sigma_3)^2 + 8\sigma_{ft}(\sigma_1 + \sigma_3) = 0 \qquad \text{if } 3\sigma_1 + \sigma_3 < 0, \tag{6.13}$$

where σ_{ft} is the uniaxial tensile strength. This is shown graphically in Fig. 6.12. This criterion predicts that the compressive strength of a brittle material is eight time the tensile strength, which is fairly close to the observed values for most brittle materials. This failure criterion agrees well with experimental data in the tension-compression zone, but it does not work too well in the compression-compression zone, where splitting will generally occur normal to the unloaded σ_2 direction.

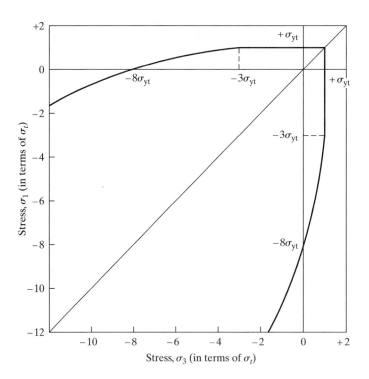

Figure 6.12
Griffith criterion for fracture under biaxial stress (tension positive).

6.2.4 Notch Sensitivity

One additional concept which must be considered is that of notch-sensitivity. A notch-sensitive material is one in which the presence of a sharp notch reduces the tensile or flexural strength beyond that caused by the mere reduction in cross-sectional area. This situation occurs in elastic-brittle materials, where the stress concentration at the crack tip is extremely high. Linear elastic fracture mechanics represents the conditions for failure of such materials by the expression

$$\sigma_{ft} = \frac{K_{IC}}{Y(\pi C)^{1/2}}.$$

(6.14)

That is, the fracture stress can be increased either by increasing the fracture toughness of the material or by decreasing the size of flaws or cracks in the material. In materials which exhibit yielding (i.e., considerable plastic deformation before failure), stress concentrations would be much smaller, and these materials would not be notch sensitive. More complex nonlinear fracture mechanics must be applied to materials exhibiting plastic deformation before failure.

6.2.5 Crack Velocity

Once the elliptical crack shown in Fig. 6.8 starts to propagate, its tip will be moving at some velocity, $V_c = dc/dt$. As the crack length $2C$ increases, the semiminor axis h of the crack will also increase. For h to increase, a volume element of material near the sides of the crack must be displaced perpendicular to the crack plane, and the rate at which this material moves limits the speed at which the crack tip can advance.

The crack velocity is usually expressed in terms of the longitudinal wave velocity, C_L. For an isotropic material, C_L is given by

$$C_L = \left(\frac{E}{\rho}\right)^{1/2}$$

(6.15)

It can be shown that the crack velocity, V, in the configuration of Fig. 6.8 is given by

$$V = \left(\frac{2\pi}{k}\right)^{1/2} \left(\frac{E}{\rho}\right)^{1/2} \left(1 - \frac{C_0}{C}\right)$$

(6.16)

where C_0 is the initial crack length. It has been found that the value $(2\pi/k)^{1/2} = 0.38$. Thus, as C becomes very large, the terminal crack velocity, V_T, can be expressed as

$$V_T = 0.38 \left(\frac{E}{\rho}\right)^{1/2} = 0.38\, C_L$$

(6.17)

That is, the terminal crack velocity is 0.38 times the velocity of the longitudinal waves in the material. This would give terminal crack velocities of about 1800 m/s in steel and about 1500 m/s for concrete.

The preceding discussion of fracture and crack velocity applies only to perfectly elastic, brittle materials. Nonlinear fracture mechanics must be applied to materials in which some localized plastic deformation is produced by the high stresses near crack tips, which gives the material some additional toughness or resistance to crack propagation.

6.3   THE DUCTILE-BRITTLE TRANSITION

It is generally convenient to characterize materials as being either ductile or brittle. Whether a material behaves in a brittle or a ductile manner depends primarily on two considerations:

1. Its atomic or molecular structures
2. The service conditions.

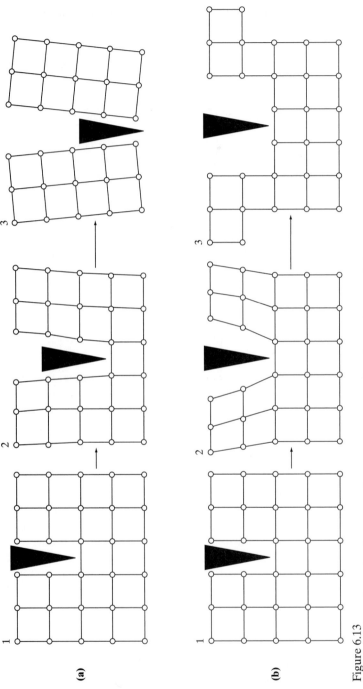

Figure 6.13
High-speed crack moving into a metal may cause atoms to break apart or merely to slide past one another. In a brittle material (a), the bonds stressed in tension fail first, the crack propagates rapidly, and the metal breaks apart. In a ductile material (b), these bonds stressed in shear fail first; the bonds break and re-form, allowing the atoms to slip, and the crack is blunted and stopped.

The effects of atomic structure can be described, at least qualitatively, in terms of the relationship between the tensile strength (σ_{ft}) and the shear strength (τ_{ft}) of the interatomic bonds. As we have seen in Sec. 6.2, a crack in a material under stress introduces stress concentrations ahead of the crack tip (Fig. 6.9), with the tensile stresses on the interatomic bonds considerably larger than the shear stresses. Now consider the regular atomic arrays shown in Fig. 6.13, with a crack moving into the atomic lattice. If the ratio of tensile stress (σ) to shear stress (τ) exerted at the crack tip is *greater* than the ratio of the tensile strength to the shear strength of the interatomic bonds, then the bonds stressed in tension break first, and the material will fail in a brittle manner, as shown in Fig. 6.13a. On the other hand, if the ratio of the tensile stress to shear stress is *less* than the ratio of tensile strength to shear strength, then failure will occur by slip; this will cause the crack to blunt and to be arrested, as shown in Fig. 6.13b. That is, we have the general relationships

$$\frac{\sigma}{\tau} > \frac{\sigma_{ft}}{\tau_{ft}} : \text{brittle fracture}$$

$$\frac{\sigma}{\tau} < \frac{\sigma_{ft}}{\tau_{ft}} : \text{yielding.}$$

For many materials, whether they behave in a brittle or ductile manner will also depend on the service conditions. The three factors that most determine whether a material will fail by yielding or by brittle fracture are the temperature, the rate of loading, and the degree of triaxiality.

1. *Temperature:* The yield stress decreases with increasing temperature; the fracture stress also decreases, but to a much smaller degree.
2. *Rate of loading:* The yield stress increases with increasing strain rate, to a larger degree than does the fracture stress.
3. *Degree of triaxiality:* The yield stresses increase markedly with an increasing degree of triaxiality; the fracture stress is relatively unaffected.

Schematically, the effects of these three parameters on the yield stress and the fracture stress are shown in Fig. 6.14a.

Rather than strength, however, it is more useful to characterize the transition from ductile to brittle behavior in terms of the *fracture energy*—that is, in terms of the amount of energy that is irreversibly absorbed in the process of fracturing.

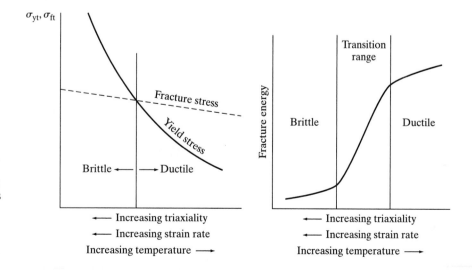

Figure 6.14
(a) Schematic description of variation of fracture strength (σ_{ft}) and yielding strength (σ_{yt}) with various factors. (b) Schematic description of the transition curve for ductile-brittle failure.

Brittle materials require much less energy for fracture than ductile ones, with temperature having a much larger effect on fracture energy than either triaxiality or rate of loading. The effects of these three factors on fracture energy are shown in Fig. 6.14b.

6.4 FRACTURE ENERGY

Fracture energy can be determined by static or impact tests, from which it is possible to calculate directly the energy involved in the fracturing process. Alternatively, the tests can allow us to define other characteristics of the fractured surface to determine whether it propagates in a ductile or a brittle manner. A static test is usually done in tension or in flexure, and the energy consumed in the fracture process can be calculated from the area under the stress-strain or load-deflection curve. The most common impact tests are carried out using a specimen with a pre-formed artificial crack (notch), to ensure that fracture occurs at a predetermined section of the specimen and that the dimensions of the fracture-inducing notch can be specified. A number of tests are available to measure the impact toughness[1] of materials. The most common one is the Charpy V-notch impact test. The principal features of the test apparatus are shown in Fig. 6.15. Typically, tests are run at a number of different temperatures, so that the temperature range over which the ductile-brittle transition

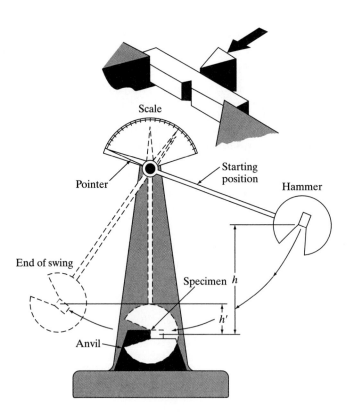

Figure 6.15
Impact testing: Charpy impact machine; and standard Charpy V-notched impact test specimen.

---

[1]The term *toughness* is sometimes used instead of fracture energy and refers to the work required to fracture a specimen. It should not be confused with the *fracture toughness, K_{IC},* defined in Sec. 6.2.2.

130   Chap. 6   Yielding and Fracture

occurs can be determined. For some materials, there is a very sharp temperature transition; for others, the transition occurs over a wide temperature range.

The toughness of the material can be expressed in three different ways: (1) *energy absorbed,* in fracturing the specimen; (2) *lateral contraction,* at the root of the notch; and (3) *fracture appearance,* in terms of the percent of the fracture surface which exhibits shear (fibrous) fracture. The Charpy test, like all other impact tests, is a completely arbitrary one. Changes in the size of the impacting hammer, or in specimen geometry, will lead to different results. In particular, as might be expected, the test is very sensitive to the size and geometry of the artificially induced notch.

6.5 EFFECT OF RATE OF LOADING

So far, we have made the tacit assumption that the mechanical properties of materials are time independent: That is, that they do not depend on the rate of loading or on the duration of loading. There are, however, some types of material behavior that are strongly time dependent. In subsequent chapters, we will consider the effects of fatigue (or cyclic loading) on material properties and the creep (or time-dependent deformation) of materials. In this section, we will consider the effects of rate of loading on the mechanical properties of materials.

The rates at which materials may be loaded can vary enormously. In a typical static test, the rate of loading is such that the material will fail in two to five minutes. A much longer time to failure, of weeks, months, or even years, occurs in static fatigue conditions; here the material fails under a constant sustained stress level, which is lower than that required to cause failure in a standard tension or compression test. Extremely short times to failure occur under dynamic or impact loading, within less than a millisecond. Therefore, to describe fully the properties of materials, they may have to be determined over a range of perhaps 12 orders of magnitude in loading rate.

The term *rate of loading* is not precisely defined. It can be used to refer to the rate at which the cross-head of a testing machine moves during a test, to the rate at which the stress is increased, or to the rate at which the strain is increased. Of these, the *strain rate* is probably the most useful, though it is often more convenient to use one of the other definitions of rate of loading.

An extreme example of a material whose properties are highly time dependent is Silly Putty®, a particular type of silicone polymer. If left sitting on a flat surface for a long period of time (i.e., sustained load under only its own weight), Silly Putty will flow like a liquid. If pulled slowly in tension, it flows like warm taffy. At higher rates of loading, as when dropped on the floor, it will bounce like a ball; that is, it will behave in an elastic manner. At very high rates of loading, however, as when struck by a hammer, Silly Putty will shatter in a brittle fashion. While the materials of interest to engineers rarely show this wide range of behavior, the effects of loading rate cannot, in general, be ignored.

6.5.1 Effect of Loading Rate on Brittle Materials

For brittle materials, such as ceramics, the ultimate strength and the elastic modulus increase with an increase in the rate of loading. Since no significant amount of yielding takes place prior to fracture, this time dependence of strength can most easily be described by extending the application of the fracture mechanics principles discussed in Sec. 6.2.

In Sec. 6.2, an expression for the terminal crack velocity was given, as $V_T = 0.38 \ (E/\rho)^{1/2}$. However, under certain combinations of applied stress and

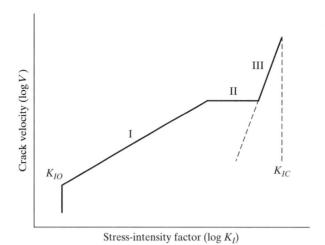

Figure 6.16
Idealized stress-intensity
factor-crack velocity diagram.

crack length, cracks will extend with velocities many orders of magnitude lower than the terminal crack velocity (slow crack growth). This phenomenon, referred to as *subcritical crack growth,* is generally controlled by a stress-induced corrosion mechanism acting at the crack tip, as the bonds between the atoms in front of the crack tip become less stable due to the high local stress field. As a result, they can be more readily broken by reaction with corrosive agents, which would not have taken place in the stress-free conditions. This is most common in glass, where the stressed silica network can be broken by reaction with water vapor. The process may be slow because it depends on factors such as the stress level, temperature, and diffusion of water molecules into the crack tip. It is this subcritical crack growth that underlies the fracture mechanics analysis of the time dependence of strength.

It has been found that, in brittle materials, there is a relationship between the stress intensity factor, K_I, and the crack velocity, V. This relationship is shown in the idealized log V versus log K_I diagram in Fig. 6.16. There are a number of features of this diagram which should be noted:

1. For at least some materials, there is a threshold value of the stress intensity factor K_{IC}, below which no crack growth occurs. For other materials, however, no such threshold has been found, which implies that for these materials, some crack growth, however slow, will occur for *any* applied stress.

2. In region I, the V-K_I relationship can be expressed as

$$V = AK_I^N \tag{6.18a}$$

    or

$$\log V = \log A + N \log K_I, \tag{6.18b}$$

    where N is the *slope* of the log V versus log K_I plot, and A is the *intercept*. In this region, the crack growth rate is controlled by the rate of reaction of the corrosive agent with the material. Region I behavior dominates the time dependence of the failure process, since for moderate loading rates the material spends most of its time in this region.

3. In region II, the crack velocity remains constant, because it is controlled by the diffusion rate of the corrosive agent to the region just ahead of the crack tip.

4. The origins of region III are not understood, but that region appears to be an intrinsic property of the material in question. Note that the slope of the log V versus log K_I curve is much steeper in region III than in region I.

5. Finally, there is a critical value of the stress intensity factor, K_{IC}, at which failure is essentially instantaneous (see Sec. 6.2.2):

$$K_{IC} = Y\sigma(\pi c)^{1/2}. \tag{6.19}$$

Subcritical crack growth is the slow growth of cracks that are too small to cause failure under the prevailing stress. This leads to a dependence of the failure stress on the loading rate, since in specimens loaded slowly, more time is available for slow crack growth than in specimens loaded rapidly. Thus, the rate of loading effect must be due, at least in part, to the growth of a crack, as governed by Eq. 6.18, until it reaches the critical value, as defined by Eq. 6.19.

6.5.2 Static Fatigue

So far, we have considered primarily the effect of the rate of loading on strength. However, there is a related phenomenon which can also be explained using the fracture mechanics approach. In a conventional strength test, the load is increased from zero to the failure load, typically at a loading rate at which failure will occur in only a few minutes. However, it has long been known that some materials, such as many ceramics, glass, concrete, and wood, will fail under sustained stresses which are significantly less than the stresses required to bring about failure in conventional strength tests. This will occur if the stresses are sustained for a sufficiently long time for a crack to grow to the critical size. This phenomenon is known as *static fatigue*. Two examples of static fatigue data, one for soda-lime glass slides and the other for timber, are shown in Fig. 6.17.

6.5.3 Effect of Loading Rate on Metals

For ductile materials, the situation is more complicated than in brittle materials, since yielding precedes fracture. For metals, an increase in the strain rate increases

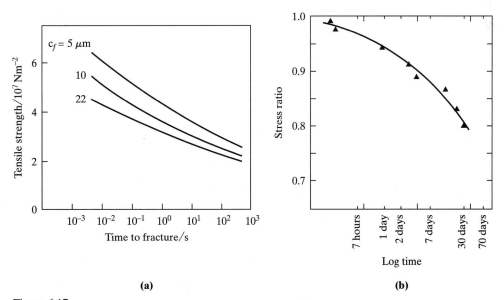

(a)   **(b)**

Figure 6.17
(a) Static fatigue curve for bend tests on soda-lime glass slides in air; c_f indicates approximate depth of flaws introduced by abrasion treatment (after R. E. Mould and R. D. Southwick, *J. Amer. Ceram. Soc.*, 42, 582); and (b) time to failure for Douglas fir beams.

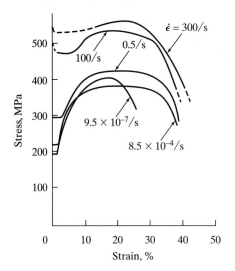

Figure 6.18
Stress-strain curves of mild steel at room temperature for various rates of strain (after Manjoine, *J. Apl. Mech. ASME Trans.,* 66, A-215, 1944).

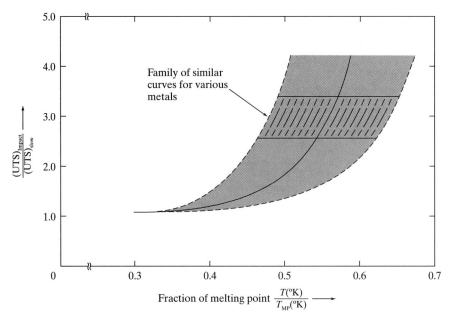

Figure 6.19
Ratio between the ultimate tensile strength (UTS) in impact and slow strain rates, as a function of temperature.

the yield strength and raises the σ-ε curve, as shown in Fig. 6.18 for mild steel. This is particularly the case for metals with a BCC (body-centered cubic) structure, such as molybdenum, tungsten, manganese, and the ordinary carbon steels; FCC (face-centered cubic) metals, such as copper, aluminum, lead, and austenitic non–nickel alloys, are much less strain sensitive. In addition, this effect becomes much more pronounced as the temperature increases. Figure 6.19 shows a family of curves for various metals, giving the ratio of impact (high rate of loading) strength to the strength measured at slow strain rates as a function of the melting point temperature. It may be seen that at temperatures below about 35% of the melting point temperature, strain rate effects are small; beyond this point, strain rate effects become increasingly pronounced. The curves for the light metals, such as aluminum, are close to the right-hand dashed curve; those for metals in which structural changes occur easily at moderate temperatures are closer to the left-hand curve. Figure 6.20 shows the relationship between the σ-ε curves for high and slow loading rates as a function of temperature. It may be seen that the relative differences in strength are greater at the higher temperature.

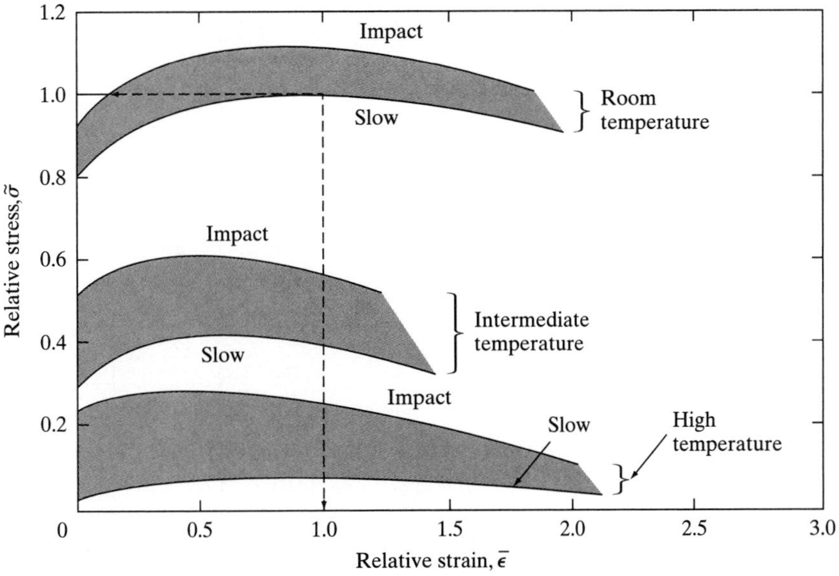

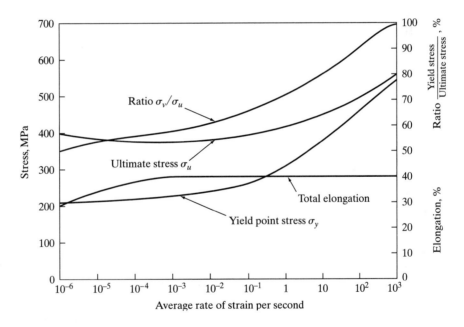

Figure 6.20
Typical stress-strain curves for slow strain rates and impact rates as they change with temperature. Relative stress and relative strain of 1.0 correspond to the stress and strain at which the ultimate tensile stress occurs at room temperature for slow straining.

Figure 6.21
Influence of rate of strain on tensile properties of mild steel at room temperature (after Manjoine, *J. Apl. Mech. ASME Trans.*, 66, A-214, 1944).

In BCC metals, there are at least five loading rate effects which are responsible for the changes mentioned previously. As the strain rate is increased,

1. The yield strength is increased.

2. The rate of strain hardening is decreased.

3. The ultimate (fracture) strength is increased, but at a lower rate than the increase in yield strength.

4. The temperature dependence of the yield strength is reduced.

5. The strain at failure is not changed very much.

Some of these effects are shown in Fig. 6.21 for mild steel.

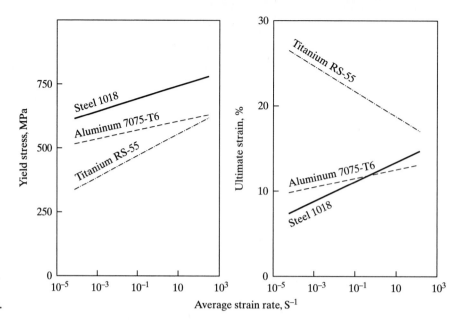

Figure 6.22
Yield strength and elongation as function of average strain rate (after R. F. Steidal and C. E. Makerov, *ASTM Bull.*, 247, 57, 1960).

However, while the yield strengths of all metals are raised with an increase in strain rate, the magnitude of this increase varies with the metal, as shown in Fig. 6.22a. On the other hand, the *ductility* may increase, decrease, or remain unchanged, as shown in Fig. 6.22b.

BIBLIOGRAPHY

BARSOM, J. M. and ROLFE, S. T., *Fracture and Fatigue Control in Structures,* 2nd ed., Prentice-Hall, Englewood Cliffs, New Jersey, 1987.

BORESI, A. P., SCHMIDT, R. J. and Sidebottom, O. M., *Advanced Mechanics of Materials,* 5th ed., John Wiley & Sons, Inc., 1993.

BROEK, D., *Elementary Engineering Fracture Mechanics,* 4th ed., Martinus Nijhoff, London, 1985.

KNOTT, J. F., *Fundamentals of Fracture Mechanics,* John Wiley & Sons, New York, 1973.

MEGUID, S. A., *Engineering Fracture Mechanics,* Elsevier Applied Science, London and New York, 1989.

PROBLEMS

1. Of the various failure theories that are available for materials exposed to multiaxial loading, how would you choose the most suitable one for a particular material or application?
2. For a material subjected to axial tension, what happens to its load carrying capacity when a transverse tensile stress is applied? What happens when a transverse compressive stress is applied?
3. What happens to the ductility of the material in question 2 when transverse stresses are applied?
4. Is it possible to relate (a) the mode of failure, or (b) the failure stress, of a structural material to its atomic structure? Discuss.

5. How do discontinuities, cracks, or other imperfections in materials affect the ultimate strength of a material?
6. Discuss the factors controlling when a material will fail in a ductile or a brittle fashion.
7. Can the behavior of a material under impact loading or other very high strain rate loading be predicted from quasistatic tests? Discuss.
8. To which types of materials can fracture mechanics *not* be applied?
9. Why do materials generally appear to be stronger when they are tested at higher strain rates?
10. Why do large test specimens appear to have lower strengths than small ones?
11. Why might a small hole drilled ahead of a slowly growing crack actually arrest the progress of the crack?

7

Rheology of Liquids and Solids

7.1 ELASTIC AND VISCOUS BEHAVIOR

The behavior of the materials discussed so far concerned their short-term response to stress. This response is the only one needed to be considered if the material is an ideal elastic solid. However, in practice, most engineering materials will exhibit an additional component in response to stress which is time dependent. This response is characteristic of viscous materials, and therefore a solid which exhibits response to stress which combines an immediate elastic component and a time-dependent viscous component is referred to as a viscoelastic material. Since most engineering materials exhibit time-dependent responses under certain conditions, one should, strictly speaking, define them all as viscoelastic. However, in practice, this term is used mainly for materials whose time-dependent response is particularly large at room temperature, such as asphalts and polymers.

The difference between elastic, viscous, and viscoelastic materials may be seen by referring to Fig. 7.1 and considering the response of the three types of materials to the same instantaneous load versus time curve (where the load is applied instantaneously at t_a and removed suddenly at t_r). For an elastic material, all the strain is instantaneous; when the external load is removed, all of the strain is recovered. For a purely viscous material, the strain increases continuously with time under load and is not recoverable. For the intermediate case of a viscoelastic material, there is an instantaneous elastic strain when a stress is applied. However, there is additional strain, which increases with time under load and is partially recoverable when the load is removed.

138

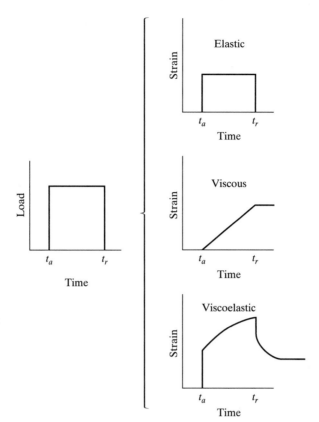

Figure 7.1
Response of three different types of materials to the load-time cycle shown, where the load is applied instantaneously at t_a and released instantaneously at t_r: (a) elastic; (b) viscous; (c) viscoelastic.

The *elastic* response for an ideal solid has already been discussed and is characterized by the relation between stress and strain (Hooke's law) as follows:

$$\varepsilon = \sigma/E, \tag{7.1}$$

where ε is the strain, σ is the stress, and E is the modulus of elasticity.

The *viscous* response can be described by an analogous relationship between stress and strain rate. That is, for a viscous material loaded in tension,

$$\dot{\varepsilon} = \sigma/\mu, \tag{7.2}$$

where $\dot{\varepsilon}$ is the rate of strain, σ is the stress, and μ is the coefficient of viscosity. Similarly, for an ideally viscous material (Newtonian fluid) loaded in shear,

$$\dot{\gamma} = \tau/\eta, \tag{7.3}$$

where $\dot{\gamma}$ is the rate if shear, τ is the shear stress, and η is the coefficient of viscosity.

If we assume further that viscous fluids are incompressible, then by analogy with Eq. 5.22 in Chapter 5,

$$E = 2(1 + v)G. \tag{5.22}$$

Assuming that Poisson's ratio is 0.5, we have

$$\mu = 2(1 + 1/2)\eta. \tag{7.4}$$

Finally, we can show that the generalized strain rate equations for Newtonian fluids have the same form as Eq. 5.25 in Chapter 5:

$$\dot{\varepsilon}_x = \frac{1}{\mu}\left[\sigma_x - \frac{1}{2}\left(\sigma_y + \sigma_z\right)\right] \tag{7.5}$$

$$\dot{\varepsilon}_y = \frac{1}{\mu}\left[\sigma_y - \frac{1}{2}\left(\sigma_x + \sigma_z\right)\right]$$

$$\dot{\varepsilon}_z = \frac{1}{\mu}\left[\sigma_z - \frac{1}{2}\left(\sigma_x + \sigma_y\right)\right].$$

7.2 SIMPLE RHEOLOGICAL MODELS

The most convenient way of depicting the behavior of viscoelastic materials is by means of mechanical models. These models may be built up by various combinations of the basic rheological elements. The three basic elements considered here are shown in Fig. 7.2. The *Hookeian element,* or spring, is perfectly elastic; all of the energy imparted to the specimen is stored as strain energy. Its stress versus. strain behavior is given by $\sigma = E\varepsilon$, where, in this case, E represents the stiffness of the spring. The *Newtonian* element, or dashpot, is perfectly viscous. All of the energy imparted to it is dissipated, and its stress versus. strain rate behavior is given by Eq. 7.2, $\sigma = \mu\dot{\varepsilon}$.

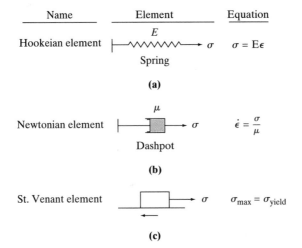

Figure 7.2
The three basic rheological elements: (a) Hookeian (spring), (b) Newtonian (dashpot), (c) St. Venant.

The *St. Venant element* represents a block that resists motion under stress by virtue of the friction between the block and the horizontal surface on which it rests. If the applied force exceeds the force of friction, the block moves. Since this would apply an *acceleration* of the block once it overcame friction, which is unrealistic, the St. Venant element is used only in conjunction with other elements; in such combinations, it represents a yield strength which is time independent. With only these three basic elements, increasingly complex rheological models may be built up by suitable combinations of the elements to simulate the viscoelastic behavior of real materials.

The simple models can be combined together in various combinations to account for the time-dependent response of solids and fluids to stress. Such models are known as rheological models, and the branch of mechanics dealing with modeling and macroscopial characterization of time-dependent response to stress is known as rheology. It covers both solids and fluids, neither of which behave in practice according to the ideal models for solid (spring-Hookeian) or fluid (dashpot-Newtonian). For civil engineering applications, there is a need to address both solids and fluids, as many of the more important construction materials are being processed on site while they are still in their fluid state (e.g. portland cement concretes and asphalt concretes).

7.3 RHEOLOGY OF FLUIDS

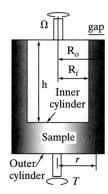

Figure 7.3
Schematic description of a rotational coaxial cylinder viscometer for measuring the rheological properties of fluids.

The evaluation of relations between stresses and strains in fluids, which is essential to characterize their behavior, is not straightforward as in the case of solids, where loads (stresses) can be applied directly and deformations (strains) can be measured by mounting gages directly on the loaded specimen. In fluids the measurements are indirect in nature, using instruments which are collectively known as viscometers. The more common viscometer is the coaxial cylinder type, in which the outer cylinder is rotating at a controlled angular velocity and the inner cylinder is stationary (Fig. 7.3). The torque required to keep the inner cylinder stationary is measured as a function of the angular velocity of the outer cylinder. If the gap between the two cylinders is sufficiently small, then for an ideally Newtonian fluid the following relation can be derived:

$$T = \eta \cdot 4 \cdot \left[\frac{1}{R_i^2} - \frac{1}{R_o^2} \right] \pi h \Omega, \tag{7.6}$$

where T is the torque, η is the viscosity of the fluid, R_i and R_o the inner and outer radii, respectively, h is the height of the cylinder, and Ω is the angular velocity. Thus, a linear correlation exists between T and Ω, analogous to the one between shear stress and rate of shear strain, $\tau = \eta \dot{\gamma}$. Thus, measured curves of torque against angular velocity in a coaxial viscometer can provide information of a similar nature to that obtained if the direct stress and strain rate could be measured in fluids. For a Newtonian fluid, the T versus Ω curve would be linear and pass through the origin; the slope of the curve can enable the calculation of the viscosity coefficient η based on Eq. 7.6. Curves of this kind are known as flow curves and are the basis for characterization of fluids.

Measured flow curves indicate that in many fluids the behavior is more complex than the one defined as the ideal Newtonian. There are four different types of fluids, three of which are non-Newtonian, as shown in Fig. 7.4. In the shear thickening fluid, the viscosity increases with increase in rate, indicating a greater resistance to flow as the strain rate increases. In shear thinning, the viscosity decreases at higher rates, suggesting that bonds between particles are being broken by the shear stress, thus allowing for easier flow. These two types of fluids are sometimes called pseudoplastic, and their flow curve can be represented by the following general equation:

$$\tau = A\gamma^n, \tag{7.7}$$

where A is a constant characteristic of the fluid and n is the index of flow; for $n > 1$,

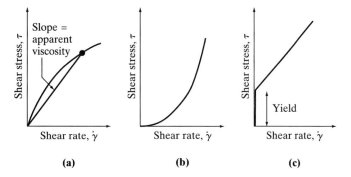

Figure 7.4
Three types of rheological behavior of fluids as exhibited by the flow curves: (a) shear thinning, (b) shear thickening, (c) yield-Bingham behavior.

the behavior is shear thickening, and for $n < 1$, it is shear thinning. Although these fluids do not have a single value of viscosity, they are often treated as Newtonian at a given shear step by defining an apparent viscosity (Fig. 7.4a).

Flow curves which exhibit yield (Fig. 7.4c) are characteristic of fluids in which initial stress must be overcome before flow can start. If the flow curve beyond the yield is linear, then the behavior can be described by the Bingham model for fluids (Fig. 7.5). The dashpot and the frictional block are in series, and no stress can be transferred to the dashpot until the frictional block yields. Bingham behavior is shown only by solid suspensions, such as cement pastes and concretes in their fresh, fluid state. The yield stress represents the mechanical breakdown of flocculated structures (see Fig. 4.15).

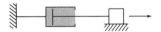

Figure 7.5
Bingham model for fluids.

The flow curve is not necessarily reversible; that is, on reducing the shear rate, the downward curve may not necessarily coincide with the upward branch and a hysteresis loop may form. This is characteristic of shear thinning fluids, in which increasing shear involves gradual breakdown of the flocculated structure, especially in particulate suspensions, where the initial mixing and shearing separates the particles and reduces the attractive forces between them. This is shown in Fig. 7.6a. If this breakup is maintained and carries into the descending branch, the shear stress required for the same strain rate is lower than in the ascending curve, as seen in Fig. 7.6a. If after completion of the cycle, when the fluid is at rest, the bonds can re-form and the next test cycle provides an identical curve, the material is said to be thixotropic (Fig. 7.6b). In a nonthixotropic fluid the second test will result in an ascending curve identical to the descending branch of the first test (Fig. 7.6c).

In civil engineering we often have to deal with suspensions rather than pure liquids (e.g., cement grouts, fresh concrete, and asphalt cement), whose flow can be described by the models reviewed previously. The presence of suspended solids affects the rheological parameters; both concentration and particle size are important. One example of a relationship of this kind is as follows:

$$\eta = \eta_0 \left(1 - \frac{\rho}{\rho_m} \right)^{-[\eta]\rho_m},$$

where η = viscosity of suspension
η_0 = viscosity of pure fluid
ρ = volume fraction of solid particles
ρ_m = the maximum volume fraction when the particles are closed packed
$[\eta]$ = constant related to viscosity of suspension with a low concentration of solids.

Figure 7.6
Illustration of the characteristics of thixotropic and nonthixotropic fluid as exhibited by flow curves: (a) first cycle-shear thinning and hysteresis, (b) second cycle in a thixotropic fluid; (c) second cycle in a nonthixotropic fluid.

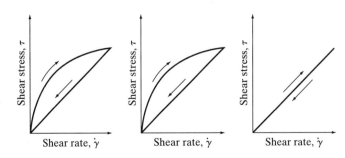

Chap. 7   Rheology of Liquids and Solids

Non-Newtonian fluids can also be described as viscoelastic materials in which the viscous component is dominating. The rheological models for describing such behavior are similar to those used for the viscoelastic solids described in the following section.

7.4 RHEOLOGY OF VISCOELASTIC SOLIDS

The rheological behavior of solids can be modeled by different combinations of the basic elements described in Fig. 7.2. With them, increasingly complex rheological models may be built to account for the behavior of real engineering materials. The simplest of the rheological models are those made up of only two elements each: the Maxwell model, the Kelvin model, and the Prandt model. These are shown in Fig. 7.7.

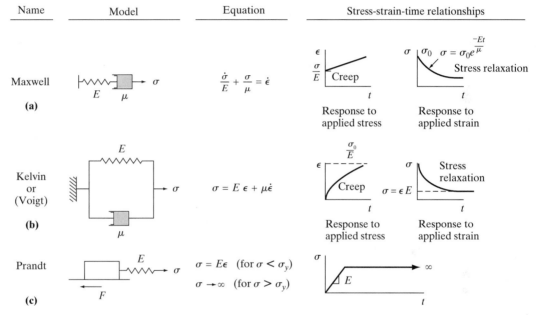

Figure 7.7
Two-element rheological models.

7.4.1 Maxwell Model

The Maxwell model consists of a spring and a dashpot in *series*. The same stress acts on both elements, and so the total strain is equal to the sum of the strains of the two elements. The extension of the spring is given by $\varepsilon_s = \sigma/E$; the extension of the dashpot obeys the relationship $\dot{\varepsilon}_d = \sigma/\mu$. Differentiating ε_s with respect to time, and summing, we get

$$\dot{\varepsilon}_s + \dot{\varepsilon}_d = \dot{\varepsilon} = \frac{\dot{\sigma}}{E} + \frac{\sigma}{\mu}. \tag{7.8}$$

Now consider the response of the Maxwell model to two limiting loading cases: constant stress and constant deformation. Under constant stress, $\dot{\sigma} = 0$ and so Eq. 7.8 becomes

$$\dot{\varepsilon}_s = \frac{\sigma}{\mu}. \tag{7.8a}$$

That is, there will be an instantaneous (elastic) strain, given by σ/E, which is recoverable, followed by a linearly increasing strain, which is irrecoverable, as shown in Fig. 7.7a. This type of behavior is often referred to as *creep* (see Sec. 7.5). On the other hand, if a strain is suddenly applied to the system and held constant, $\dot{\varepsilon} = 0$, then the stress as a function of time is given by

$$\frac{\dot{\sigma}}{E} + \frac{\sigma}{\mu} = 0. \tag{7.8b}$$

Solving, we get

$$\sigma = \sigma_0 e^{-Et/\mu}. \tag{7.8c}$$

This means that there is an exponential *stress relaxation,* as shown in Fig. 7.7a.

7.4.2 Kelvin Model

The Kelvin (or Voigt) model consists of a spring and a dashpot in *parallel*. In this case, the elongation in each element remains the same. Therefore, $\sigma_s = E\varepsilon$, and $\sigma_d = \mu\dot{\varepsilon}$, so that

$$\sigma = \sigma_s + \sigma_d = E\varepsilon + \mu\dot{\varepsilon}. \tag{7.9}$$

Under a constant stress, σ_0 we again get creep behavior, with the solution of the differential Eq. 7.9 giving

$$\varepsilon = \frac{\sigma_0}{E}\left(1 - e^{-Et/\mu}\right). \tag{7.9a}$$

That is, the strain σ_0/E, which would be obtained instantaneously in the absence of a dashpot, is instead approached exponentially. Under a constant strain, $\dot{\varepsilon} = 0$, there is some stress relaxation, and then the stress remains constant, at $\sigma = E\varepsilon$, as shown in Fig. 7.7b. If the material is given a sudden displacement and then released, there is an exponential strain relaxation, given by

$$\varepsilon = \varepsilon_0 e^{-Et/\mu}. \tag{7.9b}$$

7.4.3 Prandt Model

In the Prandt model, there is perfectly elastic-plastic behavior. Up to the yield stress, the σ-ε relationship is given by $\sigma = E\varepsilon$; at the yield stress, the deformation continues indefinitely, as shown in Fig. 7.7c.

7.4.4 Complex Rheological Models

Clearly, it is possible to build up increasingly complex rheological models to simulate more complicated types of material behavior, but they are beyond the scope of this book. However, to illustrate this approach, we describe the simplest case which is the Bingham model.

The perfectly elastic-plastic Prandt material (Fig. 7.7c) deforms infinitely once the yield point is reached, which is clearly unrealistic. This problem can be overcome by using the extended Bingham model (Fig. 7.8), which is the simplest model that

Chap. 7   Rheology of Liquids and Solids

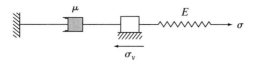

Figure 7.8
Extended Bingham model
for solids exhibiting yield.

represents the flow of a material which possesses a yield point. It consists of the three basic elements—a spring, a dashpot, and a friction block—in series. For stresses less than the yield stress, it behaves elastically. Beyond the yield point, it gives a steadily increasing strain. Its equations are as follows:

$$\varepsilon = \frac{\sigma}{E} \quad \text{for } \sigma < \sigma_y \tag{7.10}$$

$$\varepsilon = \frac{\sigma}{E} + \frac{(\sigma - \sigma_y)t}{\mu} \quad \text{for } \sigma > \sigma_y.$$

In Fig. 7.9, the result of a similar analysis for a combination of Maxwell and Kelvin elements in series is shown. This curve shows many of the characteristics observed in creep of solids.

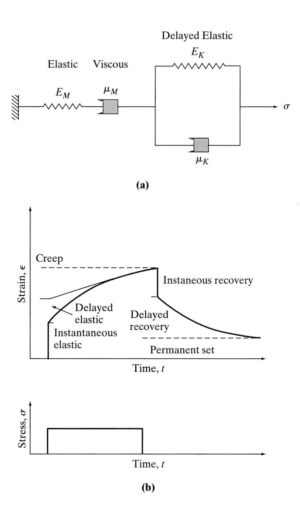

Figure 7.9
Maxwell and Kelvin models in series (Burgers model).

The viscoelastic behavior of engineering solids which is of the greatest practical significance is creep. The rheological models described in the previous section provide mathematical formulation to describe the creep curves. The characteristic rheological model and the constants of the basic rheological units of which it is composed are usually determined by matching a model to an experimental creep curve. Therefore, to understand creep behavior it is essential to address experimental creep curves as well as the mechanisms by which creep is generated in the actual material. Creep will be discussed in greater detail in each of the chapters dealing with the individual materials. In this section an overview will be given.

Generally, for all materials, three types of creep curves can be identified (Fig. 7.10b). Each of them can occur for every material, depending on the stress level and temperature. The curve is usually divided into three stages (Fig. 7.10a): the primary stage, also known as transient creep; the secondary stage, also known as the steady state (the creep rate, $\dot{\varepsilon}$ is constant); and the tertiary stage, which is terminated with fracture. If the loading stress and temperature are sufficiently low, only the primary stage would occur (i.e., the creep rate at the secondary stage would be nil, as seen in the lower curve in Fig. 7.10b). At high enough stress and temperature levels, all three stages will occur, resulting in fracture of the material at the end of the third, tertiary stage. If the stress and temperature levels are intermediate, the tertiary stage may be delayed and occur at time periods greater than the service life of the material. The phrase *sufficiently high stress and temperature* was addressed so far in qualitative terms and depends on the strength and transition temperature of the materials

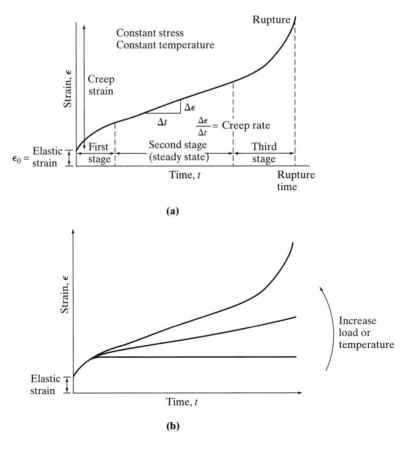

Figure 7.10
Creep curves characteristic of materials loaded under different conditions: (a) general description of the curve; (b) effect of loading and temperature conditions.

(e.g., melting temperature for metals and glass transition temperature or melting temperature for polymers). A rough estimate of the levels above which creep may become sufficiently important is ~1/3 of the transition temperature and strength. Above ~1/2 of the transition temperature and strength, creep may become critical because the material may enter the tertiary stage during its service life.

The shapes of the curves shown in Fig. 7.10b can be described on the basis of the rheological models. They can also be described by empirical relations. The latter approach is common for metals and uses the following equations:

Primary (transient) creep:

$$\varepsilon = At^n, \tag{7.11}$$

where A is a constant depending on the material, load, and environmental conditions, and n for metals is usually $1/3$.

Secondary (steady) state creep:

$$\dot{\varepsilon} = B\sigma^n \exp(-E_a/RT), \tag{7.12}$$

where B, n, and E_a are constants.

The creep in the steady-state stage is the one most important from an engineering point of view because it is the stage in which much of the creep strain is accumulated during the service life. The exponential term in Eq. 7.12 is characteristic of a thermally activated process, where the value of E_a is the activation energy. The actual processes are different for the various materials and will be briefly reviewed.

7.5.1 Creep in Metals

In metals, the main creep mechanism is the movement of dislocations. In the primary stage, their movement is gradually slowed down due to the pinning of dislocations in various sites, as described in Chapter 2. These sites could be point defects, intersecting dislocations, grain boundaries, or particles of second phase. To continue to move past these obstacles, the dislocations must acquire additional energy (to climb or jog over the obstacle). An illustration of such a process is given in Fig. 7.11. The activation energy is directly related to that of the rate of diffusion of defects such as vacancies and interstitials. This diffusion is temperature dependent, as might be expected for any thermally activated process. It becomes considerably high once the temperature increases over one-third of the melting temperature, accounting for the sensitivity of creep in metals to temperature. The accumulation of

Figure 7.11
An illustration of a dislocation climb away from obstacles, (a) when atoms leave the dislocation line to create interstitials or to fill vacancies, or (b) when atoms are attached to the dislocation line by creating vacancies or eliminating interstitials, (from D. R. Askeland, *The Science and Engineering of Materials*, PWS-Kent Publishing Company, 1985, p. 62).

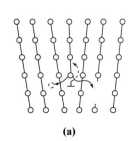
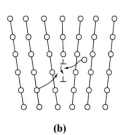

(a)   (b)

plastic deformation in the secondary creep stage may lead to necking and fracture, which occur at the end of the tertiary creep stage.

7.5.2 Creep in Polymers and Asphalts

Viscoelastic behavior (creep and stress relaxation) in polymeric and asphaltic materials is also a thermally activated process. It involves the sliding of macromolecules past each other or slow extension of individual polymeric chains when kept under load. This extension is characteristic of an amorphous polymer or the amorphous part of a polymer chain in a partially crystallized polymer. It involves movement of polymer segments of approximately 50 carbon units, when they acquire sufficient thermal energy to allow them to move or rotate past local obstacles. The probability for acquiring this energy is proportional to the exponential term in the equation describing thermally activated processes, like Eq. 7.12. In the case of polymer and asphalt materials, the activation energies are much lower than in metals, as exhibited also by their relatively low transition temperatures (which may be in the range of $100°C$ for amorphous polymers). Thus, applying the rule of thumb that creep (and, for that matter, stress relaxation) is becoming important at temperatures above $\sim 1/3$ of the transition temperature, in polymers one must expect considerable creep at levels of temperatures which are close to service conditions.

It is common to represent creep and stress relaxation data for polymers in a range of temperatures, usually by means of creep modulus or relaxation modulus curves. The creep modulus is defined as the applied stress (constant throughout the test) divided by the strain at a given time, while relaxation modulus is defined as the stress measured at given time divided by the strain (constant throughout the test). Both moduli decrease with time, and both types of curves provide a similar kind of information.

Relaxation modulus curves are shown in Fig. 7.12a. The temperature dependence of the relaxation modulus can be obtained by plotting the data obtained at a specific creep time as a function of temperature (Fig. 7.12b). The latter curve resembles the modulus of elasticity-temperature curves for polymers, as presented in Chapter 15. This similarity indicates equivalent effects of time and temperature, which are

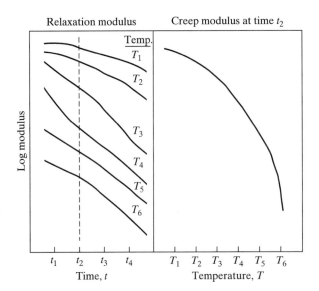

Figure 7.12
Relaxation modulus curves of polymeric materials at different temperatures (a) and the effect of temperature on the creep modulus obtained at time t_1 (b).

expected to occur in a process which is thermally activated: both an increase in temperature and an increase in time period will increase the probability of a segment acquiring sufficient energy to overcome the activation energy barrier to movement.

This similarity has led to the development of the time-temperature correspondence concept, which enables the prediction of a relaxation or creep curve at a specified temperature from the curve obtained at another temperature. This is done by means of a *shift factor* in the time scale in the relaxation and creep test. This factor is a function of the material and the temperatures of the two curves. The shift factor A_T for amorphous polymers is given by:

$$\log A_T = \frac{-C_1(T - T_g)}{C_2(T - T_g)}, \tag{7.13}$$

where A_T is the shift factor between the curve at temperature T and the transition temperature T_g, C_1 and C_2 are constants which change slightly from one polymer to the other. The curve at T_g is called the reference curve. The relations between this curve and the other curve at temperature T can be formulated as follows:

$$E(T_g, t_{ref}) = E(T, t) \tag{7.14}$$

$$t_{ref} = t/A_T.$$

With this concept it is possible to calculate the creep (relaxation) modulus-time curve at one temperature if the curve at another temperature is known. Thus, with a single curve, referred to as master or reference curve, it is possible to determine the whole family of curves at different temperatures (Fig. 7.13). This concept is usually applied to predict the long-term creep and stress relaxation expected at lower temperatures, by carrying out a short test at a higher temperature. This procedure is also common in asphalts. A similar concept has been applied in metals to predict the time to fracture in creep, using the Larson-Miller parameter, which is defined as:

$$T(C + \log t_r), \tag{7.15}$$

where C is a constant (usually on the order of 20), T is the temperature in the Kelvin scale, and t_r is the fracture lifetime. The fracture lifetime of a given metal measured at some specific stress level will vary with temperature in order that this parameter will remain constant.

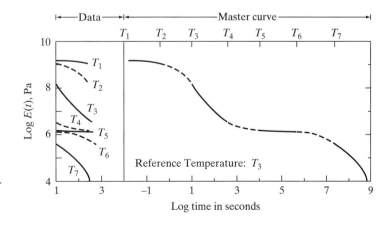

Figure 7.13
Stress relaxation modulus master curves and experimentally measured curves at various temperatures (after Aklonis and Macknight, *Introduction to Polymer Viscoelasticity*, Wiley, 1983).

7.5.3 Creep in Portland Cement Concrete and Wood

The creep mechanisms in portland cement concrete and wood are associated with the movement of water. Both materials are porous and contain water in the pores, as well as in tiny confined spaces between molecules or particles, where considerable bonding of the water to the surface of the solid is effective. The load applied results in stresses in the solid and in the water. The water responds to this stress by flowing slowly to spaces where stresses would be smaller. As a result, creep strain occurs because additional stress is imposed on the solid particles as the water moves out. The rate of this process will depend on the diffusion characteristics of the water in the pore structure (which is a materials property) as well as on the driving force to the diffusion, which is the stress applied on the water in the material. Because diffusion is also a thermally activated process, one may expect dependencies on temperature similar to the ones observed for metals and polymers, but in this case they would not be related to transition temperatures. However, a much more important environmental condition for these materials is humidity. If external drying conditions exist, they will provide an additional driving force for water movement, thus enhancing the time-dependent strains. Drying alone, without any loading, can cause contraction as water is diffusing from the pores and from the spaces between the particles, causing them to approach each other. This strain is referred to as shrinkage, and it comes on top of the creep. In practice, the two occur simultaneously because the wood or portland cement concrete are usually subjected to load and drying at the same time. Although the creep and shrinkage are the result of similar processes, their strains may not necessarily be additive, as discussed in Chapter 11.

BIBLIOGRAPHY

AKLONIS, J. J. and MACKNIGHT, W. J., *Introduction to Polymer Viscoelasticity,* John Wiley and Sons, 1983.

ASKELAND, D. R., *The Science and Engineering of Materials,* 3rd edition, PWS-Kent Publishing Co. Boston, 1984.

BANFILL, P. F. G. (editor), *Rheology of Fresh Cement and Concrete,* E&FN Spon, UK, 1991.

BAZANT, Z. P. and CAROL, I. (editors), *Creep and Shrinkage of Concrete,* E&FN Spon, UK, 1993.

ILLSTON, J. M. (editor), *Construction Materials,* E&FN Spon, UK, 1994.

Superpave Performance Graded Asphalt Binder Specifications and Testing, Asphalt Institute Superpave Series No. 1, (SP-1), Asphalt Institute, Lexington, Kentucky, USA, 1994.

TATTERSALL, G. H., *Workability and Quality Control of Concrete,* E&FN Spon, UK, 1991.

PROBLEMS

7.1 Discuss the differences in the creep processes of cementitious materials and polymers.

7.2 Dry and wet wood specimens are exposed to creep under load. In which will the creep strains be higher? Explain.

7.3 Two polymers which are similar in their general structure have a different glass transition temperature. If both are subjected to a constant load at a similar service temperature, which will experience higher creep strain? Explain.

7.4 Creep of engineering materials can follow a behavior which can be described by a Kelvin model in series with a spring or a Maxwell model.

    (a) From an engineering point of view which behavior is preferred? Explain.

    (b) What loading and environmental parameters can cause a shift in the creep behavior of a given material to be changed from one which can be described by a Kelvin model in series with a spring to a Maxwell model.

7.5 In order to predict the viscoelastic behavior of an asphalt material is it necessary to carry out a laboratory test at the actual service temperature? If not, would you recommend it out at a higher or lower temperature? Explain.

7.6 Two fresh concrete mixes with a rheological behavior which can be described by a Bingham model have the same apparent viscosity which is measured in the mixer at a specified rotation velocity.

    (a) Do the mixes necessarily have the same rheological properties?

    (b) If not, what could be the differences, and which rheological behavior would be preferred?

7.7 Discuss which rheological behavior would be preferred for a fresh mortar which is applied on a vertical surface: Newtonian or Bingham?

8

Fatigue

8.1 INTRODUCTION

Thus far, we have only considered the response of materials either to monotonically increasing loads (Chapters 5 and 6) or to sustained loads (Chapter 7). However, under certain conditions, a material may fail due to the *repeated* application of loads that are not large enough to cause failure in a single application. This phenomenon is referred to as *fatigue failure*. The existence of fatigue failure implies that, under repeated stresses, materials may undergo some internal, progressive, permanent structural changes. Fatigue failures are all the more dangerous because they generally occur suddenly, without significant prior deformations.

Fatigue failure may result either from a repetition of a particular loading cycle or from a random variation in stress. There are many examples of fatigue failure. There is hardly a student who has not whiled away a particularly boring lecture by bending paper clips back and forth until they broke; this remains the classic classroom example of fatigue failure. The fracture surface of such a paper clip is shown in Fig. 8.1. On a more serious note, fatigue failures are also common in axles, propeller shafts and blades, crank shafts, and moving machine parts. One of the most famous examples of fatigue failure was the Comet aircraft disasters of 1953 and 1954, in which three passenger aircraft crashed, with considerable loss of life. In each case, the failure appeared to be due to the growth of a crack starting from a small rivet hole in the fuselage; the repeated loading was due to the pressurizing and the relaxing of the fuselage as the aircraft ascended and descended.

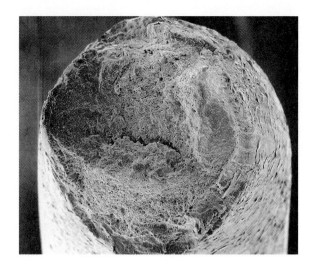

Figure 8.1
Fracture surface of an
ordinary paper clip broken
by ~6 cycles of repeated
bending (80×).

8.2 THE NATURE OF FATIGUE FAILURE

There are two basic stages involved in fatigue failure:

1. Crack initiation
2. Propagation of the crack through the material.

8.2.1 Crack Initiation

Generally, fatigue cracks originate at a free surface, at a point of high stress concentration in the material. This may often be a preexisting flaw in the material, or perhaps a human-made discontinuity, such as the root of a thread, a rivet or bolt hole, or any point at which there is a sharp change in the size or shape of the material. It appears that, at least in metals, fatigue cracks nucleate due to the mechanism of *slip*. Recall that slip occurs by the movement of dislocations (Chapters 2 and 7), which produce fine slip bands. However, at a free surface, when slip takes place, the relative displacements of the atoms along the slip planes cause "steps" to occur, of the order of a nanometer (10^{-9} m) high. Under cyclic loading, reversed slip on adjacent slip planes may lead to the formation of *extrusions* and *intrusions* at the surface, as shown in Fig. 8.2. These may act as the nucleus of a surface crack, as additional slip continues to occur along only a few slip bands rather than across a much wider region. Thus, incipient fatigue cracks may form after only 5% to 10% of the specimen's fatigue life. Subsequently, a crack will begin to grow. Initially, the crack will grow along the slip plane, but it will eventually change direction until it is growing in a plane perpendicular to the principal tensile stress (Fig. 8.2).

In nonmetallic materials, other flaws or inhomogeneities act as the weak points from which fatigue cracks propagate. For instance, in concrete, the interface between the cement and the coarse aggregate particles, which is a zone of weakness, is thought to be the region in which fatigue cracks initiate.

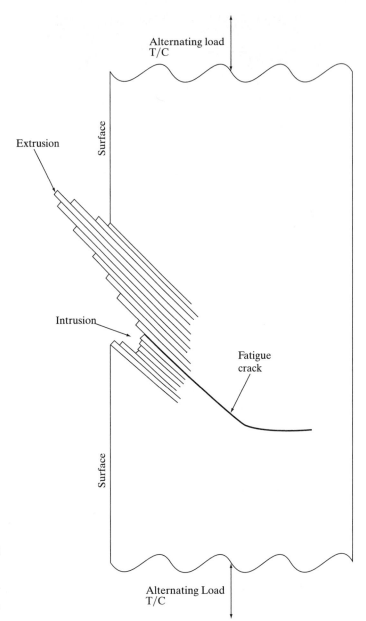

Extrusion

Surface

Alternating load
T/C

Intrusion

Fatigue
crack

Surface

Alternating Load
T/C

Figure 8.2
Schematic diagram of crack
initiation and subsequent
crack growth, first along
the slip line and then at
right angles to the principal
tensile stress.

8.2.2 Crack Propagation

Fatigue cracks will propagate under shear or tensile loading but not under compressive loading, since compression will *close* cracks rather than open them. On each tensile loading cycle, very high stresses occur at the crack tip (due to the stress concentrating effect of a sharp crack), causing the crack to propagate into the still undamaged material ahead of it. It is important to note that the crack propagates a finite distance in *each* loading cycle; this crack advance may be as much as 25 μm/cycle. The process of crack extension can be modeled (at least for metals) as shown in Fig. 8.3. Initially, the crack is sharp; under tensile loading, it both opens and advances, becom-

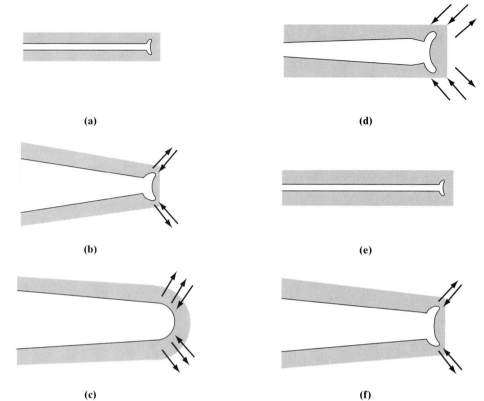

Figure 8.3
The process of fatigue crack propagation: (a) zero load; (b) small tensile load; (c) maximum tensile load; (d) small compressive load; (e) maximum compressive load; (f) small tensile load. The double arrowheads in (c) and (d) signify the greater width of slip bands at the crack in these stages of the process. The stress axis is vertical.

(a)

(b)

(c)

(d)

(e)

(f)

ing blunted as a plastic zone develops at the crack tip. During the unloading cycle, the material at the crack tip is compressed, which again sharpens the crack. In addition, the back stresses exerted on the material during unloading cause deformation markings (often called *striations* or *beach marks*) on the fracture surface, as shown in Fig. 8.4. This process is repeated until the crack reaches some critical length, at which time a sudden failure occurs, in either a brittle or a ductile manner, depending on the characteristics of the particular material.

Figure 8.4
Micrograph of fatigue fracture surface, showing the characteristic beach marks. From N. H. Polakowski and E. J. Ripling, *Strength and Structure of Engineering Materials*, Prentice Hall, 1966.

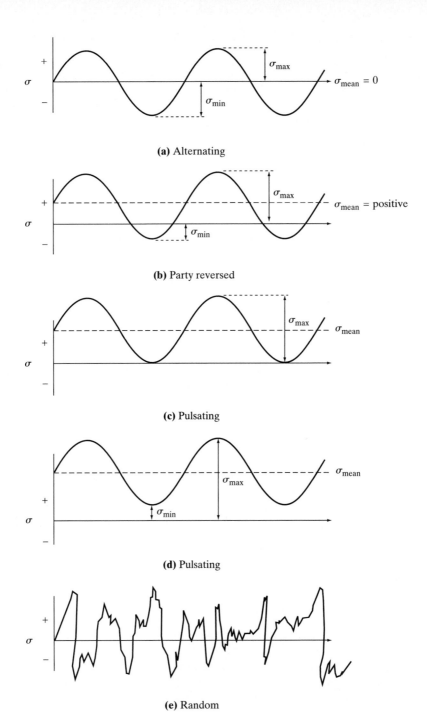

(a) Alternating

(b) Party reversed

(c) Pulsating

(d) Pulsating

(e) Random

Figure 8.5
Common types of fatigue loading.

There are, of course, infinitely many possible repeated loading spectra. However, the most common types of fatigue loading are shown in Fig. 8.5. In Table 8.1 we define several terms which are useful in describing fatigue loading. The simplest form of repeated loading, and that most often carried out during fatigue testing, is the constant-frequency, sinusoidally varying stress shown in Fig. 8.5a. For this type of loading, which represents the bending stresses in a rotating shaft, the maximum and minimum stresses are numerically equal, (i.e., $|\sigma_{max}| = |\sigma_{min}|$). Thus, the mean stress, σm, is zero. The fluctuating stress shown in Fig. 8.5b represents the case of partly reversed bending. The mean stress, $\sigma m \neq 0$, may be considered to consist of a static preload superimposed on the alternating stresses of Fig. 8.5a. A special case of fluctuating stress, shown in Fig. 8.5c, occurs when the minimum load is equal to zero; here the mean stress is one-half of the maximum stress (i.e., $\sigma m = \sigma_{max}/2$). When σ_{max} and σ_{min} are both either tensile or compressive, as shown in Fig. 8.5d for the tension case, this is referred to as a *pulsating stress*. Finally, Fig. 8.5e represents the case of a randomly varying stress, such as might occur in practice during, say, the vibration of an airplane wing in flight or traffic loading of an interstate pavement.

TABLE 8.1   Definitions Relating to Fatigue Loading.

σ_{max} = maximum stress in the loading cycle = S
σ_{min} = minimum stress in the loading cycle
σm = mean stress = $(\sigma_{max} + \sigma_{min})/2$
σa = stress amplitude = $(\sigma_{max} - \sigma_{min})/2$
σr = stress range = $\sigma_{max} - \sigma_{min} = 2\sigma a$
R = stress ratio = $\sigma_{min}/\sigma_{max}$
A = amplitude ratio = $\sigma a/\sigma m$
N = number of loading cycles to failure for a particular set of loading conditions
σF = fatigue limit, or endurance limit

Note: The usual sign convention is used: tension is + (positive), compression is − (negative).

8.4   BEHAVIOR UNDER FATIGUE LOADING

The *fatigue life,* or *endurance,* of a material refers to the number of repeated cycles of loading (N) that the material may undergo before it fails. Of course, N will vary with the particular set of loading conditions. However, the general rule is that the higher the fatigue stress level, the fewer the number of loading cycles required to cause failure. Fatigue data are most commonly presented in terms of *S-N curves*. *S-N* curves are generally drawn as a plot of the maximum stress as the ordinate versus the logarithm of the number of cycles to failure as the abscissa. (S is the symbol generally used for the maximum stress, σ_{max}; for fully reversed cyclic loading, as in Fig. 8.5a, S is also equal to the stress amplitude, σa.)

S-N curves are shown in Fig. 8.6 for the two possible types of material response to cyclic loading. Curve *a* represents the behavior of ferrous alloys and titanium. These materials exhibit a definite, or true, *fatigue limit* (or *endurance limit*), σF. That is, they may withstand infinitely many cycles of loading without failing, as long as $S < \sigma F$. On the other hand, materials such as aluminum and other nonferrous alloys, concrete, and wood behave as shown by curve *b*. These materials do not exhibit

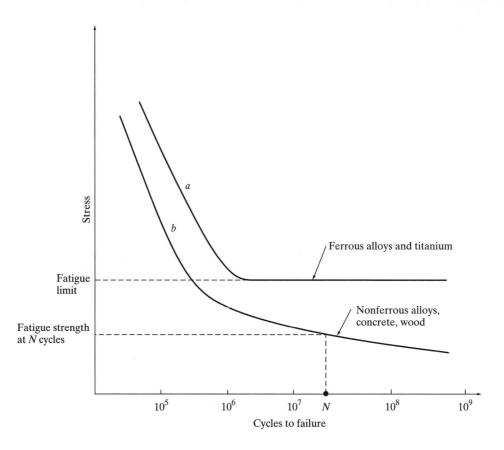

Figure 8.6
Two types of material response to cyclic loading: (a) materials with a definite fatigue limit; and (b) materials without a definite fatigue limit.

a definite fatigue limit. Rather, for any value of S, however low, they will eventually fail under cyclic loading if N is sufficiently large. For such materials it is common to define a *fatigue strength for a given number of cycles,* as shown in Fig. 8.6.

How can we explain this difference in behavior? A *perfect* Hookeian solid (i.e., a perfectly elastic material, with no imperfections or discontinuities on either the micro- or macroscale) should be able to withstand an infinite number of loading cycles, as long as the stress range is confined to within the elastic range of the material. Such a material would always exhibit the stress-strain behavior shown in Fig. 8.7a. However, as was described in Chapters 2 and 3, there is no such thing as a perfect solid. All materials contain a variety of imperfections, ranging from microstructural flaws, such as lattice defects and grain boundaries, to microcracks and capillary pores, and finally to large discontinuities, such as bolt holes. All of these flaws tend to act as stress concentrations (see Chapter 6), and under any form of loading, there is bound to be at least some rearrangement of atoms and molecules at these points of high stress concentration. Therefore, under cyclic loading, the stress-strain behavior of a real material, even when stressed only within its elastic range, will most likely be represented by the *hysteresis loop* shown in Fig. 8.7b. The area contained within the hysteresis loop can be considered to be the amount of energy (per unit volume) irreversibly lost during each cycle of loading; some of this energy goes into a rearrangement of atoms and molecules, and most of the rest into heat.

If, below some stress level, the hysteresis loop stabilizes completely (or if the material behaves perfectly elastically), this indicates that any structural changes that occur in the high stress region during each cycle of loading are reversible and do not lead to further "damage." Materials which behave in this way will have a definite

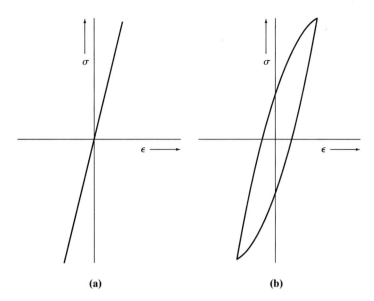

Figure 8.7
Perfect elasticity (a) and "elastic" hysteresis loop (b) exhibited by a cyclically stressed material below the fatigue limit.

(a)

(b)

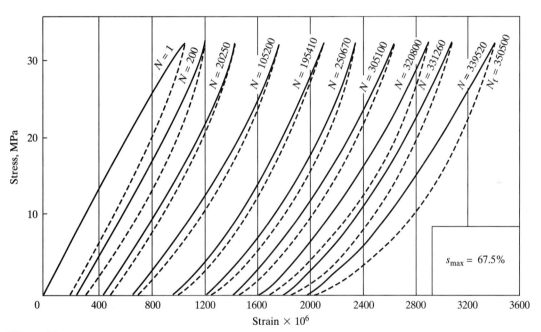

Figure 8.8
Variation of stress-strain curve with number of cycles, for concrete in cyclic compression.

fatigue limit. On the other hand, if the hysteresis loop continues to grow, however slowly, then failure will eventually occur. An example of a material exhibiting an unstable hysteresis loop is shown in Fig. 8.8. These data are for concrete cyclically loaded only in the compression range, up to about two-thirds of its ultimate compressive strength.[1] Since the loading does not extend into the tensile range, only the upper half of the hysteresis loop is shown. The progressive change in both shape and size of the hysteresis loop is a clear indication of steadily increasing damage in the specimen.

[1]While we have said that fatigue cracks do not grow in compression, in concrete, compressive loading induces secondary tensile stresses, and it is these stresses which cause crack growth.

The physical criteria which determine whether a hysteresis loop will either stabilize or continue to grow are not well understood. However, for ferrous metals, it appears that the intercrystalline regions are stiff enough to restrain the spread of slip *within* the crystals until a sufficiently high stress is reached. These materials, therefore, have a definite fatigue limit. On the other hand, for nonferrous metals, the intercrystalline regions tend to be relatively soft and are incapable of restraining the slip within the crystals even at low stresses. Therefore, these materials do not have a true fatigue limit. Heterogeneous materials such as concrete have numerous internal flaws and discontinuities which can initiate active cracks.

8.5   THE STATISTICAL NATURE OF FATIGUE

As we have already seen in the preceding chapters, all measured mechanical or physical properties of materials are subject to a certain amount of *variability*. This variability is due primarily to two factors:

  1. For real materials, containing imperfections of various kinds, no two samples of material are truly identical.
  2. It is virtually impossible to reproduce precisely the same test conditions over a large number of tests.

Fortunately, for most mechanical or physical properties, the variations in observed test data are relatively small, and an average value based on a small number of test observations will fairly represent the property in question. For instance, for wood, which is probably the most variable of structural engineering materials, bending test results on small, clear specimens would all be expected to lie within ±50% of the mean value; that is, the ratio between the highest and lowest strength would be about 3:1. For concrete cylinders prepared in a laboratory and tested in compression, all of the measured strength values would be expected to fall within ±15% of the mean value, giving a ratio of the highest to the lowest strength of about 1.35:1. For steels, the variability in tensile strength would be much smaller again, certainly within ±5% of the mean value, giving a ratio of strength of only about 1.1:1.

For fatigue data, however, the scatter in results is much larger, as shown schematically in Fig. 8.9. It is not uncommon to find tests in which the ratio of the largest N (number of cycles to failure) to the smallest N is 100. Thus, *S-N* curves by themselves may be misleading. Customarily, *S-N* curves are a plot of the maximum stress versus the *median2* fatigue life (Fig. 8.9). However, the median fatigue life is not a useful value for design purposes, since, of course, about 50% of the specimens will fail *below* this value. Therefore, to establish a satisfactory basis for choosing design values in practice, a *statistical* approach must be taken. One way of doing this is to look at the *distribution* of fatigue data. Figure 8.10 shows the distribution of the number of specimen failures under identical loading conditions versus the logarithm of the number of cycles to failure. It has been found that, if the experimental fatigue data are plotted in this way, the data will give a close approximation to the *normal,* or *Gaussian,* distribution. This means that fatigue data can be represented as having a lognormal distribution (which can be considered to be a transformation of the normal distribution function, in which the variable is expressed as the logarithm of N

[2]The median fatigue life is defined (see ASTM E208) as "the middlemost of the fatigue life values, arranged in order of magnitude, of the individual specimens in a group tested under identical conditions. In the case where an even number of specimens are tested, it is the average of the two middlemost values." For presentation of fatigue data, the use of the sample median rather than the arithmetic mean is preferred.

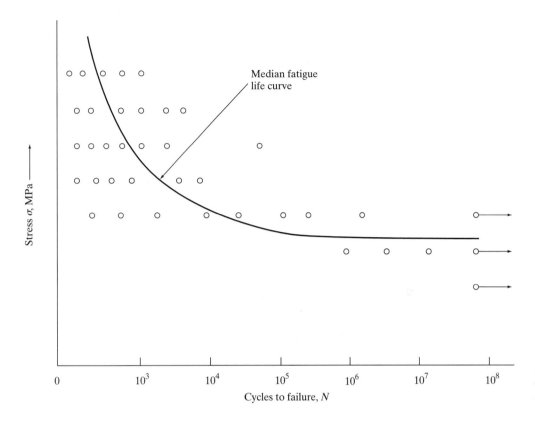

Figure 8.9
Plot of stress-cycle (*S-N*) data as might be collected by laboratory fatigue testing of a material. Adapted from J. A. Collins, *Failure of Materials in Mechanical Design: Analysis, Prediction, Prevention,* John Wiley, 1981.

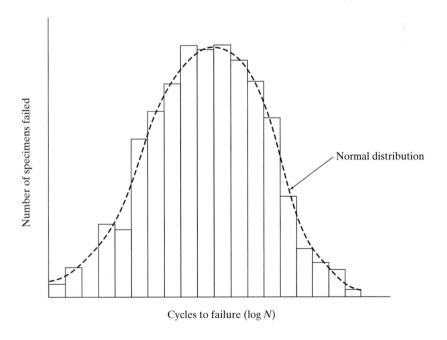

Figure 8.10
Distribution of fatigue specimens failed at a constant stress level as a function of logarithm of the number of cycles to failure (life). Adapted from J. A. Collins, *Failure of Materials in Mechanical Design: Analysis, Prediction, Prevention,* John Wiley, 1981.

rather than as N itself). The assumption of a lognormal distribution is valid only when the maximum stress is reasonably high compared to the fatigue limit; when the maximum stress is near the fatigue limit, the nature of the distribution function is uncertain. For the purposes of the following discussion, however, we will assume that the lognormal distribution function is a reasonable approximation.

8.6 THE STATISTICAL PRESENTATION OF FATIGUE DATA

The problem in the analysis of fatigue data is to take account of the scatter so that sensible design values can be chosen. There are a number of ways in which this can be done. One method would be to draw lines enclosing the entire scatter band, as shown in Fig. 8.11a. However, this method is less conservative than may appear at first glance:

1. The width of the scatter band depends on the number of specimens tested; it increases with increasing sample size.

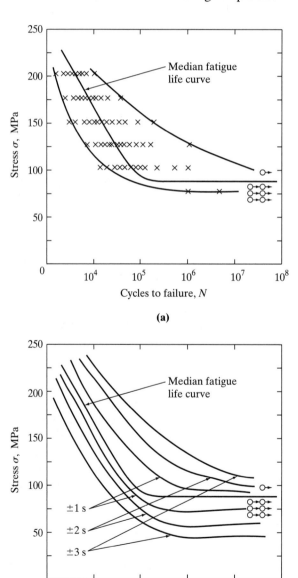

Figure 8.11
Plot of stress-cycle (*S-N*) data as might be collected by laboratory fatigue testing of any material: (a) lines enclosing the scatter band; (b) lines showing the standard deviations of the data.

2. Statistically, one would still expect at least some data to fall outside of any scatter band.

A more useful way of presenting fatigue data is to assume the lognormal distribution described earlier and then to draw a family of lines on the *S-N* curve showing the standard deviations, *s*, from the median line, as in Fig. 8.11b. Then, from the known properties of the normal distributions, one can show that

68.3% of the data will fall within ±1*s* of the median curve;

95.4% of the data will fall within ±2*s* of the median curve; and

99.7% of the data will fall within ±3*s* of the median curve.

This type of plot permits the designer to make an estimate of the fatigue stress level at which there is an acceptable probability (or risk) of failure for any particular application.

An even more useful way of presenting fatigue data, which is a variation of Fig. 8.11b, is as *S-N-P curves,* which are a family of *S-N* curves for different probabilities of failure, *P.* (Since we may define *reliability* as $R = 1 - P$, these are sometimes referred to as *R-S-N* curves.) Such a family of curves is shown in Fig. 8.12. This form of data presentation allows the designer most easily to choose an appropriate value of fatigue strength for a given application. Note that the curve for $P = R = 0.5$ represents the usual median curve.

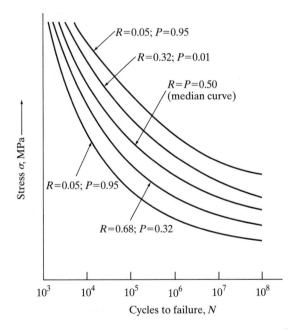

Figure 8.12
Family of *S-N-P* curves, or *R-S-N* curves, for 7075-T6 aluminum alloy. (Note: *P* = probability of failure; *R* = reliability = 1 − *P*.)

8.7 FACTORS AFFECTING FATIGUE LIFE

There are a great many variables which affect the behavior of materials under fatigue loading. We will discuss the most important ones in this section, primarily in terms of their effects on the median *S-N* curves. As shown in Table 8.2, the factors which influence fatigue behavior may be grouped broadly into three categories: *stressing conditions, material properties,* and *environmental conditions.*

TABLE 8.2 Factors Affecting Fatigue.

Stressing conditions	Type of stress
	Stress amplitude
	Mean stress
	Frequency
	Combined stresses
	Stress history
	Stress concentrations (notches)
	Size
	Rolling contact
Material properties	Type of material
	Surface conditions
	Grain size
Environmental conditions	Temperature
	Thermal fatigue
	Corrosion

8.7.1 Stressing Conditions

Type of Stress

The simplest and most common type of fatigue loading is that of fully reversed alternating stresses, with a mean stress of zero, as shown in Fig. 8.5a; most laboratory fatigue data have been measured for this case. However, fatigue may also occur under other types of loading, such as the torsional loading of coil springs (giving a state of pure shear), or repeated compressive loading of some concrete members. The behavior under these types of loading depends on the particular material, whether it is ductile or brittle, and so on. For a relatively brittle material such as concrete, for example, the fatigue strength, as a fraction of the static strength, is essentially the same for tension, compression, or bending. For ductile materials, however, this may not be true (see the following discussion of mean stress). For all forms of loading and stress conditions, the stress amplitude has the strongest influence on fatigue life, as shown by the *S-N* curves.

Mean Stress

Although most fatigue data are collected for a mean stress of zero, a nonzero mean stress often occurs in practice, due to the presence of residual stresses or to the effects of dead weight loading. In general, increasing the mean stress (i.e., going in the direction of compressive to tensile σm) for a given stress amplitude, σa, will diminish the fatigue life or will decrease the safe amplitude for a given fatigue life. This is shown schematically in Fig. 8.3. (Alternately, if we keep the maximum stress *fixed,* increasing the mean stress will lead to a larger fatigue life, since this implies a decrease in the alternating stress amplitude.) The fatigue life is affected by σm because tensile and shear stresses (which may occur as secondary stresses even under compressive loading) have the ability to enlarge voids and to propagate microcracks.

Figure 8.13 shows, schematically, the type of data that would be obtained if a series of tests could be carried out for different combinations of σa and σm, for a fatigue life of N cycles. As the mean stress approaches the ultimate tensile strength, σu, then the allowable stress amplitude approaches zero. For $\sigma m = 0$, we get the conventional fatigue strength, σF. It may be seen that $\sigma m < 0$ (i.e., compressive) has relatively small effects; $\sigma m > 0$ (i.e., tensile) has much larger effects. Since it is

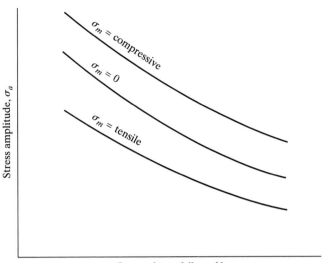

Figure 8.13
Effect of mean stress on
fatigue life.

not feasible to carry out tests on all materials for a large combination of values of σm and σa, relatively few data of the type shown in Fig. 8.14 are available. Therefore, a number of purely empirical relationships have been developed, using curve-fitting techniques, to describe the effects of a nonzero mean stress on the fatigue life. The relevant equations for the most successful of these are summarized in Table 8.3.

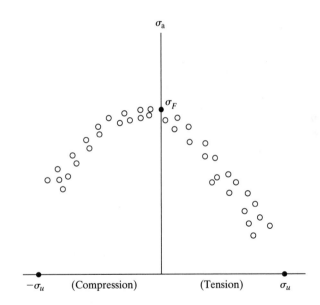

Figure 8.14
Simulated high-cycle
fatigue failure data showing the influence of mean
stress. Adapted from J. A.
Collins, *Failure of Materials
in Mechanical Design:
Analysis, Prediction, Prevention,* John Wiley, 1981.

Frequency of Loading

For stresses within the elastic range, the fatigue strength appears to increase with an increase in the frequency of cyclic loading, but the effect is small within the normal range of frequencies used for fatigue testing. At very high cycling rates, however, the fatigue strength often shows a sharp decrease, as shown in Fig. 8.15. For stresses that cause plastic deformation, a high loading frequency may cause a considerable heat buildup in the specimen, which will have a major effect on fatigue life (see Sec. 8.7.3).

Goodman (linear)	$\dfrac{\sigma_a}{\sigma_f} + \dfrac{\sigma_m}{\sigma_u} = 1$
Gerber (parabolic)	$\dfrac{\sigma_a}{\sigma_f} + \left(\dfrac{\sigma_m}{\sigma_u}\right)^2 = 1$
Elliptic	$\left(\dfrac{\sigma_a}{\sigma_f}\right)^2 + \left(\dfrac{\sigma_m}{\sigma_u}\right)^2 = 1$
Soderberg (linear, but written in terms of yield stress)	$\dfrac{\sigma_a}{\sigma_f} + \dfrac{\sigma_m}{\sigma_y} = 1$

Note: In all of these equations, σf = fatigue strength, σa = stress amplitude for a given fatigue life.

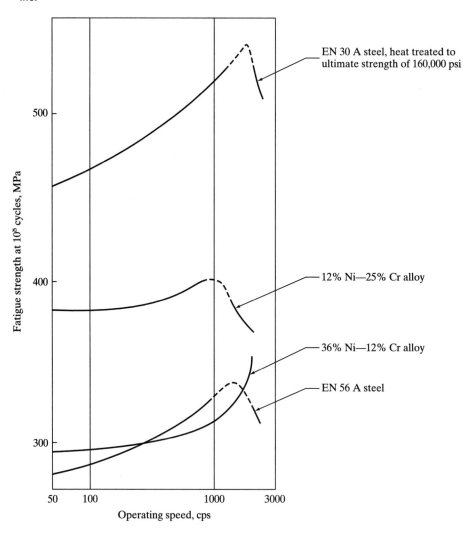

Figure 8.15
Effect of operating speed on the fatigue strength at 10^8 cycles for several different ferrous alloys. Adapted from J. A. Collins, *Failure of Materials in Mechanical Design: Analysis, Prediction, Prevention,* John Wiley, 1981.

Stress History

As has been stated earlier, the vast majority of fatigue tests are carried out by cycling the specimen at a constant stress amplitude and frequency until it fails. However, in practice, materials undergo a much more complex stress history, involving variation in both the magnitude and the frequency of the loading. Since frequency effects tend to be small, we will deal here only with the magnitude of loading. After each cycle of stress large enough to cause fatigue damage, there is at least some slight, but irreversible, structural change in the material. The *accumulation* of damage brought on by successive loading cycles thus leads to a continuing change in the *S-N* curve exhibited by the material. In other words, as shown in Fig. 8.16, a specimen which has already undergone a number of loading cycles has a different *S-N* curve from that of the undamaged, virgin material.

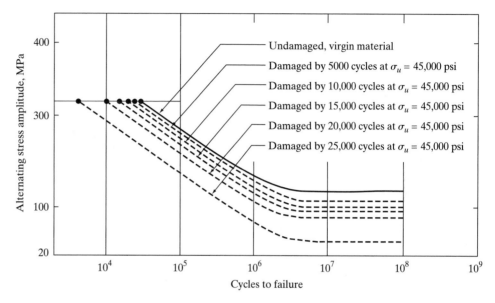

Figure 8.16
Illustration of the influence of accumulated fatigue damage on subsequent fatigue behavior of carbon steel (Note: life of virgin material at $\sigma_\alpha = 317$ MPa [45,000 psi] is approximately 30,000 cycles.)

Since the sequence of loading can affect the fatigue life of a material, the practical engineering problem that must be solved becomes one of predicting the fatigue life under variable amplitude loading. The empirical approach to this problem is based on the concept of cumulative damage. That is, it is assumed that for each cycle of loading at a particular stress amplitude, a certain amount of damage occurs. When the total damage accumulated over the entire load history reaches some critical value, failure occurs.

Over the years, many different cumulative damage theories have been proposed; we will deal here only with the oldest and simplest of these, the Palmgren-Miner hypothesis, or the *linear damage rule*. With reference to Fig. 8.17, failure would be expected to occur by the Palmgren-Miner hypothesis when

$$\sum \frac{N_i}{N_{f_i}} \geq 1, \tag{8.1}$$

where N_i is the number of stress cycles in region i, and N_{f_i} is the number of cycles that would be required for failure under the stress pattern in region i. That is, using the Palmgren-Miner hypothesis, it is assumed that the amount of damage, N_i/N_{f_i}, at any stress level is linearly proportional to the ratio of the number of cycles, N_i, to the total number of cycles, N_{f_i}, required to cause fatigue failure at that stress level.

The Palmgren-Miner hypothesis is very simple to apply and hence is widely used. It has two major shortcomings:

1. The damage that accumulates in each cycle of loading is assumed to be independent of the stress history, but this is clearly not true, as shown by Fig. 8.16.

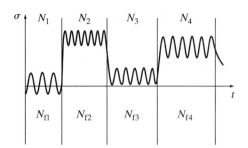

Figure 8.17
Summing damage due to
initiation-controlled
fatigue.

2. It is assumed that there is no effect of the *order* in which different stress levels are applied, but experimental evidence indicates that this too is untrue.

Nonetheless, in spite of its shortcomings, the Palmgren-Miner hypothesis at least permits the designer to make a reasonable preliminary estimate of fatigue life.

Stress Concentrations

As we have seen in Chapter 6, any geometrical discontinuities (holes, notches, cracks, abrupt changes in cross section) will introduce stress concentrations into a material. The severity of the stress concentration increases with increasing flaw size and with a decreasing radius of curvature of the flaw tip (i.e., with increasing sharpness of the discontinuity). Stress concentrations may have a serious effect on fatigue behavior; their general effect is to reduce the fatigue life. Under repeated loading, stress concentrations may initiate cracks in the localized, highly stressed volume of material near the tip of a flaw. These cracks can then propagate, as described in Sec. 8.4. To make matters worse, the relatively flat shape of the *S-N* curve means that even a small increase in load stress due to stress concentrations will bring about failure much more quickly. The reduction in fatigue life increases with the severity of the flaw; this is shown by changes in the *S-N* curves for specimens with different flaws, (Fig. 8.18).

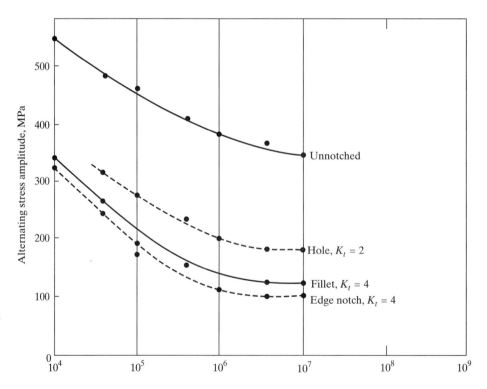

Figure 8.18
Effects of geometrical discontinuities on the *S-N* curve of SAE 4130 steel sheet, normalized, tested in completely reversed axial fatigue test. *Kt* is the theoretical linear elastic stress concentration.

There are a great deal of data regarding the values of the linear elastic stress concentration factors, Kt, for a variety of specimen and notch geometries. (Remember that Kt is defined as the ratio of the maximum stress at the root of a notch calculated from elastic theory to the nominal stress at the same point in the absence of a notch; see Eq. 6.12). Much less data of this kind exist for fatigue. However, it has been found that in fatigue, the severity of the stress concentrations is somewhat reduced. That is, if we define a *fatigue strength reduction factor, Kf*, as the ratio of the fatigue limit for an unnotched specimen to that for a notched specimen, then $Kf > Kt$, though in some cases the difference is small. This difference between Kf and Kt is thought to be due primarily to two effects:

1. Since in fatigue the highly stressed region is localized, there is a lesser probability of finding a critical flaw within it.
2. There is a notch size effect, related to the shape of the stress gradient at the root of a notch, as a function of notch radius.

For easier comparison of different materials and notch geometries, it is convenient to define a fatigue notch sensitivity index:

$$q = \frac{K_f - 1}{K_t - 1}.\tag{8.2}$$

The values of q, as determined from experiments, lie between 0 and 1. If the notch has no effect on fatigue properties, $Kf = 1$, and $q = 0$. Conversely, if the value of Kf becomes equal to that of Kt, then $q = 1.0$. It has been found that q increases with increases in the tensile strength of the material, and with increasing severity of the flaw (i.e., increasing Kt).

In general, it has been found that Kf increases with increasing specimen size (i.e., larger specimens exhibit a shorter fatigue life than smaller ones). This is due in part to the fact that in larger stressed volumes of material, there is a greater probability of finding a critical flaw, though other factors may also play a role. Thus, one should remember that laboratory-size fatigue specimens will generally exhibit higher fatigue strengths than are found in practice.

8.7.2 Material Properties

As a general rule, a finer grain size leads to improved fatigue behavior. This is thought to be due to the observation that smaller grains appear to inhibit the nucleation of fatigue cracks. Since fatigue failures generally initiate at the surface, the surface condition of the material may be important. In general, *smooth surfaces* will result in improved fatigue behavior. The *type of surface finish* can also be important. Grinding, plating, and so on may all affect fatigue life, either for better or for worse. Finally, *residual stresses* in the surface must be considered. Tensile residual stresses will reduce the fatigue life. On the other hand, compressive residual stresses, which in metals may be induced by cold-rolling or shot-peening, will increase the fatigue life.

8.7.3 Environmental Conditions

Temperature

The fatigue strength of most materials decreases as the temperature increases, but this effect is small below the temperatures at which *creep* is not significant (for metals, this is about one-half the melting point temperature). Materials that exhibit a definite fatigue limit at room temperature apparently lose this ability at high temperatures.

Stresses induced in a material by *thermal cycling* rather than by load cycling may also lead to fatigue failure; this is referred to as *thermal fatigue*. This form of loading may occur in internal combustion engines, exhaust flues, furnaces, nuclear reactors, and so on. It is a very complicated phenomenon, since it includes not only all of the loading effects described previously, but also thermal effects. The thermal stress that results from a change in temperature may be estimated from the expression

$$\sigma_{\text{thermal}} = \alpha E \, \Delta T, \qquad (8.3)$$

where α is the coefficient of thermal expansion and ΔT is the change in temperature. The higher the temperature change, ΔT, the shorter the fatigue life. For wood and concrete, thermal effects may be complicated by superimposed changes in water contents.

Corrosion

The effect of a corrosive atmosphere is to lower the fatigue strength of a material, as shown schematically in Fig. 8.19. Of course, the type of corrosive atmosphere that is important depends on the particular material; it should be remembered that even ambient air may be corrosive to some materials. Corrosion fatigue is a complex phenomenon, since simultaneous cyclic loading and corrosion create more damage than the sum of the two effects taken separately. The *pitting* of the surface that often accompanies corrosion will exacerbate the problem, since the pits will then act as stress raisers. As is the case at higher temperatures, materials that exhibit a definite fatigue limit in a noncorrosive atmosphere lose this characteristic in a corrosive one.

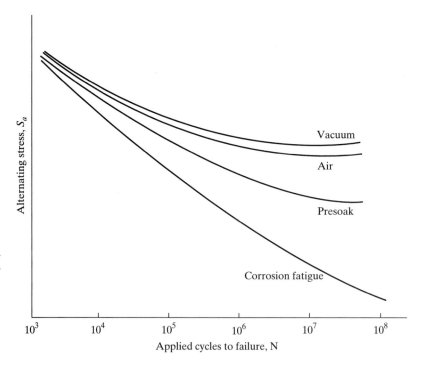

Figure 8.19
Comparison of fatigue life behavior of typical engineering alloys under various environmental conditions. Pre-soaked condition implies soaking specimens in corrosive liquid prior to testing in air (after H. O. Fuchs and R. I. Stephens, *Metal Fatigue in Engineering,* Wiley, 1980, p. 220).

Only in exceptional circumstances (aircraft assemblies, certain automobile parts) are fatigue tests carried out on full-scale components or structural assemblies, because of the specialized equipment required and because of the expense. Normally, fatigue tests are instead carried out on small laboratory specimens, and some of the difficulties in translating these laboratory tests to field performance have been outlined previously. Since fatigue tests may be carried out in bending, direct tension and/or compression, torsion, or multiaxial loading, a bewildering number of fatigue machines have been developed, many of these specific to particular materials. We will describe here only a few of the basic types of fatigue tests.

It is important to recognize that there are two basic types of machines, which may give different information:

1. *Constant load machines*: The loading cycle remains constant throughout the test; the deflection (or strain) gradually increases as the specimen sustains damage.

2. *Constant displacement machines*: The displacement cycle remains constant throughout the test; the resulting stresses may change as the specimen undergoes damage or structural alteration.

Because of the large number of cycles to failure and the number of specimens that must be tested because of the large scatter in the data, testing machines of all types are run at high speeds, typically in the range of 25 to 500 cycles per second.

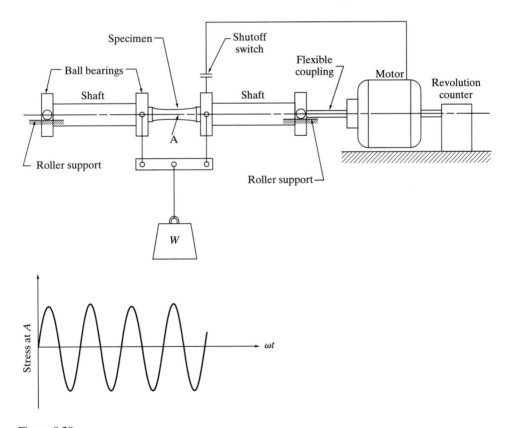

Figure 8.20
Rotating-bending fatigue testing machine of the constant bending moment type.

8.8.1 Fatigue Machines

The most common type of fatigue machine is the *constant moment rotating bending machine,* shown schematically in Fig. 8.20. The central portion of the specimen is subjected to a constant moment; as the specimen rotates, any point on the surface undergoes a sinusoidally varying, fully reversed stress cycle, from maximum compression on the top to maximum tension on the bottom. The major drawbacks of such simple machines are that: (1) they are not suitable for use with a nonzero mean stress; and (2) they require a specimen with a circular cross-section.

To overcome both of these problems, a reciprocating-bending machine, such as the one shown schematically in Fig. 8.21, may be used. This type of machine is often used for flat specimens. By appropriately positioning the specimen with respect to the mean displacement of the connecting rod, a nonzero beam stress may be applied.

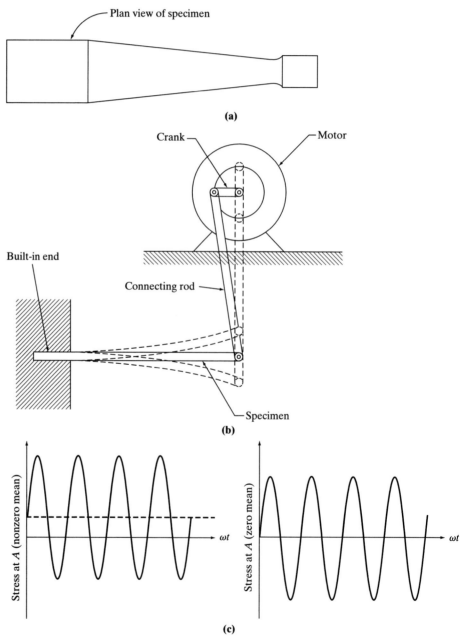

Figure 8.21
Reciprocating-bending fatigue testing machine. Part (b) from C. W. Richards, *Engineering Materials Science*, Wadsworth Publishing Company, 1961, p. 383.

Chap. 8   Fatigue

For direct tension or compression, *direct stress machines* of the type shown in Fig. 8.22 may be used. Stresses are applied by the eccentric drive. By varying the position of the fixed end of the specimen, a nonzero mean stress may also be applied. The machines shown in Figs. 8.20 to 8.22 are relatively simple. Much more complex fatigue machines exist, which require computer control to be used effectively.

8.8.2 Fatigue Test Procedures

Many different types of fatigue testing procedures are in use, depending on the type of information required. Originally, it was common to test only one specimen at each of a number of stress levels; from these data, an *S-N* curve could be con-

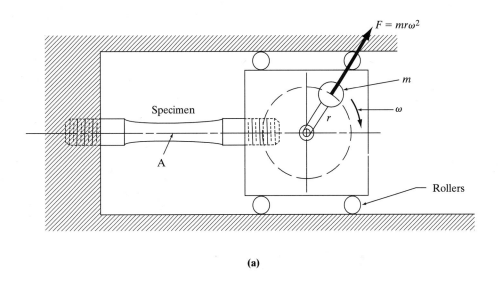

(a)

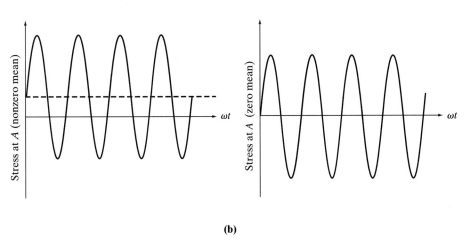

(b)

Figure 8.22
Brute-force direct-stress fatigue testing machine. Part (a) from C. W. Richards, *Engineering Materials Science*, Wadsworth Publishing Company, 19___, p. 384.

structed. However, because of the small sample size, such curves were not particularly reliable. Subsequently, it became more common to run groups of test specimens at each stress level, so that *S-N-P* curves could be constructed and better probabilistic information could be obtained. However, the above-mentioned methods are not very efficient at stresses close to the fatigue (or endurance) limit, because in this range there will always be a number of specimens that are "run out" (that is, they remain unbroken over the number of cycles of loading that can be performed). To determine the fatigue limit more accurately, the *survival method,* or *response method,* may be used. This involves testing groups of specimens at close intervals within, say, ±2 standard deviations of the estimated fatigue life. When the data are plotted on normal probability paper (stress level versus probability of surviving N cycles), the mean fatigue limit may be determined directly.

The survival method too requires a large number of specimens and is thus not very efficient. A more rapid method for finding the fatigue limit, called the *Prot method,* involves testing the specimens at *steadily increasing stress levels* until they fail. There are, therefore, no runouts. Groups of specimens are then run at different rates of stress increase. Prot suggested that the fatigue limit, E, could be calculated from the expression

$$S_\alpha = E + K\alpha^n, \tag{8.3}$$

where S_α is the Prot failure stress, S_α is the rate of stress increase, and K and n are material constants. The data are plotted as in Fig. 8.23. The object is to find a value of n that gives a linear relationship between S_α and αn. Extrapolating this curve back to a zero rate of stress increase gives the fatigue limit. Although this method is now often used, there is still some question about its reliability. Its use is limited to materials that have a definite fatigue limit and that are not sensitive to the accumulated

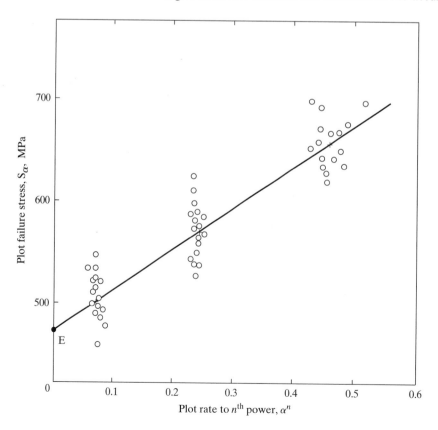

Figure 8.23
Prot test results showing the determination of fatigue limit E by plotting failure stress S_α versus αn.

damage that may have occurred at lower stress levels than the one at which failure finally occurred.

BIBLIOGRAPHY

J. M. Barsom and S. T. Rolfe, *Fracture and Fatigue Control in Structures,* 2nd ed., Prentice-Hall, Englewood Cliffs, New Jersey, 1987.

M. M. Eisenstadt, *Introduction to Mechanical Properties of Metals,* Macmillan, New York, 1971.

H. O. Fuchs and R. I. Stephens, *Metal Fatigue in Engineering,* John Wiley & Sons, New York, 1980.

PROBLEMS

8.1. What is the significance of fatigue in engineering?

8.2. Describe the process of a fatigue failure in a metal.

8.3. In design, how should one account for the very high scatter that occurs during fatigue tests? Would it be safe to design for *average* values of fatigue life?

8.4. How might you improve the fatigue life of a particular member?

8.5. Do *constant load* and *constant displacement* fatigue tests lead to the same results? Discuss.

8.6. What sort of fatigue testing would you carry out if asked to evaluate a new metal alloy for use in the wing or fuselage of a modern airplane? Discuss.

Part III

PARTICULATE COMPOSITES: PORTLAND CEMENT AND ASPHALT CONCRETES

9

Particulate Composites

9.1 INTRODUCTION

Portland cement and asphalt concrete mixtures are both particulate composite materials in that they essentially consist of a single continuous phase, (the portland cement or asphalt cement binder, respectively), and a single discontinuous particulate phase, (the aggregate). As a consequence, many features of the structure and behavior of these materials are similar. However, there are also substantial dissimilarities due to the following:

1. Differences of the relative proportion of the binders used in the two types of concrete: higher contents and some interlocking of the aggregates in the asphalt concrete (Fig. 9.16)
2. Differences in the mechanical properties of the two binders: The asphalt is much softer and more viscoelastic; therefore, there is significance for the interlocking between the aggregates to provide some additional stability.
3. Differences in the process responsible for the hardening of the initially fluid mix: a chemical reaction in the case of portland cement concrete and cooling or evaporation of solvent or water in asphalt concrete.

As a result of these differences, these two materials serve different purposes of construction: asphalt concrete, which is weaker and more flexible, is extensively used for paving roads, where the function of the asphalt layer is to redistribute the concentrated loads applied by moving vehicles so that the loading on the subgrade below will be distributed more uniformly. Portland cement concrete, which is stronger and much less viscoelastic in nature, is used as the material for load-bearing components such as columns, beams, and slabs.

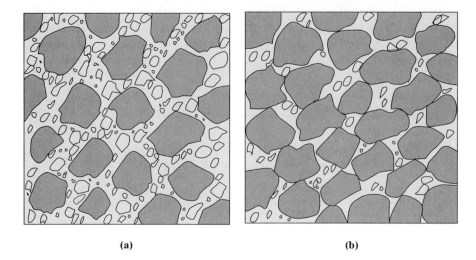

Figure 9.1
(a) Schematic structure of portland cement; and
(b) schematic structure of asphalt concrete.

(a)　　　　　　　　　　　**(b)**

These differences are sufficiently large that it is not possible to present a unified treatment of the two composites; therefore, they will be treated separately in Chapters 11 and 12, following an examination of their common component, the mineral aggregates, in Chapter 10. However, before proceeding, it is worthwhile to describe some of the concepts of the mechanics of particulate composite materials and briefly review their implications for the two concretes in view of their differences in structure and properties. Simple models may be developed to illustrate the principles of behavior of a particulate composite under an external perturbation. Figures 9.2a and 9.2b show the simplest models, in which the composite elements are in series or parallel to the external perturbation. We will consider the elastic response to an applied stress since this is the perturbation of most general interest to a civil engineer. Thereafter, we will discuss failure mechanisms in relation to interfaces between the phases.

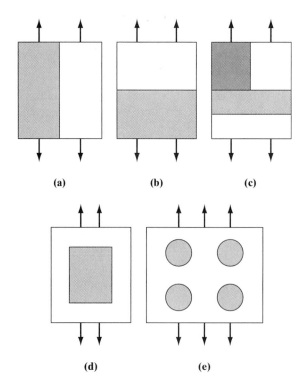

(a)　　　　　**(b)**　　　　　**(c)**

(d)　　　　　**(e)**

Figure 9.2
Various types of composite models: (a) parallel;
(b) series; (c) Hirsch's;
(d) Counto's; and
(e) aggregate.

9.2.1 Elastic Behavior

When a stress is applied to the parallel model, the modulus of elasticity of the composite can be calculated by Eq. 9.1:

$$E_c = E_p V_p + E_m V_m, \tag{9.1}$$

where E_p and E_m are the moduli of the particulate and matrix phase, respectively, and V_p and V_m are their corresponding volumes (e.g., $V_p + V_m = 1$). The derivation of Eq. 9.1 assumes equal strain in both components, which implies a perfect bond between the two; each component therefore carries a different stress. Equation 9.1 is commonly known as the law of mixtures. Soft materials in a hard matrix behave in a manner approximating to Eq. 9.1. In the series model the assumption is made that the two components carry the same stress, but do not have equal strains. In this case Eq. 9.2 can be derived:

$$\frac{1}{E_c} = \frac{V_p}{E_p} + \frac{V_m}{E_m} \tag{9.2}$$

$$E_c = \frac{E_p \cdot E_m}{E_m \cdot V_p + E_p \cdot V_m}. \tag{9.2a}$$

In this case there is no bond between the two components and the model applies best to hard particles contained in a soft matrix. These models also apply to the calculation of bulk and shear moduli. The law of mixtures has been commonly applied to calculate other composite properties, such as tensile strength, electrical resistance, and dielectric constant (from an applied electric field) and thermal expansion or conductivity (from applied temperature). However, no particulate composite conforms exactly to the parallel or series models. Equations 9.1 and 9.2 represent the upper and lower bounds for the modulus of the composite (Fig. 9.3), and more complex composite models have been suggested, which are more realistic representations of particulate composites. For example, Hirsch's model (Fig. 9.2c) is a combination of the parallel and series models, and E_c is calculated as follows:

$$\frac{1}{E_c} = X \left(\frac{1}{V_p E_p + V_a E_a} \right) + (1 - X) \left(\frac{V_p}{E_p} + \frac{V_a}{E_a} \right), \tag{9.3}$$

$$\underset{\text{Parallel}}{} \underset{\text{Series}}{}$$

where X is a measure of the contribution of each model to the actual behavior ($0 < X < 1$). Hirsch considered the stress distribution developed between the particulate and matrix phases in arriving at his model, and X can be interpreted as the degree of bonding between the two components; it is usually taken to be 0.5 for portland cement concrete. The influence of V_p on modulus of concrete is illustrated in Fig. 9.4 for aggregates of different moduli.

A major drawback of the series model and Hirsch's model is the fact that they predict a zero modulus for porous materials.[1] Counto's model (Fig. 9.2d) was

---

[1] In this case, $E_p = 0$. The parallel model reduces to $E_c = E_m \times V_m = V_m(1 - V_p)$, where V_p is now the porosity. This expression underestimates the influence of porosity on modulus. Commonly, $E_c = (1 - V_p)^n$, where n ~ 3.

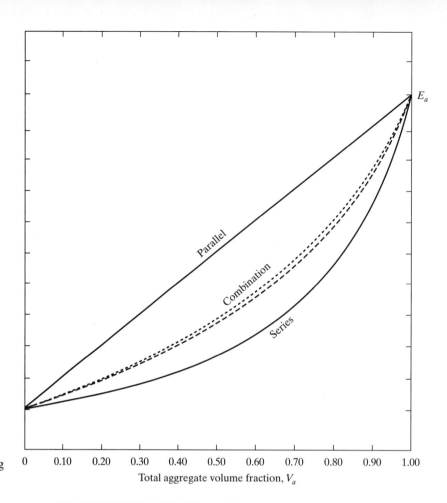

Figure 9.3
Calculation of modulus of
elasticity for concrete using
different models.

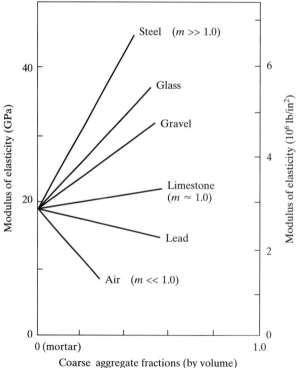

Figure 9.4
Effect of aggregates on the
modulus of elasticity of
concrete (adapted from
O. Ishai, *J. Amer. Concrete
Inst.*, Vol. 59, No. 9, 1961,
pp. 1365–1368).

developed to overcome this objection, but it results in a rather clumsy equation:

$$\frac{1}{E_c} = \frac{1 - V_p^{1/2}}{E_m} + \frac{1}{\left(\dfrac{1 - V_p^{1/2}}{V_p^{1/2}}\right)E_m + E_p} \cdot \qquad (9.4)$$

A more convenient approach is the use of the aggregate model (Fig. 9.2e), which is a symmetric array of nonbonded, noninteracting particles, and predicts E_c according to the following equation:

$$E_c = \left[\frac{(1 - V_p)E_m + (1 + V_p)E_p}{(1 + V_p)E_m + (1 - V_p)E_p}\right]. \qquad (9.5)$$

All of these more complex models are still incapable of handling particles of varying sizes, random dispersion, and variable degrees of bonding. Equations 9.3. to 9.5 give very similar predictions for realistic values of E_m and E_p for solid components (see Fig. 9.3).

9.2.2 Failure in Particulate Composites

Failure under an applied stress is greatly influenced by the nature of the composite. The influence of the dispersed particulate phase on the plastic deformation of metals through the "pinning" of dislocations has already been discussed in Chapters 2 and 6. Here we look briefly at the effects in brittle materials. Consider the case of hard particulate filler in a soft, weak matrix, which is the situation in ordinary portland cement concrete. A crack that develops and starts to propagate through the matrix will encounter a particle in its path and will be deflected. In the process the crack loses energy and eventually is fully arrested. Thus, several cracks are likely to form before complete failure of the composite occurs. The situation is aggravated by the fact that the interfacial zone is usually the weakest part of the composite. In some cases, the filler particles will not bond well to the matrix or they may adversely influence the microstructure in its immediate vicinity (Fig. 9.5a), giving a weaker bond. In all composites the elastic mismatch between the hard particles and soft matrix creates a stress concentration at the interface (Fig. 9.5b), whose magnitude is dependent on the modular ratio m ($m = E_p/E_m$). In these cases, cracks are often created and localized at the interface before propagating through the matrix. The multiple cracking leads to nonlinear stress-strain behavior and a certain degree of

Figure 9.5
Influence of particulate inclusion on (a) matrix microstructure creating an interfacial zone; and (b) stress concentration across the interface. [The term $k(\sigma)$ is the factor by which the applied stress is increased internally; ϵ_p, ϵ_m are the representations of strain in particle and matrix, respectively.]

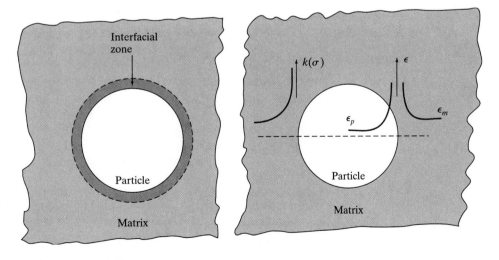

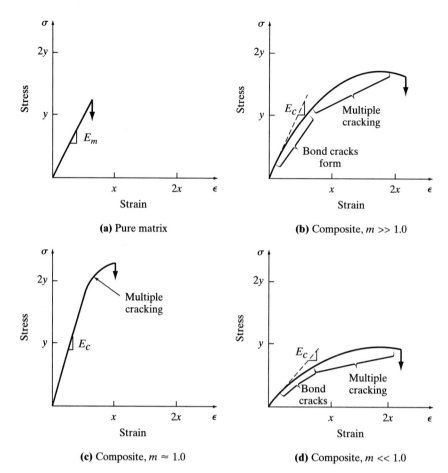

Figure 9.6
Stress-strain relations in a
brittle particulate composite: (a) pure matrix,
(b) composite, $m >> 1.0$,
(c) composite, $m \approx 1.0$,
(d) composite, $m << 1.0$.

(a) Pure matrix

(b) Composite, $m >> 1.0$

(c) Composite, $m \approx 1.0$

(d) Composite, $m << 1.0$

strengthening (Fig. 9.6b). When the modular ratio is almost equal to one, stress concentrations will be nearly absent and the tendency for crack arresting will be greatly
reduced unless a very weak interface is present. The stress-strain response is now almost linear (Fig. 9.6c). When soft particles are imbedded in a hard matrix (i.e.,
$m < 1$), multiple cracking will again occur but the overall strength of the composite
is decreased (Fig. 9.6d).

The influence of the bond between the filler and the matrix is clearly illustrated
in Fig.9.7. Additions of small quantities (less than 20%) of filler actually decrease the
strength of the material. The introduction of a weak interface (either due to poor
bond or stress concentrations) provides flaws from which cracks will more readily
propagate to failure. Increasing the amount of filler above 20% provides for a
strengthening due to crack deflection and raises the applied stress required for failure. For this to occur, the aggregate must be sufficiently strong and rigid, which is the
case for the alumina and SiC aggregates in Fig. 9.7.

When the matrix is a viscoelastic material, such as a polymer or asphalt, the situation becomes complicated because stress concentrations introduced by the elastic
mismatch between filler and matrix will be redistributed due to local yielding. Although the filler will raise the elastic modulus, it will carry very little stress after redistribution and thus the effective yield strength of the viscoelastic material will be
lowered, as seen in Fig. 9.8.

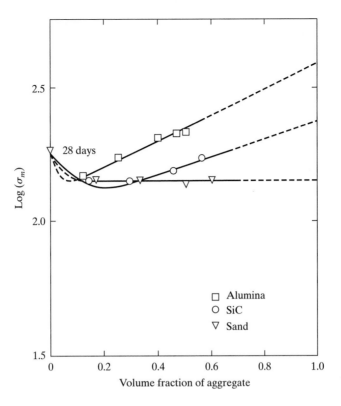

Figure 9.7
Influence of aggregates added to a very high-strength cement paste on the strength of the particulate composite (after E. Dingsoyr, T. Mosberg, and J. F. Young, "Influence of Aggregates on the Strength and Elastic Modulus of High Strength Mortars Containing Microsilica," pp. 211–218 in *Very High Strength Cement-Based Materials,* J. F. Young, ed., *Proc. Materials Research Society Symp.*, Vol. 42, 1985).

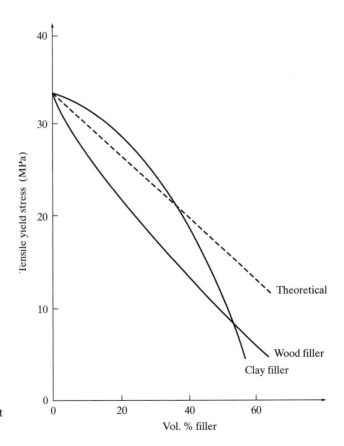

Figure 9.8 Influence of fillers on the yield stress of high-density polyethylene. The theoretical curve is that expected if the filler acts as diluent and does not support stress.

Sec. 9.2   The Mechanics of Particulate Composites   185

The mechanical properties of hardened portland cement paste (the binder in portland cement concrete) and asphalt cement (the binder in asphalt concrete) are very different. Hardened cement paste is a relatively strong and hard solid material and is almost perfectly brittle. Asphalt cement, on the other hand, is relatively weak and deforms in a viscoelastic manner. Therefore, to produce useful construction materials, different proportions of the binder and the aggregate must be used in portland cement and asphalt concrete mixtures. The differences in composition and structure between these two materials are shown schematically in Fig. 9.1.

In portland cement concrete, the portland cement paste occupies about 30% by volume of the mixture, and the aggregate the approximately 70% remaining. Thus, the aggregate particles are truly dispersed in a continuous matrix of the paste, with little interparticle contact. In these proportions, the aggregate acts to modify the brittle behavior of the hardened paste. In asphalt concrete, on the other hand, the aggregate particles occupy approximately 90% of the volume of the mixture, and the asphalt cement only about 10%. Most of the particles are in direct contact (i.e., intelocked) with, at most, a thin film of asphalt cement between the rest. In this case, the binder is effectively confined to the void spaces between the interlocked aggregate particles. It is this interparticle contact, rather than the properties of the binder itself, that imparts stiffness and stability to asphalt concrete mixtures.

9.4   INTERFACIAL PROPERTIES

The properties of a composite material are not determined solely by the relative proportions and properties of the constituent phases; the properties of the interfacial region, or interface, between the two phases may also be very important. Here again there are considerable differences between portland cement and asphalt concretes.

In an asphalt concrete, there is no chemical interaction between the aggregate and the asphalt cement. This leads to an intimate contact between the two phases, with no interfacial discontinuities. The lack of any chemical interaction also means that the properties of the asphalt cement at or near the aggregate interface are the same as those of the "bulk" asphalt cement. Thus, for asphalt concrete, the only interfacial property of significance is the adhesion between the asphalt cement and the aggregate.

For a portland cement concrete, however, the situation is more complex. First, with a few aggregates there is some chemical reaction between the portland cement and the aggregate. Second, and probably of greater significance, less efficient packing of the cement particles near the aggregate surface ("wall effect") and the "bleeding" of water that occurs in fresh concrete (i.e., the tendency for water, the lightest component in the mixture, to try to rise to the surface) often leads to the formation of water pockets beneath the coarse aggregate particles. Even after the concrete hardens, there remains around the aggregate a matrix zone which is more porous than the bulk matrix. Hence the structure of the hydration products in this zone, which is about 20 μm thick, is different from that in the bulk paste: The interfacial material has a more porous structure and is, therefore, weaker. Indeed, the interfacial zone is now considered to be the "weak link" in concrete, and one way to produce higher-strength concrete is to modify the interfacial region by the use of appropriate mineral admixtures and processing techniques.

Both asphalt concrete and portland cement concrete undergo a "fresh" state, in which they behave as viscous fluids. In the case of asphalt concrete, the fluidity is achieved either by heating the asphalt cement (the mix hardens when cooled) or by using the cement in a liquid form, which subsequently hardens when the solvent evaporates. For portland cement concrete, the mixing water provides a fluid mix until the hydration of the cement (i.e., chemical reaction between cement grains and water) begins to supply structural rigidity. For both materials, this initial fluid state is transient, lasting for only several hours, but is necessary to permit the materials to be mixed, placed, and compacted into their final form. However, the properties of these materials in the hardened state are generally of much more interest.

The characteristic deformational behavior of a composite material is normally that of the phase which is continuous. Therefore, portland cement concrete is expected to be very brittle, reflecting the brittle nature of the hydrated cement paste. On the other hand, asphalt concrete is expected to be highly viscoelastic, since the asphalt cement is a viscoelastic material. In reality, however, the behavior of these materials is more complex than this simple approach suggests. While the portland cement concrete is still a brittle material, it is much less so than the hardened cement paste (or the aggregate itself) because of the interaction between the cement and the aggregate at the interfacial zone. This may be seen in Fig. 9.9: Concrete has an intermediate stiffness, as would be expected from composite theory, but its stress-strain behavior is nonlinear, whereas that of both the aggregate and the hardened cement paste is essentially linear until failure. Similarly, asphalt concrete exhibits viscoelastic behavior, but much less so than the asphalt cement itself. This is due to the fact that the particle-to-particle contact within the aggregate phase effectively governs the deformational behavior at the low asphalt cement contents normally used in paving mixtures. Only at high asphalt cement contents, or high temperatures, does the characteristic viscoelastic behavior of the asphalt cement govern.

Thus, although both types of concrete consist of aggregates dispersed in a continuous matrix of binder, their properties are different (Fig. 9.10). In practice, the behavior or properties of either material cannot be predicted with any great

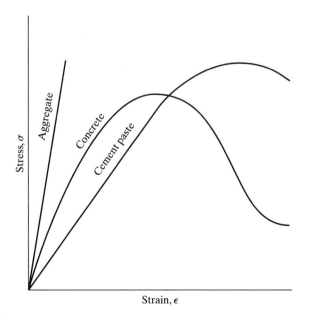

Figure 9.9
Stress-strain behavior of hardened portland cement paste, aggregate, and portland cement concrete.

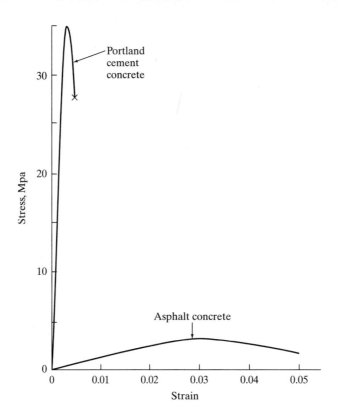

Figure 9.10
Typical stress-strain behavior in compression of asphalt cement and portland cement concretes.

accuracy using composite materials theory, and so empirical methods must be used. These will be described in greater detail in Chapters 11 and 12.

BIBLIOGRAPHY

S. Popovics "Concrete-Making Materials" McGraw-Hill, NY 1979.

SCI Committee 221 "Guide for the Use of Normal Weight Aggregates in Concrete" Doc. 221R–89. Amer. Concr. Inst., Farmington Hills, MI 1989.

Concrete Manual 8th Ed., U.S. Bureau of Reclamation, Denver, CO, 1975.

S. Mindess and J. F. Young, "Concrete" Prentice-Hall, NY 1981.

PROBLEMS

9.1 Calculate the predicted value of E for a composite where $E_m = 12$ GPa, $E_p = 62$ GPa and $V_m = 30\%$ using the following models.
   (a) Parallel ;
   (b) Series,
   (c) Hirsch x = 0.3, and
   (d) Aggregate

9.2 Repeat the calculations in Problem 1 where $E_m = 2$ GPa, $E_p = 62$ GPa and $V_m = 0.1$.

9.3 Lightweight concrete contains aggregates with a high internal porosity. Sketch the expected stress-strain curve for such a concrete and compare it to that for normal weight, normal strength concrete.

10

Aggregates

10.1 INTRODUCTION

Mineral aggregates are the predominant civil engineering material in terms of volume of use. They are used alone (in road bases and various types of fill) and with cementing matrices (such as portland cement or asphalt cement) to form composite materials or concretes. We are most interested in the use of aggregates as a component of portland cement and asphalt concretes, and this will form the basis of the discussion in this chapter. Mineral aggregates constitute about 70–80% by volume of portland cement concretes and 90% or more of asphalt concretes. As a consequence, aggregate characteristics have a direct influence on many of the properties of these concretes.

Aggregates are defined as inert, granular, inorganic materials which normally consist of stone or stonelike solids. (In some specialty concretes, organic materials are used and some aggregates do react slowly with hydrated portland cement). The large majority of the aggregates used in practice are (1) natural, in that the material of the particles has not been changed artificially during their production (although the aggregate itself may have been subjected to processes such as crushing, washing, and sieving); and (2) normal weight (i.e., with a specific gravity or relative density of about 2.6). Examples of such aggregates are natural sands and gravels and crushed stone.

Synthetic or artificial aggregates may be:

1. Byproducts of an industrial process, such as blast-furnace slag.
2. Products of processes developed to manufacture aggregates with special properties, such as expanded clay, shale, or slate, which are used for light-weight aggregates.

3. Reclaimed or waste construction materials, such as recycled portland cement concrete.

Such aggregates are not commonly used, but may be employed where justified on the basis of economic considerations or where their specific properties are required. However, the use of reclaimed or waste construction materials as aggregates should substantially increase in the future as natural aggregates become less plentiful and consequently more expensive.

The properties of aggregates are determined by testing appropriate samples, usually in accordance with standard methods or procedures. Because aggregates are variable in quality, proper sampling procedures must be followed to ensure that the sample to be tested is representative of the aggregate as a whole.

10.2   COMPOSITION AND STRUCTURE

Natural aggregates are derived from rocks. These may consist of a single mineral (e.g., limestone), but more often they contain several different minerals (e.g., granite). They are classified according to origin into three major types—igneous, sedimentary, and metamorphic—which may be further subdivided according to chemical and mineral composition, internal structure, and texture.

Igneous rocks form by cooling from a melt (i.e., molten rock matter) either above or below the earth's surface. Phase separation during cooling usually occurs, and these rocks consist of crystals or minerals in a crystalline (or sometimes glassy) matrix. The rate of cooling determines the crystal size and hence the rock properties: fine-grained rocks with smaller grain sizes are generally stronger (see Fig. 10.1 and Table 10.1).

Sedimentary rocks are formed by the consolidation of deposits of the products of weathering and erosion of existing rocks at the earth's surface. Although some additional crystallization may take place during the formation process, it is mainly one of reduction of porosity, as very little sintering occurs. These rocks tend to be weaker than the igneous ones, since mechanical compaction is not as efficient as crystal growth in producing strong microstructures. These rocks generally reflect any anisotropic behavior of their component minerals.

Metamorphic rocks are formed by the application of intense heat and pressure to sedimentary rock deposits. The resulting rocks are less porous (i.e., they are more dense) and have a stronger matrix because recrystallization and grain growth can occur during the formation process. They are therefore stronger and less anisotropic than sedimentary rocks (see Fig. 10.2).

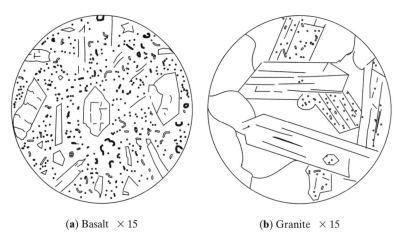

Figure 10.1
Fine-grained basalt (a) is stronger than coarse-grained granite (b).

(a) Basalt × 15   **(b)** Granite × 15

TABLE 10.1   Properties of Rocks with Different Microstructures

Rock Type	Example	Specific Gravity	Porosity (% Vol.)	Compressive Strength (MPa)	Modulus of Elasticity (GPa)	Remarks
Igneous	Basalt	2.6–3.0	0.1–1.0	50–200	30–70	Fine grained
	Granite	2.6–3.0	0.5–1.5	100–250	5–50	Coarse grained
Sedimentary	Shale	2.0–2.7	10–30	10–100	5–25	
	Limestone	2.3–2.8	5–20	35–250	2–70	
	Sandstone	2.2–2.7	5–25	20–175	5–50	Composite structure
Metamorphic	Slate	2.6–2.9	0.1–0.5	100–200	10	From shale
	Marble	2.6–2.8	0.5–2.0	100–250	40–100	From limestone
	Quartzite	2.6–2.7	0.5–5.0	100–300	10–70	From sandstone

Figure 10.2
Microstructures of limestone (a) and marble (b) illustrating differences resulting from metamorphic process. Photographs courtesy Dept. of Earth and Ocean Sciences, University of British Columbia.

10.3.1 Geometric Properties

Particle Size and Grading

Aggregates are granular or particulate materials, and the usefulness of an aggregate is to a considerable extent related to the size and gradation, or size distribution, of the particles of which it is composed.

The size of an aggregate particle is taken as some measure of the diameter of the particle. In practice, the size of a particle, regardless of its shape, is taken as d_i if that particle will just pass through a square opening of this size. Because the particles in a sample of aggregate may have an infinitely large range of sizes, a series of sieves or screens with different size openings is used to obtain some estimate of the distribution of the particle sizes. The size of a particle which passes through a sieve with an opening of size d_i but is retained on the next smaller sieve with an opening of size d_{i-1} is then taken as $d_i - d_{i-1}$, or the particle is said to belong to the $d_i - d_{i-1}$ size fraction. For example, all particles passing through a sieve with a 40 mm opening but retained on a sieve with a 20 mm opening belong to the 40–20 mm size fraction. The maximum-size-aggregate (MSA) is defined as the smallest sieve opening through which the entire aggregate sample passes.

The distribution of particle sizes in an aggregate sample, called the grading or gradation, is usually expressed as the cumulative percentage of particles that are smaller or larger than each of a series of sieve or screen openings. These distributions are usually presented in a graphical form, known as a grading curve, where the cumulative percentage passing or retained on a particular sieve is plotted against the sieve size.

There are several types of aggregate gradations. The most commonly used is the continuous or dense gradation, in which the aggregate contains every size fraction between the maximum and minimum particle sizes. Aggregates which contain particles of only one size fraction are called open or uniformly graded, and those which are missing particles of one or more size fractions are called gap graded. Schematic representations and grading curves for continuous, uniform, and gap-graded aggregates are shown in Fig. 10.3. Together with the MSA, the gradation determines the specific surface area (i.e., surface area per unit weight or volume) of the particles in an aggregate sample. For a densely (continuously) graded aggregate, the specific surface area increases as the MSA decreases.

Aggregate gradations in which the space between the aggregate particles (i.e., the voids space) is a minimum and the density of the aggregate is therefore a maximum are important in several applications. Theoretical or ideal maximum density gradations (Fig. 10.4a), have been developed on the basis of the packing characteristics of spherical particles (Fig. 10.4b).

Aggregates are normally divided into two classes on the basis of their size: fine and coarse. The division is arbitrarily made at the No. 4 (4.75 mm) or 5.00 mm particle size; smaller particles comprise the fine aggregate, and larger particles, the coarse aggregate. Aggregate specifications usually contain grading limits for fine and coarse aggregates separately; typical grading limits for fine and coarse aggregates used in portland cement concrete mixtures are given in Tables 10.2 and 10.3, respectively.

The coarseness or fineness of an aggregate gradation can be indicated by an empirical factor called the fineness modulus, which is obtained by adding the total percentages of a sample of the aggregate retained on each of a specified series of sieves and

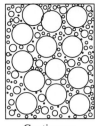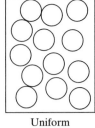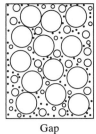

| Continuous | Uniform | Gap |

(a)

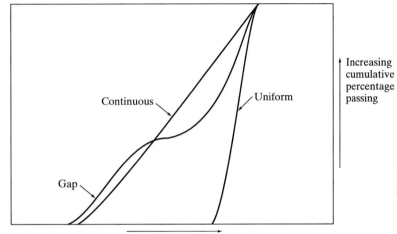

Continuous

Uniform

↑ Increasing cumulative percentage passing

Figure 10.3
Schematic representations of different aggregate gradations (a) and corresponding grading curves (b).

Gap

Increasing particle size/sieve opening

(b)

TABLE 10.2   ASTM C-33 Grading Limits for Fine Aggregates

Sieve Size	% Passing
9.50 mm (3/8 in)	100
4.75 mm (No. 4)	95–100
2.36 mm (No. 8)	80–100
1.18 mm (No. 16)	50–85
600 μm (No. 30)	25–60
300 μm (No. 50)	10–30
150 μm (No. 100)	2–10

TABLE 10.3   ASTM C-33 Grading Limits for Coarse Aggregates

Sieve Size	% Passing (Nominal Maximum-Size-Aggregate)			
	37.5 mm ($1\frac{1}{2}$ in)	25.0 mm (1 in)	19.0 mm (3/4 in)	12.5 mm (1/2 in)
50 mm (2 in)	100			
37.5 mm ($1\frac{1}{2}$ in)	95–100	100		
25.0 mm (1 in)	—	95–100	100	
19.0 mm (3/4 in)	35–70	—	90–100	100
12.5 mm (1/2 in)	—	25–60	—	90–100
9.50 mm (3/8 in)	10–30	—	20–55	40–70
4.75 mm (No. 4)	0–5	0–10	0–10	0–15
2.36 mm (No. 8)	—	0–5	0–5	0–5

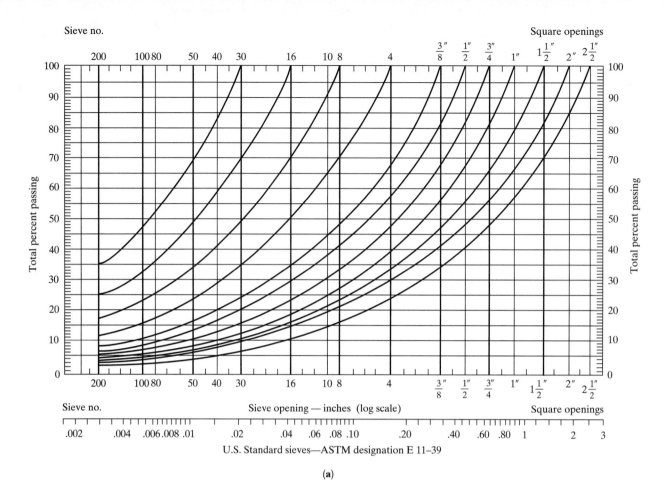

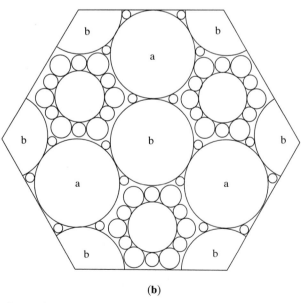

(b)

Figure 10.4
Densely graded aggregate: (a) Fuller-Thompson maximum density (minimum voids content) gradation curves; (b) schematic description of the concepts of dense grading.

dividing the sum by 100. The standard sieve sizes of this series are 0.15 mm (No. 100), 0.30 mm (No. 50), 0.60 mm (No. 30), 1.18 mm (No. 16), 2.36 mm (No. 8), 4.75 mm (No. 4), 9.50 mm (3/8 in), 19.0 mm (3/4 in), 38.1 mm ($1\frac{1}{2}$ in), 76.0 mm (3 in), and 150 mm (6 in). For a fine aggregate, the cumulative percentages of material retained on sieves below the No. 4 (4.75 mm) or 5.00 mm size are used to determine the fineness modulus, and this value normally falls between 2.3 and 3.1, increasing as the coarseness of the aggregate increases. An example of the calculation of the gradation, in terms of cumulative or total percent passing, and the fineness modulus (FM) for a fine aggregate is given in Table 10.4. Aggregates having the same fineness modulus

TABLE 10.4 Example of Sieve Test Results and the Calculation of Grading Curve and Fineness Modulus

Sieve Size (mm)	No.	Weight Retained (g)	% Retained	Cumulative % Retained	Cumulative % Passing
4.75	4	0	0	0	100.0
2.36	8	241.9	11.9	11.9	90.6
1.18	16	388.9	19.1	31.0	74.0
0.60	30	505.5	24.9	55.9	51.7
0.30	50	543.4	26.7	82.6	25.1
0.15	100	340.8	16.8	99.4	8.4
Pan*		11.3	0.6	100.0	0

$$\text{FM} = \frac{11.9 + 31.0 + 55.9 + 82.6 + 99.4}{100} = \frac{280.8}{100} = 2.81.$$

*Pan refers to the amount of material remaining in the pan which sits under the set of sieves.

may not necessarily be identical in their grading curves.

The strength of asphalt concrete mixtures is largely due to the mechanical interlock between the aggregate particles in the mixture, and it is therefore important to ensure that the size and the gradation of the particles are such as to promote aggregate interlock. Although dense gradings are required for high strength mixtures, the voids contents of the Fuller-Thompson maximum density gradations are too low to accommodate sufficient asphalt cement or binder, and in practice, the gradings used in an asphalt concrete mixture will depart slightly from them. For portland cement concrete mixtures, however, interaction between the aggregate particles interferes with the workability of the material in the fresh state, and less dense aggregate gradings must be used. In addition, the specific surface area is smaller in these less dense gradings (although the voids space is larger), so less portland cement paste is required to coat the aggregate particles and the cost of the mixture may be lower. However, depending on the maximum size aggregate and certain properties of the mix, some minimum amount of fine aggregate is required to ensure that a portland cement concrete has sufficient cohesiveness and plasticity.

Optimum aggregate gradations, in terms of grading limits, are specified by various organizations for different end uses. Commercially available aggregates may not conform to such specifications, and hence two or more aggregates having different gradations may have to be combined or blended to produce a specified gradation.

Particle Shape and Surface Texture

Particle shape and surface texture are also important in determining the amount of interaction that occurs between aggregate particles. However, these properties have not been adequately defined in a quantitative sense, and therefore their influence on the properties of concrete mixtures cannot be evaluated precisely. Both particle shape and surface texture are a result of the processing operations to which the aggregate has been subjected, as well as its mineral composition and crystalline structure.

Particle shape is related to two relatively independent characteristics: angularity and sphericity. Angularity is a function of the relative sharpness of the edges and corners of the particle. It can be defined numerically, but it is normally expressed descriptively: rounded as opposed to angular, for example. Sphericity is a function of the ratio of the surface area of the particle to its volume. It can also be defined numerically, but again it is normally expressed in descriptive terms on the basis of the relationship between the particle dimensions: equidimensional, flaky, or elongated, for example. Particle surface texture is related to the relative degree to which the surface is polished or dull, smooth or rough. This property is also assessed descriptively or qualitatively.

Since the degree of mechanical interlock between aggregate particles increases as the angularity of the particles increases, aggregates used in asphalt concrete mixtures require a minimum amount of crushed particles with fractured or broken faces to achieve satisfactory levels of strength or stability. The rough surface texture associated with a crushed aggregate also increases interparticle friction and therefore the strength of the mixture. Furthermore, although a rough surface texture is more difficult for asphalt cement to wet than a smooth, glassy texture, once wetted it offers more adhesion. Workability requirements dictate the use of rounder, smoother aggregates for portland cement concrete mixes. However, rough textured surfaces improve the cement paste-aggregate bond in the hardened concrete, and therefore its mechanical properties.

10.3.2 Physical Properties

Porosity and Voids Content

Porosity is defined as the volume inside individual aggregate particles that is not occupied by solid material (i.e., intraparticle volume). It does not refer, therefore, to the volume of interparticle voids in a packing of aggregate particles, which is the voids content. Porosity as volume percent is defined numerically as

$$p = (V_p/V_b) \times 100, \tag{10.1}$$

where V_p = volume of pores inside the aggregate particles, and
V_b = total bulk volume of the porous aggregate particles.

The pores in the particle are of two types: interconnected and discontinuous. The interconnected pores are open to the surface of the aggregate particle and can be penetrated and filled with water or other liquids. The porosity associated with the interconnected pores is called the effective porosity, and in practice it is assumed that this effective porosity is equal to the total porosity of the aggregate particles.

Many of the properties of the aggregate are significantly affected by its porosity. As the porosity increases, the bulk density and hence the strength, elastic modulus, and abrasion resistance of the aggregate decrease (see Table 10.1). Aggregates with low porosity are desirable for use in portland cement concretes, since water will be absorbed from the mixture by the aggregate particles if their porosity is high, and less will be available to provide workability. In asphalt concrete mixtures, a minimum porosity of about 0.5% is desirable for adhesion purposes, since the area of contact between the aggregate and the asphalt cement is increased.

The voids content in a sample of aggregate is the volume of the spaces between the individual aggregate particles, the interparticular volume. As a volume percent, voids content is defined numerically as

$$v = (V_v/V_t) \times 100 \qquad (10.2)$$

where V_v = total volume of voids between the aggregate particles in the sample, and
V_t = total volume of the sample, including the voids between the aggregate particles.

It is an important parameter in proportioning both portland cement and asphalt concrete mixtures. In asphalt concrete technology, voids content is referred to as the voids-in-mineral-aggregate (VMA).

Absorption, Moisture Content, and Permeability

Aggregate particles can absorb water, since they contain some porosity, and can retain a thin film of water on their surface. Consequently, aggregates can exist in one of the following four moisture states, as shown in Fig. 10.5:

1. **Oven-dry (OD).** All moisture is removed from the aggregate by heating in an oven at 105°C to constant weight (overnight heating usually suffices). All pores are empty.
2. **Air-dry (AD).** All moisture is removed from the surface of the aggregate particles, but the internal pores are partially full.
3. **Saturated-surface-dry (SSD).** All of the aggregate pores are filled with water, but there is no film of water on the surface of the particles.
4. **Wet.** All of the aggregate pores are completely filled with water and there is a film of water on the surface of all of the particles.

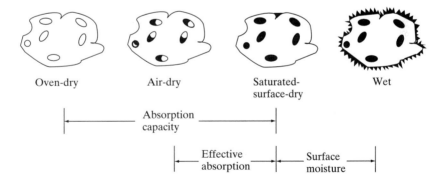

Figure 10.5
Moisture states of aggregates.

Of these four states, only two—the oven-dry and the saturated-surface-dry—correspond to specific moisture contents and can therefore be used as references. The absorption capacity of the aggregate is the maximum amount of water it can absorb, and it is the difference in water content between the oven-dry and saturated-surface-dry moisture states. It is expressed as a percentage of the oven-dry weight as follows:

$$\text{absorption capacity } (\%) = \frac{W_{\text{SSD}} - W_{\text{OD}}}{W_{\text{OD}}} \times 100, \qquad (10.3)$$

where W_{SSD} and W_{OD} represent the weight of the aggregate sample in the saturated-surface-dry and oven-dry states, respectively.

The effective absorption represents the amount of water required to bring an aggregate from the air-dry to the saturated-surface-dry state, expressed as a percentage of the saturated-surface-dry weight as follows:

$$\text{effective absorption}(\%) = \frac{W_{SSD} - W_{AD}}{W_{SSD}} \times 100, \qquad (10.4)$$

where W_{AD} represents the weight of the aggregate sample in the air-dry condition.

The surface moisture is the water on the surface of the aggregate particles—that is, the difference between the wet and saturated-surface-dry states—and can also be expressed as a percentage of the saturated-surface-dry weight:

$$\text{surface moisture }(\%) = \frac{W_{WET} - W_{SSD}}{W_{SSD}} \times 100, \qquad (10.5)$$

where W_{WET} represents the weight of the aggregate sample in the wet condition.

The moisture content of an aggregate in any state (in a stockpile awaiting use, for example) can simply be expressed as

$$\text{moisture content }(\%) = \frac{W_{AGG} - W_{OD}}{W_{OD}} \times 100, \qquad (10.6)$$

where W_{AGG} is the weight of the aggregate in the stockpiled condition.

If the moisture content is greater than the absorption capacity, then the aggregate is wet (i.e., there is surface or free moisture on the aggregate particles), and if it is less, then the aggregate is air-dry and will absorb additional moisture. The moisture content of an aggregate and its absorption capacity are very important parameters in proportioning portland cement concrete mixtures, since they determine the amount of water that must be added to the mix to provide the required level of workability. Most normal-weight aggregates have absorption capacities between 0.5 and 2.0%, with higher values indicating high-porosity aggregates, which may have potential durability problems.

Permeability, or the susceptibility to passage or penetration by fluids, is also a function of the porosity of the aggregate. This property has a significant effect on the resistance of the aggregate to deteriorating influences, particularly those associated with the presence of water in the aggregate particles.

Specific Gravity

The specific gravity, or relative density, of an aggregate is the ratio of its mass (or weight in air) to the mass of an equal volume of water at stated temperatures. Three types of specific gravity are commonly described, and differ on the basis of the moisture content of the aggregate; they are:

$$\text{bulk specific gravity}, G_b = \frac{W_{OD}}{W_{SSD} - W_{SW}} \qquad (10.7)$$

$$\text{bulk specific gravity (saturated-surface-dry)}, G_{bssd} = \frac{W_{SSD}}{W_{SSD} - W_{SW}} \qquad (10.8)$$

$$\text{apparent specific gravity}, G_a = \frac{W_{OD}}{W_{OD} - W_{SW}} \qquad (10.9)$$

where W_{SSD} and W_{OD} represent the weight of the aggregate sample in the saturated-surface-dry and oven-dry states, as above, and W_{SW} represents the weight of the saturated sample in water.

The 'true' specific gravity of the aggregate, G_t, is the ratio of its oven-dry weight to the volume of solids in the aggregate particles excluding all pores. The relationship between the various specific gravity values is

$$G_b < G_{bssd} < G_a < G_t.$$

10.3.3 Strength and Toughness

The strength and other related characteristics of aggregate particles are dependent upon the properties of the constituent minerals and the bonding between the grains, as well as the porosity of the particles, and hence can vary within wide limits (Table 10.1). Unfortunately, there is no truly satisfactory test for measuring these properties. The Los Angeles abrasion test (ASTM[1] C-131 and ASTM C-535) has been commonly used in the past to obtain some measure of them, but it is an indirect, empirical test method which does not provide any fundamental information about the aggregate. This test measures the degradation of aggregates of standard gradings resulting from a combination of abrasion or attrition, impact, and grinding in a rotating steel drum containing a specified number of steel spheres.

Strength and toughness are required to prevent particle breakdown during manufacture, placement, and compaction of asphalt concrete paving mixtures and under traffic loading. The breakdown forces are probably more severe in open-graded, low-binder-content mixes than in dense-graded mixtures. Therefore, when strength and toughness are poor, use of the aggregate in a dense-graded mixture is preferable. Aggregates are also subjected to abrasion under traffic loads, and thus their abrasion resistance is of particular importance for portland cement or asphalt concretes used in pavement surfaces.

Aggregate strength is of less importance in portland cement concrete mixtures since, with the exception of certain specialty concretes, the aggregate is generally stronger than the portland cement paste and the particles are not in direct contact.

10.3.4 Other Properties

Surface Chemistry

The surface properties of an aggregate are of little consequence in their potential use in portland cement concrete mixtures, because other factors are primarily responsible for bonding with the cement paste. However, these properties are important to the use of an aggregate in asphalt concrete mixtures, since the surface of the particles may be electrically charged, and this will affect the adhesion or bonding between the aggregate and the asphalt cement. Hydrophobic or water-repellent aggregates, such as limestone and dolomites (basic in nature), have a positive surface charge and their surface energy is such that they are more easily wetted by an asphalt cement than by water; whereas hydrophilic or water-attracting aggregates, such as silicates (acidic), have a negative surface charge and are more easily wetted by water. Even if hydrophilic aggregates are wetted initially by asphalt

---

[1]American Society for Testing and Materials.

cement in the absence of water, the aggregate-cement bond is liable to breakdown if water is able to enter the mixture and permeate the interface (for more details, see Chapter 12).

Surface Coatings

The surfaces of aggregate particles are sometimes naturally coated with materials such as clay, silt, calcium carbonate, iron oxide, opal, and gypsum. These coatings may be detrimental to the performance of the aggregate in a portland cement or asphalt concrete mixture since they interfere with the bond between the surface and the binder. They vary widely in thickness, hardness, and adhesion to the aggregate surface. If they do not break down during aggregate processing (crushing, washing, etc.), they may break away from the aggregate after mixing and placement of the concrete, thereby having a negative effect on the properties of the mixture.

Durability

Durability is the resistance to deteriorating influences which may reside in the aggregate itself, or which are inherent to the environment in which it is exposed. Physical durability is the resistance to physical processes which can break down or otherwise damage the aggregate particles, and chemical durability is the resistance to chemical processes. The large majority of rocks used as aggregate are durable or stable under most conditions to which they are exposed. However, several different kinds of instability are known:

1. Unstable volume changes (i.e., expansion due to water absorption or to chemical reactions such as oxidation, hydration or carboration).
2. Freeze-thaw deterioration (i.e., expansion and fracturing of aggregate particles due to excessive hydraulic pressures generated by the increase in volume associated with liquid-water phase change during freezing of water). The freeze-thaw problem is related to the expansion problem in item 1 because the susceptibility of an aggregate to both is dependent on the porosity and permeability of its particles. This topic will be discussed in greater detail in Chapter 11.
3. Mechanical degradation or breakdown of aggregate particles, which may occur during handling and stockpiling operations as well as during use, as a result of low strength or toughness.
4. Chemical degradation, such as the alkali-aggregate reaction in concrete (see Chapter 12, p. 271).

Deleterious Substances

Deleterious substances can occur in aggregates in relatively small amounts (generally less than 1%) and can cause harmful effects. Examples of such substances include the following:

1. Absorbent particles, such as shale, leached chert, or porous flint. These particles are usually frost susceptible and particularly cause problems in portland cement concrete.

2. Clay lumps and "friable" particles, which crumble or are broken up easily, especially when saturated.

3. Coal or wood particles, which act as weaker unstable inclusions.

4. Organic impurities, which occur mainly in sand and gravel deposits due to contamination from the overburden. In portland cement concretes, water-soluble organic matter can interfere with the hydration reactions and retard the strength development of the concrete.

5. Flat or elongated particles. These particles can cause problems in both portland cement and asphalt concretes because their interaction with other particles leads to loss in workability and/or compactability.

Standard test methods are available to detect these materials and determine their approximate amount and probable effect (e.g., ASTM C-40 and ASTM C-142).

BIBLIOGRAPHY

SANDOR POPOVICS *Concrete-Making Materials,* Hemisphere Publishing Corp./McGraw-Hill Book Company, 1979

L. COLLIS and R. A. FOX, Eds., *Aggregates: Sand, Gravel and Crushed Rock Aggregates for Construction Purposes.* The Geological Society of London, 1985

H. G. SCHREUDERS and C. R. MAREK, Eds., *Implications of Aggregates in the Design, Construction and Performance of Flexible Pavements.* American Society for Testing and Materials, 1989. ASTM STP 1016

P. KLIEGER and J. F. LAMOND, Eds., *Significance of Tests and Properties of Concrete and Concrete-Making Materials.* American Society for Testing and Materials, 1994. ASTM STP 169C

L. DOLAR-MANTUANI, *Handbook of Concrete Aggregates*, Noyes Publications, 1984

S. MINDESS and J. F. YOUNG *Concrete*, Prentice Hall, Englewood Cliffs, New Jersey, 1981

Concrete Manual, 8th ed. Bureau of Reclamation. Denver, Colorado, 1975.

PROBLEMS

1. What is an aggregate? What is the difference between a natural and a synthetic (artificial) aggregate?
2. How are (i) igneous, (ii) sedimentary and (iii) metamorphic rocks formed? What is the general relationship between the structure and properties of these three types of rock?
3. An aggregate particle is said to belong to the d_i-d_{i-1} size fraction in the aggregate sample; what does this mean?

4. What is the difference between a continuous (dense) aggregate gradation and an open (uniform) aggregate gradation?
5. What is the fineness modulus of an aggregate? Calculate the fineness modulus of the following fine aggregates.

Sieve Size	Cumulative % Retained	
	A	B
4.75 mm	0.3	0.1
2.36 mm	13.6	9.2
1.18 mm	32.2	25.0
600 μm	52.5	43.3
300 μm	77.7	72.5
150 μm	93.8	92.0
75 μm	98.6	98.3

6. Why are continuous or dense graded aggregate not used in (i) asphalt concrete mixtures or (ii) portland cement concrete mixtures?
7. Why do aggregate specifications for asphalt concrete mixtures typically require a minimum amount of crushed particles with fractured or broken faces?
8. What is aggregate porosity and how does it affect the strength and toughness of the aggregate?
9. Three of the important physical and geometrical properties of aggregates with respect to their use in portland cement or asphalt concrete mixtures are (i) porosity, (ii) particle shape, and (iii) maximum-size-aggregate (MSA). For each of these properties:
   (a) define the property;
   (b) explain how the property is expressed or described quantitatively; and,
   (c) explain the significance of the property with respect to the use of the aggregate in either a portland cement or an asphalt concrete mixture.
10. Can the absorption capacity of an aggregate be used as a measure of its total porosity? If not, why?
11. 1500 g of an oven-dry aggregate absorbs 55 g of moisture to reach the SSD condition. Calculate the absorption capacity of the aggregate.
12. Calculate the SSD weight of 1200 g of oven-dry gravel, when its absorption capacity is 1.4%.
13. Stockpiled sand and gravel are brought to the oven-dry state. 500 g of sand loses 35 g of water, while the 1500 g of gravel loses 18 g of water. If the absorption of sand and gravel are 1.2% and 1.6% respectively determine:
   (a) the stockpiled moisture state of each aggregate, and
   (b) the surface moisture or effective absorption of each aggregate.
14. The weight of a sample of coarse aggregate was measured under various conditions with the following results:

oven-dry weight in air	1076.3 g
air-dry weight in air	1083.3 g
saturated-surface-dry weight in air	1089.2 g
weight of saturated sample in water	681.5 g

For this aggregate sample, calculate the:

    (a) air-dry moisture content;

    (b) absorption capacity;

    (c) effective absorption;

    (d) bulk specific gravity;

    (e) saturated surface-dry bulk specific gravity; and

    (f) apparent specific gravity.

15. Why are aggregate strength and toughness important parameters in asphalt concrete pavement mixtures?

16. Why are hydrophobic aggregates more preferable than hydrophilic aggregates for use in asphalt concrete paving mixtures?

17. Why are surface coatings (clay, silt, etc.) potentially detrimental to the use of an aggregate in a portland cement or asphalt concrete mixture?

18. Identify and briefly describe three physical or chemical processes that result in the breakdown or deterioration of an aggregate.

19. Identify three deleterious substances that occur commonly in natural mineral aggregates, and explain why each of these substances is considered to be detrimental to the use of the aggregates in which it occurs.

11

Portland Cement Concrete

11.1 INTRODUCTION

Portland cement concrete is a particulate composite consisting of a continuous binder phase (the cementitious matrix) and a dispersed particulate phase (the aggregates), as shown in Fig. 9.1. The aggregate grading varies over a range, starting from fine material smaller than 5 mm in size (fine aggregate) to a maximum coarse aggregate size usually in the range of 20 to 40 mm, as outlined in Chapter 10. The aggregates can be considered for a first degree of approximation as inert fillers because they seldom interact chemically with the matrix.

In normal concretes the aggregates are much stronger than the cementitious matrix, and therefore most of the physical and mechanical properties of the concrete are controlled by the continuous cementitious phase. Yet the presence of the aggregates is essential in the concrete because they serve as an inexpensive filler (the cost of the portland cement in the cementitious matrix is an order of magnitude greater than the aggregate) and reduce the deformation of the concrete, both the initial and the time-dependent deformations (i.e., creep and shrinkage). The latter influence is the result of the modulus of elasticity of the aggregate, which is considerably higher than that of the matrix.

Having established the cementitious phase as the more important one in concrete, it is essential to understand its structure in relation to mechanical, chemical, and physical properties, to provide the necessary background to deal with the concrete itself. Therefore, the treatment of the concrete (Sec. 11.3) will follow the discussion of the cementitious phase (Sec. 11.2), and it will be subdivided into three parts: early age concrete, chemical admixtures, and hardened concrete and the durability of concrete (Sec. 11.4). The final section in this chapter will deal with the concepts of mix design (Sec. 11.5).

The cementitious phase is made of a mixture of portland cement and water, known as the cement paste. The paste is initially fluid, but as time passes it becomes hard and strong, as expected of a solid that serves as a construction material. The change from the fluid to the hardened state is the result of chemical reactions between the cement grains and water, which is referred to as the hydration reaction. The hydration products are responsible for the properties of the hardened paste. A simplified description of the hydration process is shown in Fig. 11.1, demonstrating some of the basic characteristics of this system:

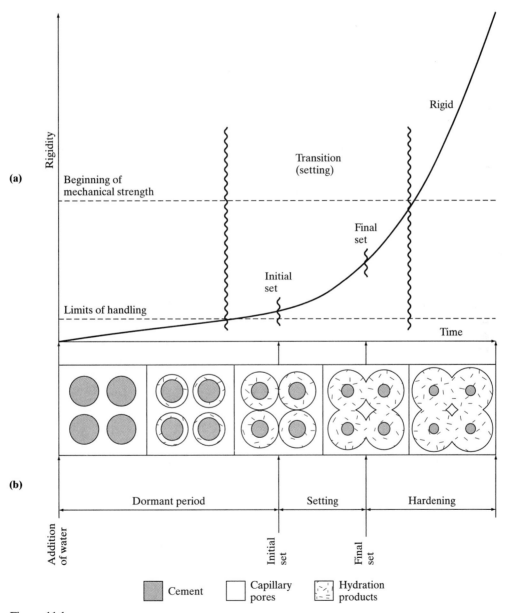

Figure 11.1
Schematic description of the hydration process in a cement paste: (a) setting and hardening (after S. Mindess and J. F. Young, *Concrete,* Prentice Hall, 1981, p. 223); and (b) structure formation (after I. Soroka, *Concrete in Hot Environments,* E&FN SPON, 1993, p. 26).

1. The paste is fluid as long as the individual grains are separated from one another.

2. The hydration products occupy a greater volume than the original cement grain and therefore the advent of hydration is accompanied by a pore-filling effect (i.e., reduction in the porosity).

3. During the early stages of hydration, when sufficient contact has formed between hydration products, the paste gains sufficient rigidity to lose its fluidity; this is referred to as setting.

4. The continuation of the hydration process after the setting results in the generation of strength due bonding interactions between the hydration products, which are facilitated by the close proximity of the hydration products as the porosity decreases. At this stage the rigidity increases to such a degree that at the time referred to as final setting the paste is already a solid, although of a very low strength.

5. The progress of hydration from the final setting time is accompanied by measurable strength increase; this is referred to as the hardening stage.

In practice the process is much more complex, because the cement grains vary in size and composition and the hydration products are not uniform (their composition and microstructure vary considerably with time and with location in the paste). A detailed account of these characteristics will be provided in the following sections; they will be discussed in terms of the engineering properties of the paste.

11.2.1 Composition and Hydration of Portland Cement

Portland cement consists of four major phases, all of which can react with water to produce different types of hydration products. The composition of these phases and their content in a typical normal cement are given in Table 11.1.[1] It can be seen that they are all made of oxides, which are abundant in the earth's crust. The raw materials for the production of such compounds are limestone ($CaCO_3$), quartz (SiO_2), and clays, which are the source of Al_2O_3 and Fe_2O_3. The compounds in the portland cement are obtained in a process involving quarrying, crushing, and mixing of these raw materials, after which they are calcined and burned at a rotary kiln in a temperature of about 1450°C. The product of the firing process (called clinker) is in the

TABLE 11.1   Composition of Portland Cement

Compound	Formula	Wt. Fraction (%)
Tricalcium silicate	Ca_3SiO_5 $(C_3S)^a$	55
Dicalcium silicate	Ca_2SiO_4 (C_2S)	20
Tricalcium aluminate	$Ca_3Al_2O_6$ (C_3A)	10
Tetracalcium aluminoferrite (ferrite phase)	$Ca_4Al_2Fe_2O_{10}$ (C_4AF)	8
Gypsum	$CaSO_4 \cdot 2H_2O$ $(C\bar{S}H_2)$	5

[a]Shorthand notation is commonly used: The formulas are written as combinations of oxides which are given the following symbols: CaO = C; SiO = S; Al_2O_3 = A; Fe_2O_3 = F; SO_3 = $\bar{S}$; H_2O = H. The combining ratio is written as a subscript. Hence $Ca_3SiO_5 = 3CaO \cdot SiO_2 = C_3S$, $Ca_6A_2lO_3 = 3CaO \cdot Al_2O_3 = C_3A$; etc.

[1]A shorthand notation is used in the chemistry of cement. It will be used throughout this chapter, using the definitions given in Table 11.1.

form of hard granules of about 10-mm size, which are then interground with approximately 5% gypsum to a fine powder with particle size in the range of about 1 to 100 μm (average size about 10 μm). This final product is portland cement. The chemical composition and the fine size of the particles make them sufficiently reactive to undergo in the hydration reactions required to develop the properties of the cementitious phase.

The principal strength-producing compounds are the calcium silicates, which together make up 75% to 80% of the cement. Their hydration is very similar and can be represented (in shorthand notation) by:

$$2C_3S + 7H \rightarrow C_3S_2H_8 + 3CH;[2] \qquad \Delta H = -500 \ J{\cdot}g^{-1} \qquad (11.1)$$

$$2C_2S + 5H \rightarrow C_3S_2H_8 + CH; \qquad \Delta H = -250 \ J{\cdot}g^{-1}. \qquad (11.2)$$

The negative enthalpy (heat of hydration) values indicate an exothermic reaction. Overall, we can write

$$\text{Calcium silicates } (C_3S + C_2S) + \text{water} \rightarrow$$
$$\text{Calcium silicates hydrate } + \text{ calcium hydroxide,} \qquad (11.3)$$

where the amount of calcium hydroxide (CH) depends on the relative proportions of tricalcium and dicalcium silicates (C_2S and C_3S) in the cement. The calcium silicate hydrate consists of extremely fine particles in the colloidal size range, and it is an amorphous material which does not have the exact composition indicated in Eqs. 11.1 and 11.2. Therefore, it is often called C—S—H to avoid the implication of a specific formula.

The C_3A compound, without the presence of gypsum, would react vigorously with water to produce in a short time a large amount of hydration products, which can cause setting within a few minutes, thus impairing the fluidity of the paste. From a practical point of view, this is undesirable because the concrete should remain in the fresh state for at least a few hours. However, in the presence of gypsum, the C_3A reacts to form ettringite, which builds up as a layer on the C_3A particle and prevents rapid hydration:

$$C_3A + 3C\bar{S}H_2 + 26H \rightarrow C_6A\bar{S}_3H_{32};[3] \qquad \Delta H = -1350 \ \text{J}{\cdot}\text{g}^{-1}. \qquad (11.4)$$

$$\left(\begin{matrix} \text{tricalcium} \\ \text{aluminate} \end{matrix} + \text{gypsum } + \text{water} \rightarrow \text{ettringite} \right)$$

The ettringite layer is stable as long as there is sufficient gypsum available. Once it is depleted, ettringite and C_3A react further as follows:

$$C_6A\bar{S}_3H_{32} + 2C_3A + 4H \rightarrow 3C_4A\bar{S}H_{12}. \qquad (11.5)$$

$$\left(\text{ettringite } + \begin{matrix} \text{tricalcium} \\ \text{aluminate} \end{matrix} + \text{water} \rightarrow \text{monosulfoaluminate} \right)$$

The overall reaction is therefore

$$C_3A + C\bar{S}H_2 + 10H \rightarrow C_4A\bar{S}H_{12}; \qquad \Delta H = -350 \ \text{J} \cdot \text{g}^{-1} \qquad (11.6)$$

and monosulfoaluminate is the stable phase in mature concrete.

In principle, the ferrite phase (C_4AF) will form the same reaction as C_3A, but since it is much less reactive it is likely to combine with very little of the

---

[2]In shorthand notation H = H_2O; thus CH = $CaO{\cdot}H_2O$ or $Ca(OH)_2$.
[3]In shorthand notation, H = H_2O, C = CaO, $\bar{S}$ = SO_3;
thus $C\bar{S}H_2$ = $CaO{\cdot}SO_3 \cdot 2H_2O$ = $CaSO_4 \cdot 2\ H_2O$.

gypsum. So unless the C_3A content is low, a better representation of the C_4AF reaction is

$$C_4AF + 2CH + 14H \rightarrow C_4(A, F)H_{13} + (A, F)H_3;^4 \qquad (11.7)$$

$$\text{ferrite} + \text{calcium hydroxide} + \text{water} \rightarrow \begin{array}{c} \text{tetracalcium} \\ \text{aluminate hydrate} \end{array} + \text{ferric-aluminum hydroxide.}$$

The calcium hydroxide comes from the hydration of tricalcium silicate. Tetracalcium aluminate hydrate is structurally related to monosulfoaluminate, while the ferric-aluminum hydroxide is amorphous.

The chemical reactions of the four phases occur at different rates (Fig. 11.2), and the strength generated by the hydration products is different (Fig. 11.3). It can be seen that the aluminates and ferrite phases, although chemically reactive, produce very little strength. Strength development is almost entirely due to the hydration of the calcium silicates, and the contribution of each depends on their relative content and rates of hydration. C_3S, which hydrates at a higher rate and generates strength at accelerated rates compared to C_2S, is the major component responsible for the strength generation in the hydration of the calcium silicates is the C—S—H.

When considering hydration reactions, attention should be given to the processes that take place within the first 24 hours because they determine the state of fluidity of the fresh paste, its setting, and the beginning of hardening, all of which are of great practical significance for the application of concrete on site. Since all the reactions are exothermic (i.e., they are accompanied by liberation of heat), it is easy to follow the rates of reactions by monitoring the rate of heat liberation at the early stages. On the basis of such curves, it is possible to divide the hydration process into five stages, as outlined in Fig. 11.4. The first stage is a very short one, and the heat liberated there is due mainly to the wetting and early dissolution of the cement grains. The induction period follows is characterized by negligible reaction rates, which allows the paste to maintain its fluidity, as seen schematically in Fig. 11.1. The onset of rapid reaction rates at the beginning of Stage III is mainly due to onset of hydration of the C_3S phase. About the time of the initial rise in rates, there is already sufficient buildup of hydration products to lead to initial setting, as shown

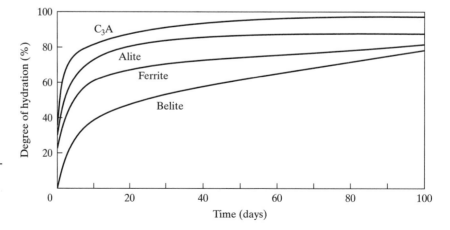

Figure 11.2
Rates of hydration of the four main chemical components of portland cement (after S. Mindess and J. F. Young, *Concrete*, Prentice Hall, 1981, Fig. 3.3, p. 28).

[4](A, F) means that F substitutes for A in the formula.

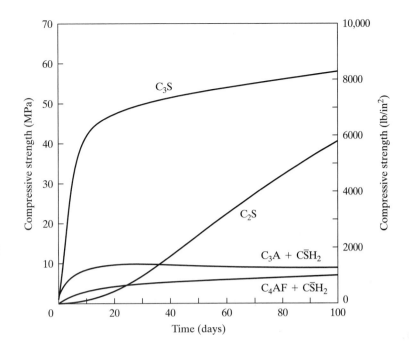

Figure 11.3
Strength development over time of the four main chemical components in portland cement (after S. Mindess and J. F. Young, *Concrete*, Prentice Hall, 1981, Fig. 3.4, p. 29).

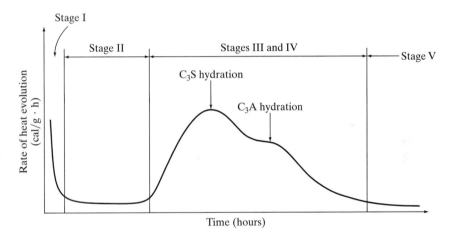

Figure 11.4
Rate of heat evolution during the hydration of portland cement (after S. Mindess and J. F. Young, *Concrete,* Prentice Hall, 1981, Fig. 4.4, p. 85).

schematically in Fig. 11.1, while final setting has occurred before hydration has reached the peak marking the end of Stage III. The arrow marking the C_3A hydration in Fig. 11.4 is indicative of the time when gypsum is depleted and the C_3A now begins a more rapid reaction according to Eq. 11.5. This occurs generally during the slowing down of the hydration of C_3S in Stage IV, but sometimes it is found during Stage III. The gypsum content is adjusted in the cement so that this reaction would occur well beyond the dormant stage to prevent flash set, which can cause premature stiffening during the induction period. Finally, in Stage V, all of the reactions proceed more slowly as the hydration products fill in the pore space between cement grains.[5]

[5]Note that the slower reactions of C_2S and C_4AF do not contribute significantly to the heat liberation curve, except in stage I.

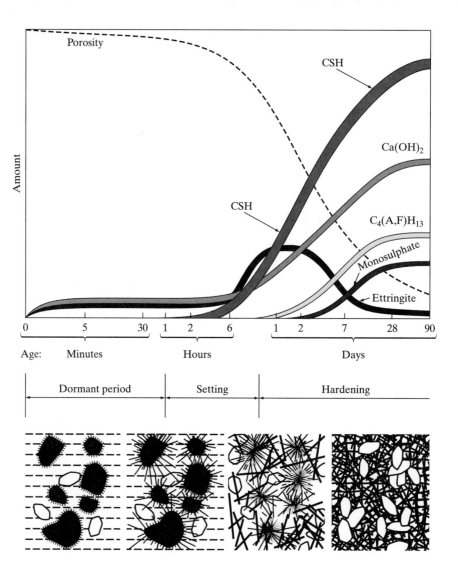

Figure 11.5
Schematic description of the reaction products and microstructure which develop during the hydration of portland cement (after F. W. Locher and W. Richartz).

When referring to the early hydration reactions, attention should be given to the relatively high rates of heat evolution that occur. In practice, this results in an increase in the temperature of the concrete, which can rise to a considerable extent within the first day or so above the ambient temperature. This rise is particularly affected by the C_3S and C_3A, which are characterized by highly exothermic reactions (high negative ΔH values in Eqs. 11.1 and 11.4), and occur at a high rate (Fig. 11.2). The increase in temperature can be favorable at low ambient temperatures because it can accelerate reaction and thereby the rate of strength gain. However, in massive structures this temperature builds up, and its nonuniform distribution as the structure is later cooled to ambient temperature can generate thermal stresses, which may lead to cracking.

11.2.2 Microstructure and Properties of Hydration Products

A schematic description of the reaction products developed over the course of hydration and of their microstructure is provided in Figs. 11.5 to 11.7. The main characteristic features observed are the reduction in porosity accompanied by the

Figure 11.6
Scanning electron micrograph of hardened Portland cement paste after 7-days curing. A cement particle coated with C-S-H (bottom left) is surrounded by ettringite needles growing into the pore space. A few platelets of monosulfoaluminate can be seen in the upper left. A large crystal of calcium hydroxide is at the far right × 3500. (Photo courtesy of Dr. Paul Stutzman).

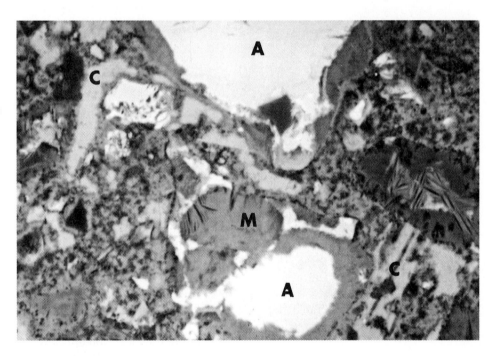

Figure 11.7
Backscattered SEM picture of a cut and polished hardened portland cement paste after 28 days curing. The bright areas (A) represent the remnants of unreacted cement particles surrounded by a dense coating of C-S-H. A pocket of monosulfoaluminate (M) lies adjacent to one cement particle. The ground mass consists of calcuim hydroxide (light gray, labeled C), C-S-H (dark gray) and porosity (black) × 2500 (Photo courtesy of Dr. Paul Stutzman).

formation of a continuous matrix of C—S—H, in which are imbedded unhydrated cement grains, microscopic crystals of monosulfoaluminate, and relatively massive crystals of calcium hydroxide. These latter are several orders of magnitude greater than the average C—S—H "particle," and they grow around many cement grains, with their surrounding layers of C—S—H. Not all of the space is filled with solids; the resultant porosity is of great importance and will be discussed separately.

C—S—H

This material is extremely finely divided to the extent that the exact form of the particles cannot be ascertained with certainty, even by electron microscopy. It is best regarded as an amorphous colloid,[6] which means that its surface properties dominate its behavior. Since there is no regular atomic arrangement governing its internal structure, it can show wide variations in composition and morphology. The structure of this material will reflect the conditions under which it forms, and it can act as an efficient space filler. Many different models of C—S—H have been proposed; Fig. 11.8 gives a schematic representation which incorporates known features about this material.

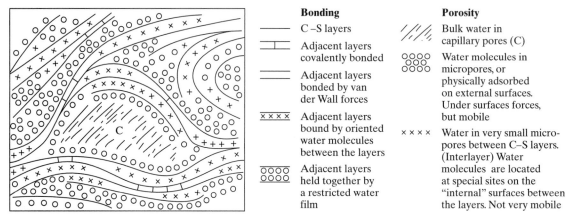

Bonding

——	C–S layers
⊥	Adjacent layers covalently bonded
——	Adjacent layers bonded by van der Wall forces
xxxx	Adjacent layers bound by oriented water molecules between the layers
0000	Adjacent layers held together by a restricted water film

Porosity

/////	Bulk water in capillary pores (C)
0000	Water molecules in micropores, or physically adsorbed on external surfaces. Under surfaces forces, but mobile
x x x x	Water in very small micropores between C–S layers. (Interlayer) Water molecules are located at special sites on the "internal" surfaces between the layers. Not very mobile

Figure 11.8
Schematic description of the structure of C—S—H.

The structure can be regarded as a poorly organized layer structure, of the kind found in clays, and as such it has a very high energy, characteristic of an unstable arrangement. The bonding between sheets ranges from weak bonds involving hydrogen bonding with water between the sheets, to strong covalent bonds involving siloxane (Si—O—Si) linkages. As water is removed from the system, the sheets will tend to move closer together (collapse), increasing the bonding between surfaces. Because of this collapse, it is difficult to get an accurate determination of the water content, density, or surface area of the material. These measurements should, in practice, be made on C—S—H in the saturated or near-saturated condition, but this is not possible; the values given in Table 11.2 are the best estimates we have. The extent of van der Waals bonding is expected to be proportional to the surface area.

The influences of C—S—H on concrete properties are summarized in Table 11.3 and will be discussed again later at appropriate places in this chapter. We can summarize here by saying that C—S—H occupies the largest volume in the paste and forms a continuous matrix binding everything together. The cohesive forces holding it together are partly covalent/ionic bonds (~65%) and partly van der Waals bonds (~35%). It is a microporous, high-surface-area material. Its structure changes depending on the temperature at which it is formed (at higher temperatures it has more covalent/ionic bonds and a lower surface area) and the water/cement ratio of the paste (at lower water/cement ratios there is less space available, and so it has a different pore structure).

[6]Hence the term C—S—H gel, which is often used.

TABLE 11.2 Solid Components of Hardened Cement Pastes

Material	Volume Fraction of Solids (%)	Density (kg m)$^{-3}$	Particle Dimensions (μm)	Surface Area (m^2 g^{-1})	Morphology and Crystallinity
C—S—H	65	2000	<1 across ~0.01 thick	400	Irregular foil, amorphous, microporous
CH	20	2250	100 across 10 thick	~0.5	Thick hexagonal plates that cleave easily and are crystalline
C$_4$A$\bar{\text{S}}$H$_{12}$ (monosulfoaliuminate)	10	1950	~2 across 0.1 thick	~2	Very thin irregular plates clustered in "rosettes"; fairly crystalline
Unhydrated residues	5	3150	~1	0.1	Remnants of original cement grains

TABLE 11.3 Influence of Paste Components on Paste Properties

Component	Strength	Deformations	Durability
C—S—H	Provides major cohesive force but is intrinsically weak because of microporosity. Dry concrete is stronger than wet (stronger van der Waals bonds).	Water loss from micro pores causes shrinkage (on drying) and creep (on loading) at room temperature.	Very insoluble. Water in micropores does not freeze and has low mobility (low permeability).
Calcium hydroxide	Contributes by reducing porosity. Cleavage may limit strength of high-strength pastes.	Is dimensionally stable. Will restrain C—S—H deformations.	Blocks capillary pores and hence lowers permeability. But slowly leached by water, which causes efflorescence and increases permeability. Carbonates, and is dissolved by acids.
C$_4$A$\bar{\text{S}}$H$_{12}$	Not significant. Reduces total porosity.	Minor effect.	Causes sulfate attack by reforming ettringite and causing expansion.
Unhydrated cement	Significant only in low-porosity pastes.	Will restrain C—S—H deformations.	Renewed hydration may cause autogeneous healing of internal microcracks.
Capillary pores	Total porosity is the major factor influencing strength.	Fine pores contribute to shrinkage and creep.	Porosity influences permeability and diffusivity. Large pores increase water flow through concrete.

Calcium Hydroxide

Calcium hydroxide is, by contrast, a well-crystallized material of definite composition, much more amenable to characterization. Its large crystals can be regarded as imbedded in the matrix of C—S—H because each crystal is distinct from its neighbors and need not form a close contact. On the other hand, the crystals grow around many of the partially hydrating cement grains and bind them together in this way, but their contribution to strength is mainly by virtue of their filling pore spaces. If the calcium hydroxide is leached out of a slab of paste by water (a slow process) the paste will not disintegrate, although it becomes weak and porous. The relatively high solubility of calcium hydroxide ($1.5 \text{ g} \cdot \ell^{-1}$) is the cause of efflorescence and as it dissolves can contribute to durability problems in the long term by increasing the porosity of concrete. Calcium hydroxide acts as a buffer to maintain a high pH inside concrete. This can be reduced by atmospheric carbonation.

Monosulfoaluminate

Like calcium hydroxide, the monosulfoaluminate helps fill space and reduce porosity and thus has an indirect contribution to strength. It may likewise have a minor influence on shrinkage and permeability by changing the distribution of pores. When sulfate ions can enter the concrete, the monosulfoaluminates can convert back to ettringite with an accompanying increase in solid volume:

$$C_4A\bar{S}H_{12} + 2CH + 2\bar{S}(aq) \rightarrow C_6A\bar{S}_3H_{32}. \tag{11.8}$$

This reaction causes internal volume expansions, which lead to internal microcracking and hence overall concrete deterioration. This is the mechanism of sulfate attack. The hydrates developed from C_4AF can also form ettringite, but to a much more limited extent.

Anhydrous Residues

These are of significance only in very low porosity pastes, where hydration is limited. Then the high bonding forces of the grains and the hydration products may improve strength and restrain shrinkage of C—S—H. (By implication, less C—S—H is also formed, and hence its contribution to bulk shrinkage is smaller.)

11.2.3 Portland Cements of Different Compositions

In view of these characteristics of the individual hydration products and their influence on the properties of the fresh and hardened paste, it is possible to assign the influences of the various phases in portland cement on the resulting properties of concrete. This is summarized in Table 11.4. By adjusting the content of these compounds it is possible to produce portland cement for specific applications, which are listed in Table 11.5. The designations are those used in the United States, specifying five different types of portland cement labeled I to V. Similar classes of cements are found in most industrial countries, but they have different designations.

High early strength (type III cement) is obtained by increasing the content of C_3S and grinding the particles to smaller sizes, both of which accelerate the production of C—S—H. Characteristics of this kind are required for early form removal, early prestressing of concrete, etc. Low heat cement (type IV) is required for massive structures (e.g., dams), where cooling of the structure (after the initial buildup of temperature induced by the exothermic reactions) creates temperature gradients between the interior and the exterior of the structure, as the latter cools at higher rates. The temperature difference can be on the order of 20 to 50°C. Low heat is achieved by reducing the content of C_3S and C_3A, both of which hydrate at a high rate and liberate considerable heat (Eqs. 11.1 and 11.6). Sulfate attack is mainly the result of the reaction of $C_4A\bar{S}H_{12}$ with sulfate ions (the source of such ions could be sea water in marine structures or salty soil in contact with the foundations of the structure). Therefore, the extent of this problem can be reduced by lowering the content of C_3A to produce sulfate-resistant cement (type V). The type II cement is intermediate between type I and type V or type IV. A comparison of the properties of concretes made with these cements is provided in Figs. 11.9 to 11.11.

TABLE 11.4   Influence of Cement Compounds on Concrete Properties

Property	Compound	Remarks
Setting behavior	C_3S	Controls normal setting
	C_3A	Can cause premature stiffening
Temperature rise during hydration	C_3S	
	C_3A	
Strength development	C_3S	Responsible for early strength
	C_2S	Contributes to long-term strength
Creep and shrinkage	C_3S, C_2S	Major contributions
	C_3A, C_4AF	Minor effects
Durability	C_3S	Leaching of $Ca(OH)_2$
	C_3A	Sulfate attack

TABLE 11.5   ASTM Types of Portland Cements and Blended Cements Used in Construction

Description	Type	Blended Cement	Composition
General purpose (normal)	I	—[a]	See Table 11.1
High early strength (rapid hardening)	III	—	High in C_3S, or ground more finely
Moderate heat of hydration	II		Moderately low in C_3S and C_3A
		I-PM	Contains 0–10% pozzolan
		I-SM	Contains 0–25% slag
Moderate sulfate resistant	II		Moderately low in C_3A ($< 8\%$)
		I-PM	Contains pozzolan low in alumina
		I-SM	
Low heat of hydration	IV[b]		Low in C_3S and C_3A
		IP, IIP	Contains 20–25% pozzolan
		IS	Contains 25–65% slag
		S	Contains $> 65\%$ slag
Sulfate resistant	V		Low in C_3A ($< 5\%$)
		IP, IIP	Contains pozzolan low in alumina
		IS, S	Contains slag

[a] In some countries, general-purpose cements may contain up to 10% of mineral admixtures.
[b] Not presently manufactured in the United States.

11.2.4 Blended Cements and Mineral Admixtures

In modern concrete technology, there is increased use of concretes in which the cementitious phase is a blend of portland cement and a mineral admixture. Its content can be in the range of 20% to 70% of the total binding material (cement + mineral admixture) in the concrete. The mineral admixture can be added as a separate ingredient to the concrete in the mixer or it can be blended with the cement in the plant, to deliver to the site a blended cement in which part of the portland cement has been replaced by the mineral admixture.

Most of the mineral admixtures used are byproducts of industrial processes, and the motivation for their application is economical and ecological, as well as technical. The two most commonly used mineral admixtures are fly ash (which is the byproduct obtained in the burning of coal for energy production) and blast furnace

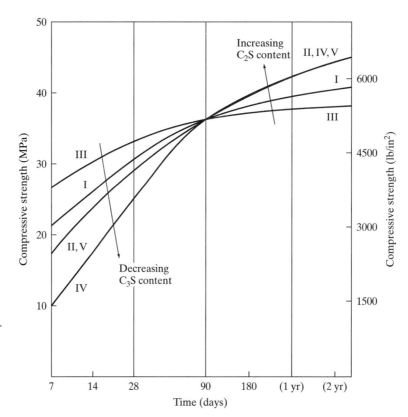

Figure 11.9
Comparison of strength development of concretes with different ASTM-type cements (after *Concrete Manual,* 8th ed., U.S. Bureau of Reclamation, Denver, Colorado, 1975).

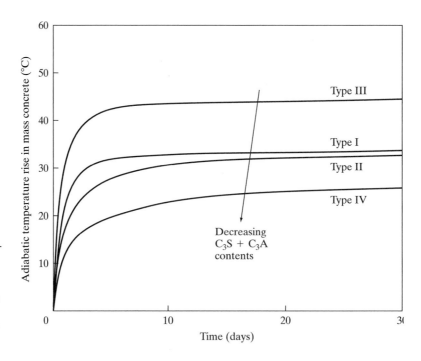

Figure 11.10
Comparison of the temperature rise from heat released during hydration of cements in mass concretes (based on data in *Concrete Manual,* 8th ed., U.S. Bureau of Reclamation, Denver, Colorado, 1975).

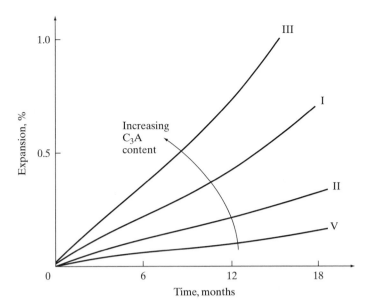

Figure 11.11
Comparison of expansion of concretes prepared with different ASTM-type cements after immersion in sulfate solutions.

slag (byproduct obtained in the production of steel). Both types confer very similar benefits to the concrete, although they react in different ways. They can lower the heat of hydration and improve the durability of concrete. In general, concretes made with these materials have a finer pore structure and hence a lower permeability or diffusivity when the same water/binder ratio is used, as compared to the portland cement alone. Addition of a mineral admixture can be used to improve workability by increasing the proportion of fine material without increasing cement contents. Classification of blended cements as specified by ASTM is included in Table 11.5.

Pozzolans

Fly ash is an example of the class of materials called pozzolans. First used by the Romans (Puozzoli is a small village near Naples where volcanic ash from Mount Vesuvius was mined), pozzolans now embrace a wide variety of materials (Table 11.6). What they all have in common is a reactive, amorphous siliceous component which can combined with the lime formed during hydration. The reaction is:

$$2S + 3CH + 5H \rightarrow C_3S_2H_8. \tag{11.9}$$

The calcium silicate hydrate is very similar to that formed from the hydration of C_3S and C_2S, although it usually has a lower amount of CaO than the formula indicates.

The pozzolanic reaction does not liberate large quantities of heat, and generally the reaction is slow, with appreciable reaction occurring only beyond 28 days. Thus, adding a pozzolan is akin to increasing the C_2S content of the cement, and the early strength will be low. However, some pozzolans can be reactive, particularly those with high surface areas (e.g., silia fume and metakaolin).

Many pozzolans also contain alumina, and this may react if it is present in the reactive siliceous component:

$$A + 4CH + 9H \rightarrow C_4AH_{13}. \tag{11.10}$$

C_4AH_{13} is similar to $C_4A\overline{S}H_{12}$ and can be involved in sulfate attack, so such pozzolans should not be used if sulfate resistance is required. Crystalline compounds (e.g., quartz, mullite, haematite) that are present in a pozzolan are considered inert

TABLE 11.6 Classification of Pozzolans

Form of Silica	Type	Examples	Typical Composition		
			SiO_2	Al_2O_3	CaO
Natural Pozzolans					
Glassy particles	Volcanic ash	Santorin earth	65	13	4
		Pozzolana	45–60	20	2–10
		Trass (rock)	55	10–20	2–5
Glassy matrix	Solidified lava	Tuffs	—	—	—
		Pumicites	—	—	—
		Rhyolites	—	—	—
Amorphous	Diatomaceous earth	—	85	5	trace
Artificial Pozzolans					
Amorphous	Calcined clays ($< 600°$ C)	Metakaolin	50–90	5–25	0–5
Glassy particles	Fly ash	Class F	50	20	0–5
		Class C	50	25	10–15
Amorphous	Silica fume	Microsilica	85–98	0–5	trace
Amorphous	Rice husk ash	—	85–98	—	trace

and thus act only as dilutents. Unburned carbon in byproduct pozzolans (fly ash, etc.) can interfere with air entrainment, and this possibility should be investigated, when frost resistance is required.

Since pozzolans are either natural or byproduct materials, they can have a wide range of compositions and reactivity. Thus a pozzolan needs to undergo extensive testing to assure its suitability for concrete.

Blast Furnace Slags

If molten blast furnace slags are cooled rapidly, glasses containing CaO, SiO_2, and Al_2O_3 are formed. Cooling in air is too slow and allows the melt to crystallize into unreactive compounds, so it must be cooled rapidly by quenching in water (granulation or pelletization). The glass thus is formed, which in contrast to a pozzolan is reactive in its own right:

$$(C—S—A)_{glass} + H \rightarrow C_3(S, A)_2H_8. \qquad (11.11)$$

It forms an aluminum-substituted C—S—H. (Some alumina, from C_3A, is also found in C—S—H formed in cement hydration, so this reaction product is again similar.) However, the slag reacts extremely slowly unless activated by the presence of $Ca(OH)_2$, as formed in cement hydration. Thus, slags are mixed with portland cement, but in these blends, more slag can be used than pozzolan because of its cementitious character. (Only about 25 wt % of pozzolan, by weight of cement, will react with all the calcium hydroxide formed in a type I cement, and any excess is an unreactive dilutant.) Otherwise, the blast furnace slag reaction is similar to the pozzolanic reaction, liberating small quantities of heat and proceeding slowly.

11.2.5 Porosity and Pore Structure

The discussion in the previous sections dealt with the composition and properties of the solid compounds in the portland cement and their hydration products. The chemistry of the system and its control is only one aspect that should be considered, and attention must also be given to the pore structure that develops during hydration because this physical characteristic is of greater significance in controlling the properties of the concrete. The simplified model in Fig. 11.1 demonstrates the pore-filling effect of the hydration products. The remnants of the spaces that were occupied by the mixing water, and which have not been filled by the hydration products, exist as capillary porosity. Yet, as we have already seen, these are not the only type of pores in the paste; there are also some pores of smaller diameter in between the aggregated hydration products (primarily C—S—H). The early model of Powers[7] accordingly subdivided the pores into two types: capillary and gel pores, as seen in Fig. 11.12. This model is a simplification of the actual pore structure of the paste because pores may exist in a range of sizes, as seen in the microstructural representations in Figs. 11.6 and 11.7 and actual evaluation of the pore size distribution in Fig. 11.13. The pores in Fig. 11.12 are subdivided into ranges according to general size classifications applicable to all porous solids:

Macropores: pores bigger than 100 nm (0.1 μm)

Mesopores: pores in the range of 2.5 nm to 100 nm (0.1 μm)

Micropores: pores smaller than 2.5 nm.

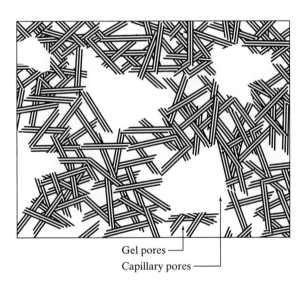

Figure 11.12
Schematic description of the paste pore structure according to Powers (after T. C. Powers, "Physical Properties of Cement Paste," pp. 577–613 in *Proc. Symp. Chem. Cement,* Washington, D. C., 1960, National Bureau of Standards, Monograph No. 43, Washington, D.C. 1962).

Gel pores
Capillary pores

The model suggested by Powers, although simplistic, can serve as an adequate working hypothesis for calculation of pore volumes, to account for some of the more important characteristics of the engineering properties of concrete. It is possible to calculate the volume fractions of the solids, the gel pores, and capillary pores in terms of two parameters: water/cement ratio (w/c) and degree of hydration (α), which represents the fraction of cement that has hydrated. The basic equations are simple and are based on experimental data obtained by Powers; they can also be derived from the hydration equations. It must be remembered, however, that the coefficients represent average values. There is a wide range in the values determined for the different cements, and they are applicable only for the temperature range of 0 to 50°C. In these equations c is the cement weight in the paste, w is the water weight in the paste, and α is the degree of hydration.

[7] T. C. Powers, Structure and Physical properties of hardened portland cement paste, *Journal of the American Ceramic Socity,* Vol. 41, No. 1, 1958, pp. 1–6

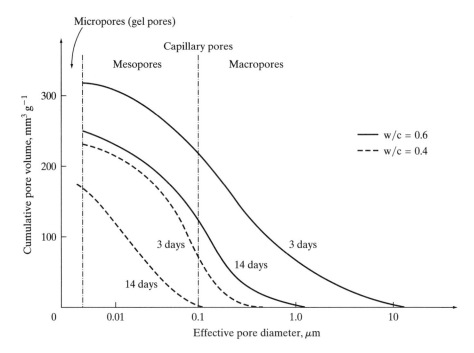

Figure 11.13
Pore size distribution of hydrated cement pastes as affected by the water/cement ratio and the time of hydration.

Volume of total hydration products $= 0.68\,\alpha c$ (in units of volume)   (11.12)
   (including gel pores),

Volume of unhydrated cement $= 0.32(1-\alpha)c$ (in units of volume)   (11.13)

Volume of capillary pores $= [w/c - 0.36\alpha]c$ (in units of volume)   (11.14)

Volume of gel pores $= 0.16\alpha c$ (in units of volume)   (11.15)

Relative volume of capillary pores in paste (capillary porosity) $= \dfrac{w/c - 0.36\alpha}{w/c + 0.32}$ (11.16)

The result of the calculations of capillary porosity as a function of w/c ratio and time are presented in Fig. 11.14 and show that the w/c ratio is a major factor controlling the porosity of hardened cement paste. Since capillary porosity controls the strength and permeability of the hardened paste (Fig. 11.15), the production of concretes of high strength and low permeability can be achieved by using mixes of low w/c ratio and assuring a high degree of hydration. The first can be achieved by adequate proportioning of the concrete ingredients that are placed into the mixer, whereas the second is obtained by making sure that proper curing is applied (i.e., sufficiently long curing in which means are taken so the concrete does not dry after it has been placed and compacted and so the water in the mix does not evaporate and is available for the hydration reaction). A low w/c ratio and adequate curing are key elements in concrete technology. A particularly low w/c ratio can be achieved in mixes with superplasticizers and very fine pozzolans like silica fume; they form the basis for high strength/high performance concretes,

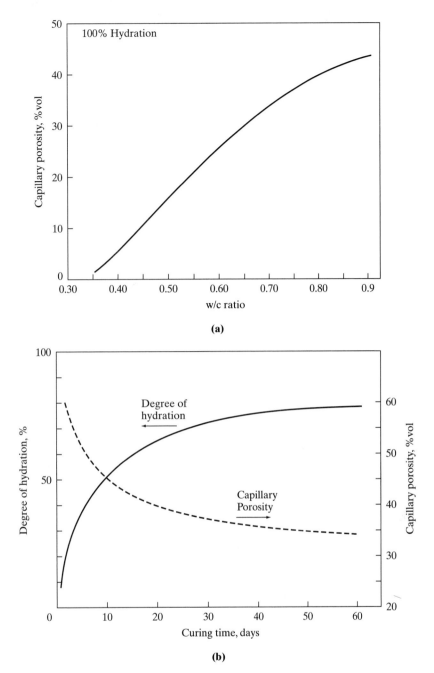

Figure 11.14
Capillary porosity as a function of w/c ratio (a) and curing time and degree of hydration (b), as calculated according to the Powers model.

which are dense and durable, with compressive strength levels in the range of 50 to 150 MPa.

Although paste porosity has the dominating influence on concrete properties, there are other types of pores found in concrete which should be considered. One type is entrapped air left after compaction (usually less than 3% by volume), or entrained air, which is added deliberately to improve the frost resistance of concrete (4% to 8% by volume). Entrained air will be discussed in Sec. 11.3.4. Very high porosity levels can be introduced to obtain lightweight and thermally insulating

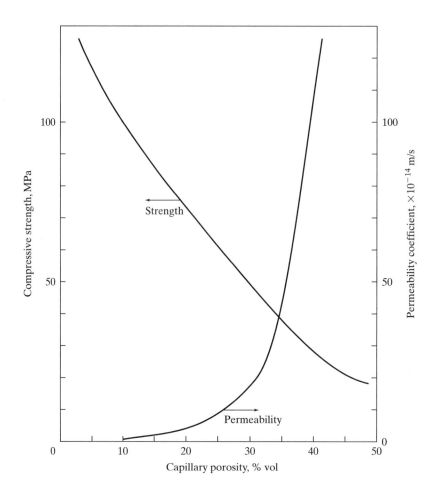

Figure 11.15
Effect of capillary porosity on compressive strength and permeability of hardened cement paste.

concrete by entraining large quantities of air or other gases (≈30% by volume or more) and by the use of porous, lightweight aggregates. In these specialized systems the property-porosity relations are different from the ones discussed here, and are beyond the scope of this chapter.

11.3   PROPERTIES OF CONCRETE

It was already demonstrated that concrete is a composite materials. The structure and properties of its two components were discussed in Chapter 10 (aggregates) and in Sec. 11.2 (cementitious matrix). In this section the overall performance of the concrete will be presented. It is essential to realize that the quality of the end product depends not only on the properties of the individual ingredients and their proportioning in the concrete, but is also very sensitive to the production process, which consists of several stages: mixing, transporting, placing, compacting, and curing. For this process to be successful, it is essential to adjust the properties of the fresh concrete to the equipment used in the production process (mixers, pumps, vibrators, etc.) as well as to the environmental conditions which exist on site. Therefore, the treatment of the concrete will begin with a section dealing with the early age properties of concrete, with particular attention to the fresh concrete workability. Since in modern concrete technology extensive use is made of *chemical* admixtures, whose main influence is on the properties of the early age concrete, a special section will be devoted to them, and only afterward the properties of hardened concrete will be discussed with reference to mechanical, physical, and durability performance.

11.3.1 Fresh Concrete

In the period prior to setting, we are dealing with a fluidlike material, usually referred to as the fresh concrete. The properties of the fresh (*or plastic*) concrete should be adjusted to the production equipment and process. This implies that the material should have adequate flow characteristics which remain stable over the whole production period, lasting for several hours. There is also the requirement that the concrete in the fresh state will remain uniform (i.e., there will be no segregation and separation between the aggregates and the paste). Such a separation is obtained due to differential settlement of the individual phases. For example, after placing and compaction, while the concrete is still fluid, the aggregates may settle down, forcing an upward movement of water and some cement particles. This effect, known as *bleeding,* can lead to the formation of a weaker concrete surface due to accumulation of water on the top surface of the concrete and reduced bond underneath reinforcing bars within the concrete (Fig. 11.16). Separation of this nature could also occur in the production process, when the concrete is induced to flow (e.g., in a concrete pump) and the different phases are not moving at the same speed, leading to separation. In view of this brief discussion, it becomes apparent that the requirements for the properties of the fresh concrete are complex and time dependent, and therefore they cannot be described in terms of one significant physical parameter. The term *workability,* which is used with respect to fresh concrete, is a qualitative one, and good workability implies good performance in all placement processes. Workability, which is adjusted for one set of conditions and defined as good, may be poor workability for a different set of conditions (i.e., different equipment, different transportation times, etc.). Therefore, it is also common to use alternative terms, such as *pumpability, compactability,* and *finishability,* which refer to the

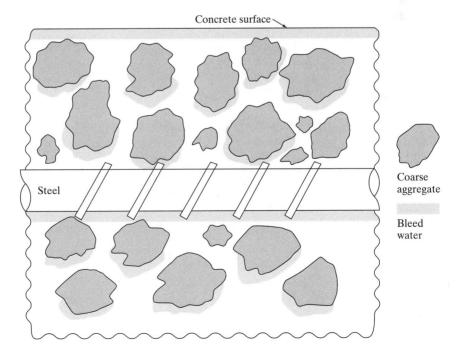

Figure 11.16
Different types of bleeding effects in concrete.

workability in conjunction with technology employed for the production of the concrete component, for which the concrete should be of adequate workability.

Although workability as a whole is difficult to quantify, it is possible to devise test methods which address some of the properties encompassed under workability. The most common and practically important are the tests for evaluation of the flow properties, treating the fresh concrete as a fluid. In this category, the rheological tests based on the concepts described in Chapter 7 can provide data of physical significance to quantify the fresh concrete. Since the aggregates in the concrete are too large to enable testing in a rotational coaxial cylinders viscometer, where the gap must be very small (Chapter 7), alternative methods were developed, using rheometers based on a concept of rotating an impeller in concrete and measuring the torque required for the rotation as a function of the rotational speed. The torque-impeller speed curve is analogous to the torque-rotational speed curve obtained from a coaxial viscometer, and with some assumptions the data from the rheometer test might be calibrated and presented as shear stress–shear rate curves. Tests of this kind suggest that the behavior of the fresh concrete follows the Bingham model (Fig. 11.17), and therefore the properties of the concrete can be described in terms of two parameters, g and h:

$$T = g + hN, \tag{11.16}$$

where T is the torque and N is the impeller rotational speed. The g and h constants represent the flow resistance (g) (analogous to yield) and the torque viscosity (h) (analogous to plastic viscosity) and are characteristics of the concrete.

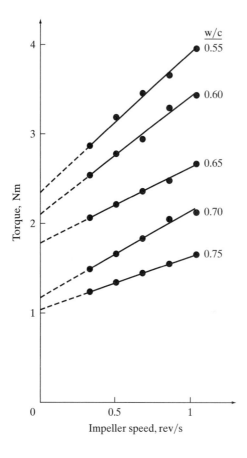

Figure 11.17
Typical results of rheological and workability testing of concrete using a rheometer (after G. H. Tattersall, "Measurements of Rheological Properties of Fresh Concrete and Possible Application to Production Control," pp. 79–97 in *Effect of Surface and Colloid Phenomena on Properties of Concrete, Proc. Symp. M,* MRS, November 1982).

The rheometer tests require special equipment and cannot be easily applied on site. Therefore, the tests used routinely and specified in standards are much simpler although less physically significant. However, proper use of such tests, with the understanding of their limitations, has proved to be successful in producing workable concrete. The most common test accepted worldwide is the slump test (ASTM C-143). It is based on filling a standard cone with fresh concrete, lifting it up, and measuring the slump (Figs. 11.18a and 11.18b). Correlation can be found between the slump value and the parameter g in the rheometer test (Eq. 11.16 and Fig. 11.19).

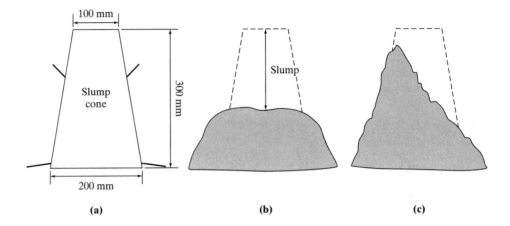

Figure 11.18
The slump cone test: (a) the slump cone; (b) slump measurement in properly slumped concrete; and (c) sheared fresh concrete.

(a)   **(b)**   **(c)**

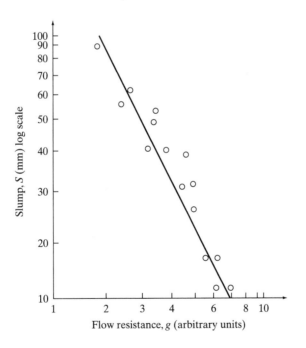

Figure 11.19
Relations between slump and g in Eq. 11.16 (after T. Scullion, "The Measurement of Workability of Fresh Concrete," M.Sc. Thesis, University of Sheffield, 1975).

The slump method is a sensitive test for concretes with slumps in the range of 10 to 180 mm. For mixes of similar composition of the solid constituents, this test can readily distinguish between concretes which are different in their water contents, as demonstrated in Fig. 11.20. Therefore, this test is referred to as one which measures the consistency of the concrete. It can be seen that the relation between slump and water content is not linear, and a small increase in the water content in the wetter

225

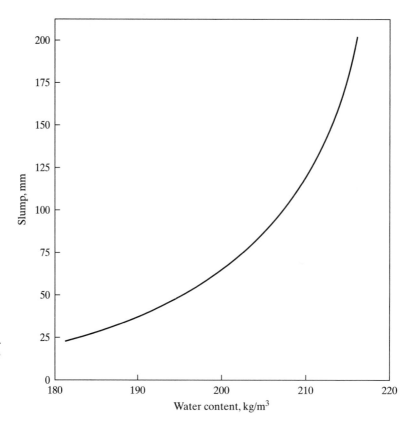

Figure 11.20
Relation between slump and the amount of mixing water (drawn from data in *Standard Practice for Selecting Proportions for Normal, Heavyweight and Mass Concrete,* ACI 2111-89, ACI Committee 211 Report, American Concrete Institute, Detroit, 1989).

mixes will result in a slump increase much greater than in the stiffer mixes. The slump of the concrete is the most common test used to characterize the properties of the fresh concrete, and it is often referred to as a measure of workability. However, the term *consistency* is the more adequate one to be used for this measure, because adequate consistency is only one of the parameters which will ensure proper workability. Ideally, the consistency measure should be supplemented by data regarding the tendency for segregation. Unfortunately, this is not quantifiable by simple tests that can be carried out on site. Yet observation of the concrete and, in particular, the mode of the slumped concrete in the slump test can provide valuable information: A concrete that in the slump test is sheared and collapsed (Fig. 11.18c), rather than sagged evenly, is indicative of a concrete that will tend to segregate and bleed. This behavior is usually the result of a lack of sufficient content of fines (usually sand) in the concrete, which are essential for generating cohesiveness in the fresh mix (i.e., sufficient level of secondary attractive forces in the fresh mix which can hold the concrete ingredients together). Such forces are generated by physical interactions between small-size particles. Mixes which are cohesive at one slump level may not be cohesive in a higher level (in particular, in the higher slump range); if the water content is too high, thus causing greater separation between the solid particles, it may reduce the attractive cohesive forces to such an extent that it may lead to segregation and excessive bleeding.

The properties of the fresh concrete change during the induction period prior to setting because there is a reduction in the free water in the mix. This is due in part to the evaporation of the mixing water and absorption into the aggregates if they are air dried when discharged into the mixer. An additional contribution comes from water consumed in the hydration process and the concomitant formation of hydration products; (which form slowly during the induction period). It shows up by reduction in the slump, and reference is made in practice to the slump loss of the concrete (Fig. 11.21). This loss must be taken into account in particular if the time period between the mixing and placing of the concrete is long; it is of greater concern in hot climates, where the evaporation is more severe. Special attention to this problem is given in ready-mix concrete technology.

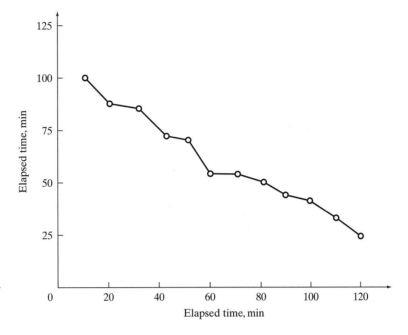

Figure 11.21
Slump loss in concrete (after D. Whiting and Dziedzic, "Behaviour of Cement-Reduced and 'Flowing' Fresh Concrete Containing Conventional Water-Reducing and 'Second-Generation' High Range Water-Reducing Admixtures," *Cement, Concrete and Aggregates,* 11(1), pp. 30–39, 1989).

11.3.2 Behavior during Setting

Evaporation of free water from the concrete will continue after the concrete has been placed and compacted and is accompanied by shrinkage. As long as the concrete is sufficiently fluid, it will be free to shrink. However, when it starts to gain rigidity during the initial setting period, the shrinkage will result in a buildup of stress in the concrete, which, if sufficiently high, will lead to cracking. This process occurs in the fresh concrete prior to the hardening stage; the shrinkage at this stage is referred to as plastic shrinkage and the cracking which might occur is termed *plastic shrinkage cracking.*

The problem of plastic shrinkage cracking will be more severe in hot, dry, and windy environmental conditions, when the rate of evaporation is high. In practice, the most efficient way to deal with plastic shrinkage cracking is to apply proper curing immediately after casting to prevent evaporation from the concrete. This is achieved by covering the concrete with plastic sheets, wet burlap, or a curing compound, which is sprayed over the surface and provides a protective film. However, practice has shown that the best method of curing is to keep the concrete continuously wet, since this not only prevents evaporation but also supplies additional water for hydration.

The stages of the initial and final setting are important in concrete technology, as they control the time during which the concrete is in the fresh state (and, from a practical point of view, we like to keep it as long as possible) and the time at which the concrete starts to gain useful strength (which we would like to keep as short as possible). The setting of concrete can be monitored by means of a simple test in which the resistance to penetration of a standard probe is measured as a function of time. Test results obtained by the method outlined in ASTM C-403 are shown in Fig. 11.22. The standard defines initial and final setting in terms of the time required to achieve a certain stress level. Although this stress level is somewhat arbitrary, it provides an adequate indication of the state of the concrete. Initial set occurs close to the end of the induction period and signals the beginning of an abrupt increase in yield strength. Since the initial and final setting are the result of the continuation of the same hydration process, these time periods are related, and it is not possible to accelerate or delay one of them without similarly affecting the other (as is demonstrated in Fig. 11.22 for the effect of temperature).

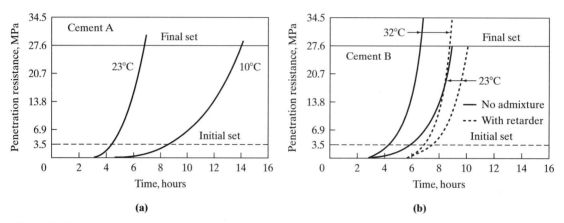

Figure 11.22
Setting curves of concrete determined according to ASTM C-403 (after J. H. Sprouse and R. B. Peppler, ASTM STP 169B, 1978, pp. 105–121).

11.3.3 Chemical Admixtures

Materials that are added in small amounts to the concrete (usually less than 1% by weight of the cement) and dissolved in the mixing water are known generically as chemical admixtures. They are different from the mineral admixture that were discussed in Sec. 11.2. The latter are added in much larger quantities and are active components in the cementitious phase, whereas the chemical admixtures can be considered as modifiers. The main influence of chemical admixtures is on the properties of the fresh concrete; this influence is brought about by affecting the rates of the early hydration reactions or regulating the rheological properties of the fresh mix. Chemical admixtures consist of a wide variety of materials; many of them are based on industrial byproducts and may contain several ingredients (Table 11.7). Classification of the different types of admixtures into five categories, based on their influence on concrete properties, is provided by ASTM C-494: water-reducing admixtures (type A), retarding admixtures (type B), accelerating admixtures (type C), water-reducing and retarding admixtures (type D), and water-reducing and accelerating admixtures (type E). Some chemicals may belong to more than one category. High-range water reducers (superplasticizers) fall under ASTM C-494 as type F and G if they are used to reduce water content, and under ASTM C-1017 if used to produce concretes of enhanced workability (high slump, flowing concrete). Air entraining admixtures are covered by ASTM C-260.

TABLE 11.7   Classification of Chemical Admixtures

Admixture Class	Active Chemical Component	Source
Air-entraining agents	Sodium abietate Sulfonated hydrocarbons	Decomposition of pine wood Petrochemical industry
Water-reducing agents Plasticizers	Lignosulfonates Starch derivatives Hydroxycarboxylic acids (e.g., gluconic acid)	Paper pulp production Food and agriculture processing Food and agriculture processing
Superplasticizers	Formaldehyde condensates of (a) sulfonated naphthalenes (b) sulfonated melamines Acrylic co-polymers	Synthesized chemicals
Retarding admixtures	(See plasticizers) Phosphonates	Synthesized chemicals
Accelerating admixtures	Soluble calcium salts [e.g., $CaCl_2$, $Ca(NO_2)_2$, Ca formate]	Synthesized chemicals or chemical byproducts
Flash-setting admixtures	Na_2SiO_3 (water glass) $NaAlO_2$, Na_2CO_3	Synthesized chemicals

Water-reducing Admixtures

These are surfactants (see Chapter 4) which adsorb at the solid-water interface and prevent the cement grains from flocculating in water (Fig. 11.23). More water is available for "lubricating" the mix and the flow properties are enhanced (hence the term *plasticizer* is often used), as can be seen by an increase in the slump of the concrete. An alternative use of these admixtures is to reduce the water content of the mix while keeping the flow properties (slump) constant. The reduction of the w/c ratio of the mix confers many benefits to the concrete: improved strength, decreased permeability, and increased durability.

Unfortunately, water-reducing admixtures often retard the setting severely. Although this can be offset to some extent, it limits the amount of admixture that can be used, so that the reduction in water content is limited to about 5% to 10%. Newer synthetic admixtures (high-range water reducers, superplasticizers) can be used in larger quantities, and a reduction of 20% to 30% in water content is possible. These admixtures give complete dispersion of the cement grains to achieve maximum flow properties, while plasticizers only partially achieve this.

The use of the water-reducing admixtures increases the slump of the concrete but does not necessarily reduce the rate of slump loss. To counteract this, the addition of these admixtures (in particular, superplasticizers) is often delayed to the stage just prior to the discharge of the concrete from the mixer. In other instances a retempering procedure is applied, involving the addition of two dosages of the admixture in a specified time sequence with remixing. A newer generation of admixtures allows high slump to be maintained for up to 2 hours.

Set-retarding Admixture

These admixtures, as their name implies, delay the setting of the concrete by slowing down the early hydration reaction and the rate of early strength development. They are used in hot weather conditions to offset the acceleration of setting caused by the higher temperature. Although they are effective in delaying setting, they may not

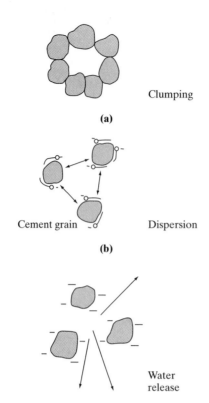

Clumping

(a)

Cement grain   Dispersion

(b)

Figure 11.23
Schematic description of the action of water-reducing admixtures to prevent flocculation which occurs normally (a), by adsorbing on the surface of cement grains, causing them to disperse (b), and releasing more water for a lubricating effect (c).

Water release

(c)

necessarily slow down the rate of slump loss. Under such conditions a combination of retarding admixture and retempering may provide an adequate solution.

Set-accelerating Admixture

These admixtures, as their name implies, accelerate the rate of setting by accelerating the rates of the early hydration reactions. This effect influences not only the hydration reaction during the setting period, but it lasts also into the hardening stage, thus leading to higher early strength gain. Their use is beneficial for concrete formulations applied for repair and for achieving high early strength for formwork removal. In cold weather conditions such admixtures are used to offset the retarding effect of low temperatures. In the past, the more common admixtures for this purpose were based on various chloride salts. However, because the presence of chlorides promotes the corrosion of steel in concrete, chloride-free admixtures should preferably be used.

Air-entraining Admixtures

These are also surfactants which act at the air-water interface, causing water to foam during mixing (Fig. 11.24). The fine foam remains stable until it is "locked" into the cement paste during hardening. The tiny bubbles act as reservoirs to accommodate water expelled from the capillary pores during freezing (see Sec. 11.3.4). Air entrainment also improves the workability of the concrete by generating a mix with improved flow properties (higher slump) and reduced segregation. The presence of air, however, leads to some reduction in strength, which may be in the range of 10% to 20%, depending on the content and dispersion of the air bubbles.

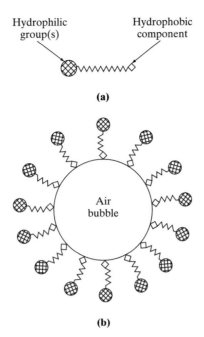

Figure 11.24
Schematic representation of air entrainment by surfactant: (a) surface active molecule, (b) stabilized air bubble (after S. Mindess and J. F. Young, *Concrete,* Prentice Hall, 1981, p. 174).

Most of the modern concretes are produced with admixtures, and in some instances the concrete would contain a combination of several chemical admixtures. Their application, although beneficial, should be exercised with care to assure compatibility between the different admixtures used and between them and the portland cement. Since many of the admixtures affect more than one property of concrete, their overdose may lead to unwarranted influences. Additional advantages can be gained by combining chemical and mineral admixtures. For example, additions of silica fume with a superplasticizer allow the production of high-strength/high-performance concretes with a low water:cement ratio and with good workability.

11.3.4 Properties of Hardened Concrete

Strength

The strength of concrete is its most important engineering property because it not only reflects its mechanical quality, but also provides an indication of long-term performance. Stronger concrete is usually denser, less permeable, and thus better resistant to deleterious environmental influences. Concrete is a brittle material, like most ceramics, and is much stronger in compression than in tension, as discussed in Chapter 6 and seen in Fig. 11.25. Therefore, concrete structures are made of reinforced concrete, where in a typical component subjected to flexure, the steel bars accommodate the tensile stress and the concrete accommodates the compressive stress and provides protection to the steel bars to prevent their corrosion. Therefore, the discussion in this section and the following one will be confined to behavior under compression.

It was already stated that in this particular composite the properties are largely governed by the cementitious matrix. It was already demonstrated in Sec. 11.2 that the strength of the paste is dominated by the capillary porosity, which is a function of the w/c ratio and degree of hydration α. The influence of the two was combined in an empirical relation proposed by Powers:

$$\sigma = 100(1 - P_{cap})^3 = 100\left(\frac{0.68\alpha}{w/c + 0.32\alpha}\right)^3 MPa \ . \tag{11.17}$$

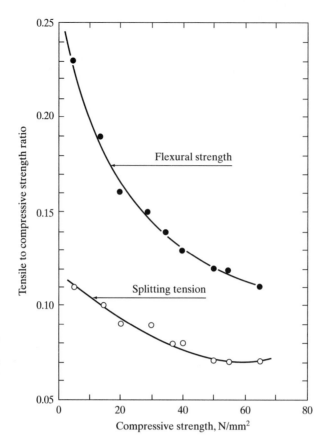

Figure 11.25
Relations between flexural, tensile, and compressive strength of concretes of different strength levels (after W. H. Price, "Factors Influencing Concrete Strength," *Proc. Amer. Concr. Inst.* 47(8), pp. 417-432, 1951).

The coefficients in Eq. 11.17 are the average of values for cements tested by Powers; his equation can also be represented in terms of constants that can be empirically determined for different cements.

A relation of this kind indicates that the strength of the concrete would depend on the w/c ratio (Fig. 11.26) as well as on the period of moist curing in water (Fig. 11.27b), since the value of α can only increase as long as moisture is available for hydration. It also depends on the temperature of curing (Fig. 11.27a); hydration occurs faster initially at higher temperature. (The dashed line in Fig. 11.26 at the lower w/c ratio range reflects a situation that might occur in low w/c ratio concretes, where the fresh concrete is too stiff and cannot be compacted efficiently.)

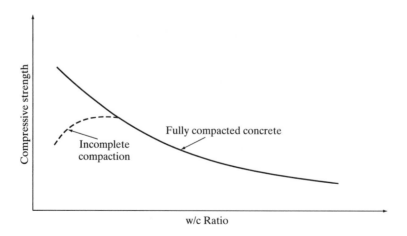

Figure 11.26
Relations between compressive strength and w/c ratio for well-cured concrete.

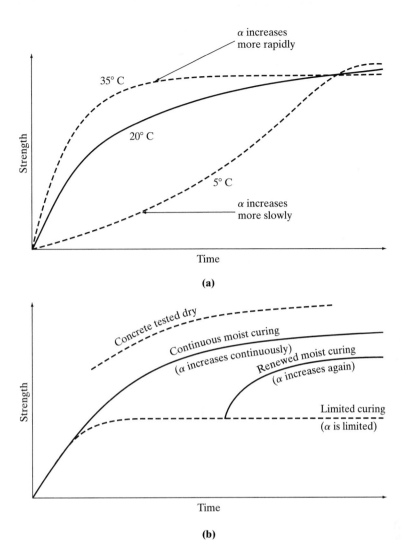

Figure 11.27
Compressive strength development as a function of concrete curing: (a) effect of temperature; (b) effect of moisture conditions.

The relation between strength and w/c ratio was first suggested by Abrams using an empirical equation of the following type:

$$\sigma = A/B^{w/c}, \tag{11.18}$$

where A and B are constants characteristic of different cements. Eqs. 11.17 and 11.18 both represent similar strength-property relationships.

The compressive strength levels also depend on the moisture state of the concrete during testing; concrete which has been dried prior to testing will be stronger because the removal of water allows stronger van de Waals bonds to develop between the surfaces of C—S—H. The differences in compressive strength between the wet and dry state are about 10%.

Stress-Strain Curve

The aggregates in normal strength concrete have only a small influence on strength, yet their presence substantially affects the stress-strain curve of the concrete. The curve is curvilinear over the whole range of applied stress (Fig. 11.29a), although at low stress levels it is very nearly linear. It has been shown by direct observations and other measurements that the shape of the curve is due to progressive internal cracking of the concrete because of its composite structure. In normal weight concrete the different response of the strong, rigid aggregate and the softer and weaker paste matrix will lead to stress concentrations at the paste-aggregate interface due to differential strain, which tends to develop between the two. This stress concentration leads to the progressive development of tiny cracks at the interface, as demonstrated by computer model simulation (Fig. 11.28b). They form at low stresses of about 25% of the strength and cause the deviation from linearity in the stress-strain curve. This is in contrast with the largely linear stress-strain curve of the individual constituents of the concrete, which are homogeneous and do not possess the interfaces leading to nonlinearity (Chapter 9). These cracks, known also as bond cracks, are stable (i.e., they remain localized and do not propagate further if the load level is maintained). As the stress is increased above 50% of the strength, the bond cracks start to propagate slightly into the matrix, and matrix microcracking also starts to develop. When the stress increases over 75% of the strength, the paste microcracks become unstable and continue to propagate and to merge with existing bond cracks into continuous cracks until failure occurs. The higher the stress is above the 75% level, the more rapid is the propagation of the cracks. If a constant load is maintained between 75% and 100% of the strength, failure will eventually occur, because the unstable cracks are given sufficient time to propagate catastrophically. This phenomenon is known as static fatigue and is shown in Fig. 11.29 in terms of stress-strain-time curves. This

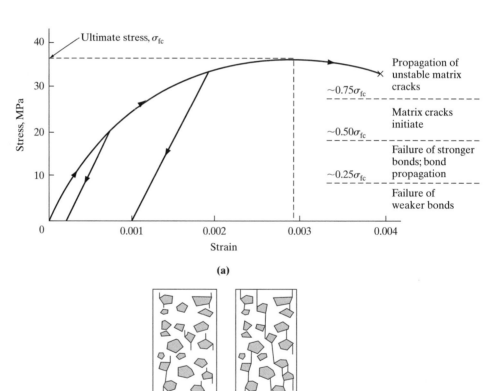

Figure 11.28
Behavior of concrete in compression: (a) stress-strain curve; (b) modeling showing progressive cracking under low (left) and higher (right) load (after Y. V. Zaitsev and F. H. Wittmann "Simulation of Crack Propagation and Failure of Concrete," *Materials and Structures,* 14(83), pp. 357–365, 1981).

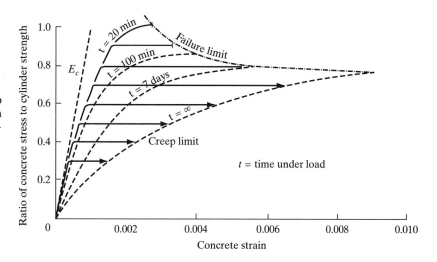

Figure 11.29
Static fatigue curves of concrete in compression: Effect of raising applied load to levels indicated and holding constant until failure. Figure indicates time to failure. Dashed stress-strain curves indicate the approximate curve to be expected if the load has been slowly raised over these time periods (after H. Rusch, "Researches Toward a General Flexural Theory for Structural Concrete," *J. American Concrete Inst.*, 57, 1960, pp. 1–28).

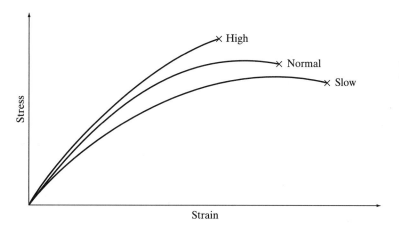

Figure 11.30
Effect of rate of loading on the stress-strain curve of concrete in compression.

influence is also manifested in the effect of rate of loading on the stress-strain curve (Fig. 11.30): At slower rates the strength would be lower and the curvature bigger, because more time is given to the unstable cracks to propagate.

The extent of cracking that occurs before failure can be estimated from the degree of curvature of the stress-strain curve. When the specimen is unloaded from intermediate stress levels, the stress-strain curve is nearly linear because further cracking does not occur. The positive intercept on the strain axis of the unloading curve represents the nonelastic strain due to microcracking (Fig. 11.28a). The fractured surface of the concrete, after it had been loaded to failure, shows a rough texture. This roughness means that some stress can be transferred across the cracked surface by mechanical interlock so that some postcracking strength is possible, and hence the stress-strain curve can often be recorded a little way past the ultimate stress. The extent to which this occurs depends on the conditions of test.

The shape of the stress-strain curve depends on the extent of microcracking, which is a function mainly of the compatibility between the aggregates and the matrix: If they are of similar strength and modulus of elasticity (porous lightweight aggregates in a normal strength paste or normal aggregates in a high-strength matrix), the stress-strain curves would tend to be more linear in nature (Fig. 11.31) because there is less stress concentration developed between the two. For normal aggregates the non-linearity in the shape of the stress-strain curve will increase with reduction in the strength of the paste matrix. Factors which increase the aggregate-paste bond

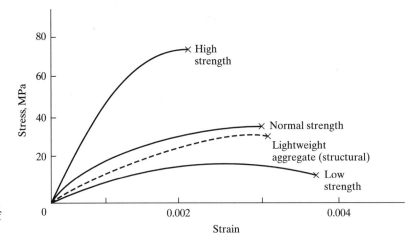

Figure 11.31
Influence of the properties of the paste and aggregates on the stress-strain curve of concrete.

will increase the linearity of the stress-strain curve because the formation of bond cracks will be more effectively suppressed. Factors of this kind include the angular shape and rough-textured surface of the aggregates.

Modulus of Elasticity

The determination and definitions of moduli of elasticity in materials with curvilinear stress-strain curves were given in Chapter 5. Since concrete is loaded in practice to a range of about 40% of its strength, it is common to define its modulus of elasticity in terms of its secant modulus of elasticity, which is taken to this range. The initial tangent modulus can be estimated from nondestructive tests, like ultrasonic pulse velocity measurements:

$$E = V^2/\rho, \tag{11.19}$$

where E is the initial tangent modulus of elasticity, V is the wave velocity, and ρ is the density.

Figure 11.31 indicates that the secant modulus of elasticity will depend on strength. Empirical equations have been developed to estimate modulus of elasticity, E, from the compressive strength, σ. Theses equations are of the type

$$E = A(\sigma)^{1/n}, \tag{11.20}$$

where A and n are constants. The equation recommended by the American Concrete Institute is

$$E = 4.73\sigma^{1/2}, \tag{11.21}$$

where E is in GPa units, and σ in MPa units.

However, for lightweight aggregate concrete a correction for the lower density, ρ, must be made. For example,

$$E = 0.043\rho^{3/2}\sigma^{1/2} \text{ GPa}. \tag{11.22}$$

When $\rho = 2300 \text{ kg} \cdot \text{m}^{-3}$ (normal weight concrete), Eq. 11.22 becomes Eq. 11.21. This correction for weight is needed because lightweight aggregates have a lower modulus than normal aggregates. However, not all normal-weight concretes can be described exactly by Eq. 11.22, because the concrete modulus of elasticity is sensitive to the modulus of the aggregate, as might be expected from the rule of mixtures:

$$E = E_a V_a + E_p V_p,\qquad(11.23)$$

where E_a, E_p, V_a, and V_p refer to the moduli and volume fractions of aggregate and paste, respectively (see Fig. 9.4 for an example).

Time-dependent strains: creep and shrinkage

THE NATURE OF CONCRETE CREEP AND SHRINKAGE

Creep and shrinkage are discussed together because there is growing evidence that these two time-dependent strains are caused by the same internal process, which involves movement of water. In the case of shrinkage, the driving force for water movement is environmental conditions causing diffusion of water outward (i.e., water is being lost). In the case of creep, the driving force is stress, which causes water to move from one location to the other within the concrete (i.e., no water is being lost). Thus, creep strains will be generated in saturated concrete which is prevented from drying; this is known as basic creep.

Creep strains are always opposed to the applied stress, so that when saturated concrete is under compressive load (basic creep) contraction in the direction of load will occur and is additive with elastic strains. The Poisson ratio is about the same in creep as at initial loading, so lateral expansion will increase. In contrast, loss of moisture on drying (free shrinkage) involves equal contractions in all directions (volumetric decrease). In practice, we are mainly interested in the time-dependent strains in the axial loading direction, which usually involves concrete in compression. Therefore, both creep and shrinkage strains lead to contraction. In practice, creep and shrinkage occur at the same time as the concrete is exposed to natural drying environment during its loading. The total time-dependent strain in the compressive loading direction under such conditions is slightly greater than the sum of free shrinkage and basic creep (Fig. 11.32). In the lateral direction, creep and shrinkage are partially compensatory. It is important to know what is meant by *creep strain*;

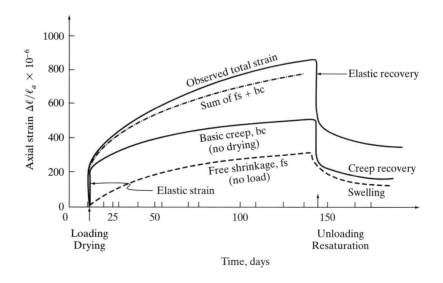

Figure 11.32
Creep and shrinkage strains observed in concrete under compression.

many engineers include in it all the time-dependent strains measured when concrete is under load without distinguishing between creep and drying shrinkage.

In the structure, creep and shrinkage show up as an increase with time in axial deformation (e.g., column under compressive load), an increase in deflection (e.g., beam in flexure), or stress relaxation (e.g., a prestressed component, where the pre-stress decreases with time as the concrete contracts due to shrinkage and creep). A time-dependent shrinkage can lead to cracking if a concrete component is being re-strained and tensile stresses develop as it tries to contract (e.g., a concrete slab on grade, where control joints are needed to localize cracks in the structure at points where they can be easily sealed). Since the shrinkage and creep strains over time can be as big as or greater than the elastic strains and deformations, they should be taken into account in the design of concrete members.

Another important aspect of creep and shrinkage is the fact that the strains are not wholly recoverable when the load is removed and the concrete resaturated. Con-crete undergoes an appreciable irrecoverable strain component when first loaded to a certain stress level or dried to a certain relative humidity. Thereafter, changes are nearly reversible if the concrete is cycled between these limits, unless it is exposed to more stringent conditions. The response of concrete to its loading environment can be complex, and the remainder of this section will be devoted to establishing general principles which can be used to make knowledgeable predictions about concrete performance.

MECHANISMS OF CREEP AND SHRINKAGE

These phenomena have their origins in the response of the cement paste to various imposed stresses. The microstructure of the paste determines the nature of the stresses and also the response of the material to the stresses (the strain).

Creep   Let us consider first the simplest condition: basic creep, in which the concrete is loaded without drying. In this case no moisture is lost from the paste while it is under load, but the water can be redistributed within the microstructure. The applied load will be carried to different extents by different components of the paste. The hydration products will carry stress in proportion to their volume fractions, but the capillary pores will only transmit very low stresses at best. Thus, the C—S—H component is under the highest stress and since a large fraction of its volume is micropores, the water therein can be under high stress. This water is no longer in equilibrium with its surroundings and will want to diffuse to regions of lower stress, which are the capillary pores. As this occurs, the surface forming the micropores can move closer together, partly because of the external stress exerted on the C—S—H as a whole and partly because this water was actively keeping the surfaces apart (disjoining them). Thus, the process can be considered as a gradual densification of the C—S—H structure through a viscoelastic response. Because the C—S—H is a random array of particles, two surfaces cannot move independently of their immediate surroundings. We can envision a rather complex process involving slipping of surfaces past one another. Creep, therefore, depends on the ability of water movement by diffusion within the paste, which is mainly a function of pore structure and the ease of slippage of C—S—H particles, which is a function of bonding. The van der Waals bonding is smaller when water separates the particles, while chemical bonding increases over time.

Drying Shrinkage   Now consider the case of moisture loss from the paste when water is lost from the capillary pores; menisci are created, and capillary stresses are developed (see Chapter 4, Eq. 4.16), acting isostatically on the paste. As the relative humidity decreases, menisci are created in smaller and smaller pores, leading to larger and larger stresses. Shrinkage due to capillary stress have been measured in porous glasses but is not large enough to account for much of the shrinkage of cement paste. Two additional processes are taking place. Under internal capillary stress C—S—H particles will move closer to each other and may undergo slip by the same processes that are active in creep. At the same time, moisture is lost from the micropores of C—S—H to maintain hygral equilibrium. This loss of water continues slowly even when complete removal of capillary water (below 40% rh) has removed the imposed capillary stresses, and it provides another mechanism which is associated with shrinkage due to a reduction in the distances between surfaces of C—S—H particles. This aspect has been described as removal of a disjoining stress which acted to swell the paste on inbibition of water, and it has the same effect on the micropores that capillary stresses have on the mesopores.

FACTORS AFFECTING CREEP AND SHRINKAGE

Influence of Aggregate   The process described in the previous subsection originates within the paste fraction. Except in rare cases, concrete aggregates do not change dimensions on drying. Thus, the shrinkage and creep of mortar and concrete will be less than that of pure paste, and we would expect a relationship of the type between concrete strain, ε_{con}, and paste strain, ε_p:

$$\varepsilon_{con} = \varepsilon_p (1 - V_a)^n. \tag{11.24}$$

If $n = 1$, then Eq. 1.24 becomes the rule of mixtures, where the effect of aggregates is merely one of dilution, and the equation implies no strain in the aggregate. However, when $n > 1$ the equation indicates that the aggregates act as rigid inclusions, actively restraining paste deformations through elastic interactions. The value of n for concrete has been expressed in terms of elastic constants of the aggregates and is about 1.8. Thus, the creep of concrete depends on aggregate type, being smaller when the aggregates have a higher modulus of elasticity. Some comparisons are given in Fig. 11.33 for concretes drying under load.

Influence of Paste Properties   Using the preceding mechanisms, we can readily understand the factors affecting paste deformations. These are simply (1) reduction in the ability of C—S—H particles to slip and move relative to each other due to changes in their nature of bonding; and (2) reduction of capillary porosity.

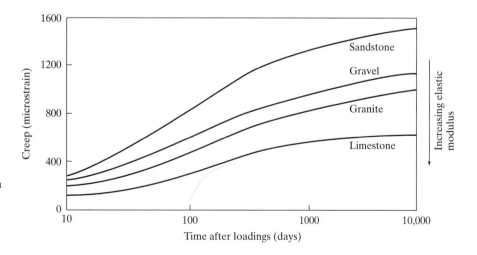

Figure 11.33
Effect of aggregate type on concrete creep (after G. E. Troxell, J. M. Raphael, and R. E. Davis, *Proc. ASTM,* 58, pp. 1101–1120, 1958).

Changes in C—S—H which reduce creep and shrinkage are caused by prolonged hydration or hydration at elevated temperatures (which increase the chemical bonding); by pre-drying before loading, which limits the amount of water which will diffuse from C—S—H while under load; and by carboration, which also will increase the chemical bonding. Reduction in porosity, due to increased hydration and low w/c ratio, reduces creep (Figs. 11.34 and 11.35). Admixtures which reduce water content should also reduce shrinkage and creep.

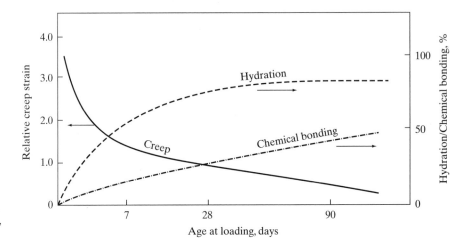

Figure 11.34
Effect of age of loading on the creep strain, hydration, and polymerization.

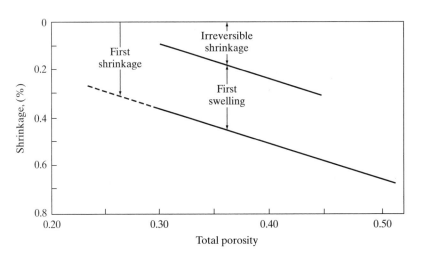

Figure 11.35
Effect of water/cement ratio on the shrinkage strains of portland cement paste (after R. A. Helmuth and D. H. Turk, *J. Port. Cem. Assoc. R&D Labs.*, 9, pp. 8–21, 1967).

Effect of Relative Humidity   The extent to which concrete can lose moisture is dependent on the relative humidity of the surroundings. For concrete to dry out, moisture must move from the interior to the surface, where it can evaporate. Moisture transport through a porous body is a diffusion-controlled process since water must move through narrow pores. In good-quality concrete of low porosity, the water moves with greater difficulty through the smaller pores. Initially, water moves by bulk liquid diffusion, but once menisci have formed, movement must be by vapor transport and diffusion coefficients are smaller still (see Table 11.8). Most of the water has to move through a partially dry region near the surface and so is governed by vapor diffusion. However, when high surface to volume ratios exist,

**TABLE 11.8 Estimated Diffusion Coefficients
Governing Water Transport in Drying Concrete ($m^2 s^{-1}$)**

	Low Porosity	Medium Porosity	High Porosity
Wet	10^{-11}	10^{-10}	10^{-8}
Dry	10^{-12}	10^{-11}	10^{-9}

initial rates of water loss, and hence shrinkage, will be appreciably higher. For example, the rate of water loss, and hence shrinkage, will be greater from an I-shaped section having the same cross-sectional areas as a solid rectangular section.

Estimates of Creep and Shrinkage Values   Detailed mathematical modeling of the processes accounting for creep and shrinkage involve complex and difficult mathematics involving combinations of Maxwell and Kelvin models (Chapter 7). Although accurate prediction is seldom required for most structures, some allowance for these deformations must be made. Usually a simple mathematical equation is used to describe the strain-time curve, and correction factors are used to take account of the most important factors: relative humidity, age of concrete, dimensions, and concrete composition. Relations of this kind may form part of a code, such as the American Concrete Institute (ACI) building code.

11.4 DURABILITY OF CONCRETE

The ultimate success of a concrete structure has to be measured by the years of maintenance-free service it provides. If a correct prediction has been made of the environmental conditions to which the concrete will be exposed (including applied loads), there is no reason for concrete not to last for many decades without the need for costly repair. We now have the knowledge and experience to protect concrete from possible harmful effects; this requires the selection of sound materials and the production of high-quality concrete on a regular basis. Unfortunately, ignorance and economic pressures to reduce initial costs often prevent this desired outcome. High-quality concrete is more expensive to produce, and its benefits accrue only in the long term. Life-cycle cost analyses usually indicate that the final cost of repairing avoidable damage is many times greater than extra initial costs would have been.

Incipient damage may have begun within the concrete for many years before the first visible signs are apparent. Once its integrity is compromised, concrete becomes vulnerable to other kinds of degradation that it might have withstood earlier. Thus, damage occurs exponentially and concrete that has apparently survived satisfactorily for 15 years may need to be completely replaced 5 to 10 years later.

Degradation of concrete can be divided into chemical attack and physical attack, as shown in Table 11.9. One of the most widespread forms of concrete deterioration is cracking and spalling due to corrosion of the steel reinforcement, particularly when promoted by chloride salts (sea water or deicing salts). This topic is covered in Sec. 11.6.

11.4.1 Permeability and Diffusivity

For concrete to be attacked rapidly, aggressive agents must be able to enter the concrete easily because surface attack is slow. Thus, the ease with which water can flow into and through concrete (carrying with it dissolved reactants) is of great

TABLE 11.9   Common Forms of Degradation of Concrete

Type	Description	Cause	Component(s) Involved	Symptoms
Chemical	Alkali-aggregate	Reaction of siliceous aggregates by alkali ions	Aggregate	Coarse "map-cracking" with viscous fluids erupting
	Sulfate attack	Reaction of paste components with sulfates	Paste	General cracking and softening
	Acid attack	Dissolution by acids	Paste (aggregate)	General etching of surface
	Rebar corrosion	Rusting of steel	Reinforcement	Cracks with rust stains above location of reinforcement
Physical	Frost attack	Freezing of water in pores	Paste	General scaling and spalling at surface
	D-cracking	Freezing of water in pores	Aggregate	Fine crack pattern roughly parallel to joints in pavements
	Fire damage	Decomposition of hydration products and development of internal stresses	Paste (aggregate)	Cracking and spalling
	Thermal cracking Shrinkage	Internal stresses from restrained contractions	Paste (aggregate)	Localized cracking.

importance. This is measured by the permeability coefficient, K_p, determined by D'Arcy's law:

$$\frac{dq}{dt} = K_p \cdot \frac{\Delta h}{x} \cdot A, \qquad (11.25)$$

where dq/dt is the rate of flow, A is the area, x is the thickness through which flow occurs, and Δh is the pressure head of fluid ($\Delta h/x$ is the pressure gradient through the sample). Since permeability is largely controlled by the volume and size of the large capillary pores, the use of low w/c ratios and provision for extended moist curing can lower the permeability drastically (Fig. 11.36). Dense, well-cured concrete has a permeability commensurate with that of low-porosity rocks, even though the

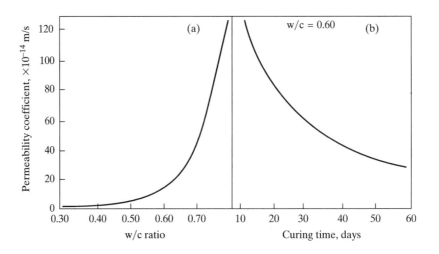

Figure 11.36
Effect of water/cement ratio (a) and curing (b) on permeability.

measured porosity of the paste is much higher. The reason for this is considered to be due to the fact that when the capillary porosity is low, there is no longer a continuous pore system through which bulk flow can occur. The pores are isolated from each other by C—S—H, so water can only flow by much slower molecular diffusion through its micropores. The importance of good curing and low w/c ratios is readily seen from Fig. 11.36. The addition of a pozzolan or blast furnace slag produces more C—S—H and hence a more discontinued pore structure.

Aggressive agents can penetrate into the concrete by a diffusion-controlled process. This could be diffusion of ions in solution (e.g., Cl^- ion diffusion) or gases (e.g., CO_2, O_2) which diffuse through vacant pores or through pore solution in saturated pores. The diffusion of these species can be described by Fick's law:

$$\frac{dc}{dt} = -D\frac{dc}{dt},$$ (11.26)

where dc/dt is the rate of diffusion, dc/dx is the concentration gradients, and D is the diffusion coefficient characteristic of the material.

The influence of the paste structure on the diffusion coefficient is generally similar to that of the permeability coefficient, with the w/c ratio and curing being of prime importance, as demonstrated in Table 11.10 in the case of w/c ratio.

TABLE 11.10 Effect of w/c ratio on Cl^- Ion Diffusion in Pastes

w/c Ratio	Diffusion Coefficient m²/s
0.4	2.60×10^{-12}
0.5	4.47×10^{-12}
0.6	12.35×10^{-12}

After C. L. Page, N. R. Short, and A. El-Tarra, "Diffusion of Chloride Ions in Hardened Cement Paste" *Cement Concr. Res.,* Vol. II, No. 3, 395–406, 1981.

11.4.2 Composition of Pore Solutions

One consequence of hydration is that the hydration products must be in equilibrium with the aqueous solution in the capillary pores (i.e., the pores are filled not with pure water, but with an ionic solution). The pore solution is that in equilibrium with $Ca(OH)_2$, giving a pH of 12.5. However, the presence of soluble alkalis, released during hydration, increases the concentration of hydroxide ions, raising the pH to above 13.0. The concrete is thus a highly alkaline medium. This can have implications for chemical durability, as discussed next.

11.4.3 Chemical Attack

Alkali-Aggregate Reaction

Certain types of rock contain reactive forms of silica, which can react with the soluble alkalis that enter the pore systems from the cement paste, or may, on occasion, come from external sources (e.g., deicing salts). The most common is the alkali-silica reaction, which is outlined in Table 11.11. Step 1 is controlled by the alkali content of the cement. Step 2 is probably rate determining and depends on the exact form of silica in the rocks. Petrographic examinations of aggregates combined with laboratory and field data show that a wide variety of amorphous or poorly crystalline silicas can react at very different rates (Table 11.12). Step 3 is the critical reaction causing extensive damage to the concrete; the gel goes from an amorphous solid to

TABLE 11.11   Description of the Alkali-Silica Reaction

Reaction Step	Reaction	Result
1	Release of alkali ions from the cement during hydration	Increases the concentration of hydroxide ions in the pore solution
2	Initial hydrolysis of siliceous fraction of the aggregate in the highly alkaline pore solution $[K(Na)OH + SiO_2 \rightarrow [K_2O(Na_2O)—SiO_2—H_2O]$ (amorphous alkali silicate gel)	Destroys aggregate integrity
3	Swelling of alkali silicate gel by inbibition of water,	Causes localized swelling internal pressure, and cracking
4	Liquefaction of alkali silicate gel by further inbibition of water	Expulsion of liquid gel through the cracks

TABLE 11.12   Forms of Silica Involved in Alkali-Aggregate Reactions

Silica	Examples	Time to Obvious Distress
Alkali-silica glass	Waste container glass	~1 year
Hydrolyzed amorphous silica	Opal, flint	~5 years
Aluminosilicate glass	Rocks of volcanic origin—rhyolites, dacites, pumicite	5–10 years
Microcrystalline silica	Weathered quartzites	10–20 years
Layer silicates (expansion of layers)	Phyllites	30–40 years

a semisolid to a liquid phase (Step 4) as water is taken into the structure. The last step does no further damage but provides a useful diagnostic symptom. Steps 3 and 4 are sensitive to the presence of moisture and will not occur if the concrete is kept dry.

Control of this reaction must be taken at Step 1, either by avoiding reactive aggregates or by keeping the alkali content of the cement low (Na_2O + 0.65 $K_2O \leq 0.60$). A useful strategy is to use a pozzolan in the concrete. Pozzolans contain alkali-reactive silica in a finely divided form, which reacts with the alkalis very rapidly. Thus, deleterious expansions are avoided because the alkali silicate gel is spread in small quantities through the matrix. The worse case is to have relatively small amounts of the reactive aggregate in fairly large pieces scattered throughout the paste. These act as "sinks" for the mobile hydroxide ions and become sites of high expansion.

This type of distress is being observed more frequently since the level of alkalis in portland cement has been increasing in recent years and there is greater use of marginal aggregates.

Sulfate Attack

Sulfate attack occurs when sulfate ions penetrate into the concrete from the surrounding environment (groundwater or sea water). We can divide the chemical processes into three steps, as given in Table 11.13. As the dissolved sulfate ions permeate into concrete, they react to form gypsum. This concentrates and fixes the sulfate in the concrete so that Step 3 can occur. The formation of ettringite is accompanied by volume expansions, which cause internal stresses and cracking.

Sulfate attack can thus be prevented by suppressing any one of these three steps. The permeability is decreased by lowering the w/c ratio (Fig. 11.37), by providing adequate moist curing, or by using a pozzolan or blast furnace slag. These mineral admixtures also reduce the amount of calcium hydroxide and suppress Step 2. Finally, using a sulfate-resistant cement with a low C_3A content (Type V $\leq$ 5% C_3A or Type II $\leq$ 7% C_3A) will prevent the damaging formation of ettrigite (Fig. 11.37).

TABLE 11.13   Sequence of the Sulfate Attack

Reaction Step	Chemical Reaction	Relevant Factor in Concrete	Control Strategy
1	SO_4^{2-}(external) $\rightarrow$ SO_4^{2-}(internal)	Permeability	Low w/c, pozzolan addition, slag addition
2	$Ca(OH)_2 + SO_4^{2-} \rightarrow CaSO_4 \cdot 2H_2O + 2OH^-$ (gypsum)	$Ca(OH)_2$ content	Pozzolan addition, slag addition
3	$C_4A\overline{S}H_{12}$ + gypsum $\rightarrow C_6A\overline{S}_3H_{32}$ (ettringite)	C_3A content of cement	Type II or V cement

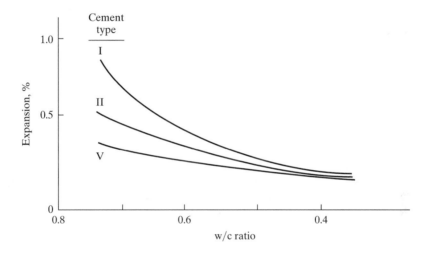

Figure 11.37
Expansion after exposure in sulfate solution for one year.

Acid Attack

Since concrete is held together by alkaline compounds (both CH and C—S—H), it will be dissolved by acidic liquids such as occur in many industrial wastes, in some natural waters, and in the vicinity of some mining operations. Attack can be severe where the acidity is high (pH < 4) and water flow is occurring. If the acid forms an insoluble calcium salt, then the precipitation of this salt within the capillary pores can slow down or prevent further attack.

11.4.4 Physical Attack

Freezing and thawing

Cement paste is highly susceptible to cycles of freezing and thawing because of its large pore volume and small pore sizes. As the temperature drops below 0°C, water can begin to freeze at about −5°C. Not all the water freezes at once, since the smaller the pore, the lower must be the temperature before ice formation can begin.

Once ice crystals form, the accompanying 9% volume expansion compresses the remaining water in the pore (Fig. 11.38). The water will exert hydraulic stress on the solid. Buildup of stresses from many capillaries will cause localized fracture within the paste. This can be avoided if the water is able to move out of the paste to relieve the stress and freeze somewhere where there is plenty of space. It has been shown that the water must not have to move more than 0.2 mm. The only way this can be done is to distribute free space throughout the paste in the form of tiny air bubbles, by the use of an air-entraining agent. The air bubbles act as "safety valves," allowing the water to freeze harmlessly without generating damaging stresses.

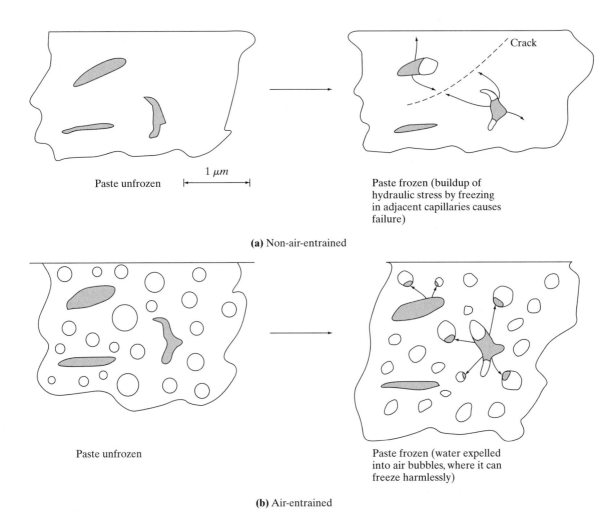

(a) Non-air-entrained

Paste unfrozen   1 μm

Paste frozen (buildup of hydraulic stress by freezing in adjacent capillaries causes failure)

Crack

Paste unfrozen

Paste frozen (water expelled into air bubbles, where it can freeze harmlessly)

(b) Air-entrained

Figure 11.38
Schematic representation of the mechanisms of frost resistance in concrete and the effect of air entrainment.

Thus, the use of an air-entraining admixture (Sec. 11.3.2) to entrain a sufficient volume of air bubbles ($\sim 6 - 8\%$), which are small enough and uniformly dispersed to achieve 0.2 mm spacing or less, is the most efficient means for achieving frost durability (Fig. 11.39).

This problem is common to all porous materials, but there are differing degrees of susceptibility depending on the pore structure. A large pore volume and

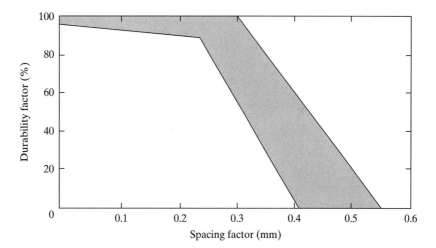

Figure 11.39
Relationship between frost durability and bubble spacing factor of entrained air (after S. Mindess and J. F. Young, *Concrete,* Prentice Hall, 1981, p. 173).

small pore diameter is the worst situation, but damage only occurs if the material is saturated or nearly so. If it is partially dried, many of the pores become empty and can play the same role as entrained air bubbles.

Certain aggregates can also have undesirable pore structures and can cause frost damage if frozen when wet. Such aggregates are widespread in the midwestern states. The high water tables that occur in many of these states mean that the aggregate remains in a near-saturated condition, particularly at the bottom of the pavement and near joints. The aggregates and surrounding pastes are slowly damaged by repeated freezing and thawing even if the concrete is air-entrained. The damage is first apparent after several winters (8 to 12 years), in the form of fine cracks at the surface referred to as "D-cracking", symptomatic of considerable damage underneath (Fig. 11.40). Since air-entrainment does not prevent this problem, such aggregates must be avoided if the concrete is likely to be frozen in a wet condition.

Fire resistance

A major advantage in the use of concrete in construction is its ability to withstand the ravages of fire. This is due to the fact that heat can penetrate concrete only slowly, so that although concrete is damaged by high temperatures, the inner layers remain at a low temperature and are not much affected (neither is the reinforcing steel). In contrast, steel structures conduct heat rapidly and cause severe loss of structural integrity to beams and columns quickly.

The low rate of heat penetration is due to several factors:

1. Concrete has a low thermal conductivity.
2. Heat is consumed, evaporating the free water in the paste and aggregate.
3. Heat is consumed, decomposing the hydration products.
4. Certain aggregates (e.g., dolomite) may decompose and consume heat.
5. Decomposed layers have an even lower thermal conductivity, providing increased thermal insulation.

These processes, loss of water and decomposition, cause loss of strength and high shrinkage (Fig. 11.41), resulting in cracking and spalling. But the core of the element, including the reinforcement, can remain relatively unaffected if exposure to high external temperatures is limited to just a few hours.

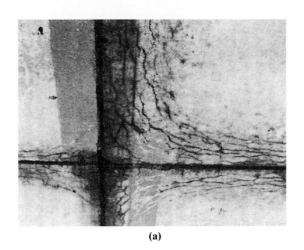

(a)

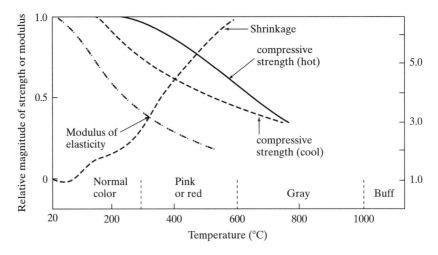

(b)

Figure 11.40
D-cracking in a concrete pavement: (a) actual damage, (b) schematic.

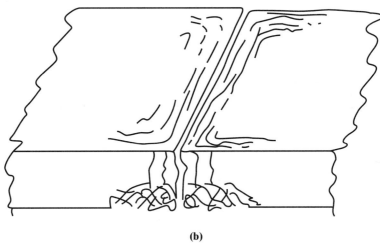

Figure 11.41
Effect of exposure to high temperature on concrete properties (after S. Mindess and J. F. Young, *Concrete,* Prentice Hall, 1981, Fig. 19.6, p. 530).

11.5.1 Corrosion Mechanism

Structural concrete is usually designed in combination with reinforcing steel bars due to the low tensile strength of the concrete. The concrete can provide excellent protection to the steel to suppress its tendency to corrode when exposed to the natural environment. However, in spite of this favorable protection potential, we frequently come across durability problems in reinforced concrete structures which are the result of the corrosion of the steel in concrete. This is usually associated with poor concrete practice, either in the design or the construction stages. It is essential to understand the mechanisms by which concrete protects the steel and thus take all the necessary steps to assure that the protective potential of the concrete is maintained.

Corrosion is essentially an electrochemical process, and its principles are discussed in some detail in Chapter 13. It involves the formation of a cathode and an anode, with electrical current flowing in a loop between the two. In the anode, the iron metallic atoms are oxidized to Fe^{2+} ions, which dissolve into the surrounding solution. The electrochemical corrosion process which may occur in steel in concrete is presented schematically along these principles in Fig. 11.42. The corrosion itself is in the anode, where there is ionization and dissolution of the metallic iron. The cathode is located at a distance, and the cathodic reaction consumes electrons and leads to the formation of OH^- ions. For the cathodic reaction to occur, moisture must be present and a supply of oxygen should be available, usually by diffusion of oxygen from the surrounding environment through the concrete cover. The causes for certain regions on the steel bar to become anodic while other are anodic are described in Chapter 13. The ions formed at the cathode and anode migrate through the aqueous solution in the pores of the paste of the concrete surrounding the steel bar. The electrical current loop consists of electrons flowing in the steel bar and ions moving in the concrete pore solution between the anode and the cathode.

The Fe^{2+} and OH^- ions which are moving in the pore solution interact chemically, close to the anode, to produce an iron oxide, which is a byproduct of the corrosion reaction, known as rust.

The corrosion of the steel in concrete produces two types of deleterious effects:

1. Reduction of the cross-section area of the steel at the anode, thus reducing the load-bearing capacity of the reinforcing bar.
2. Spalling of the concrete (Fig. 11.43), which is the result of expansion stresses created by the rust: The rust is voluminous in nature (its volume is two to six times that of the steel from which it was formed) and is deposited in a confined space between the steel bar and the concrete.

The amount of rust formed by corrosion of the outer 0.1 to 0.5 mm of steel bar is usually sufficient to cause cracking. This reduction in diameter is too small to practically influence the load-bearing capacity of the steel bar, which has a diameter greater than 10 mm. Therefore, the first damage to be concerned with in the process of corrosion of steel in concrete is the cracking and spalling in the concrete itself. The sequence of events usually involves the appearance of cracks on the concrete surface parallel to the underlying reinforcement. This is a warning sign that corrosion is occurring and maintenance treatment and repair is required.

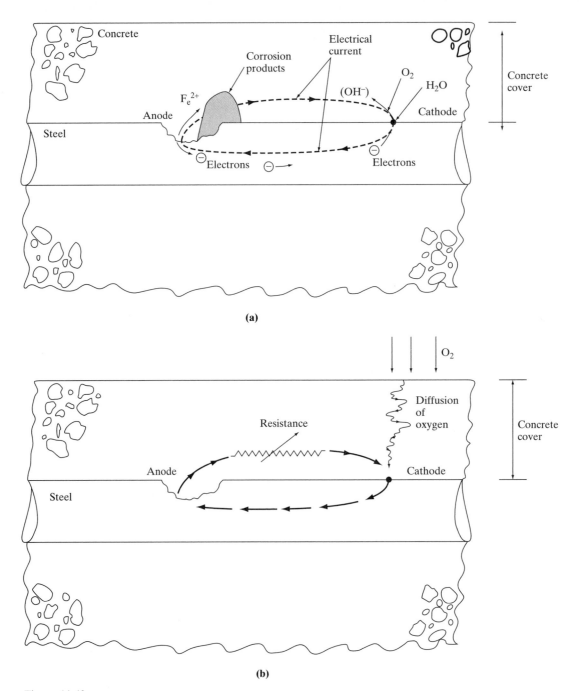

Figure 11.42
Schematic description of the process of corrosion of steel in concrete: (a) chemical reactions and charge movements, (b) parameters controlling corrosion rate—resistivity between cathode and anode and diffusion of O_2 at the cathode.

11.5.2 Corrosion Protection

Concrete can provide adequate protection to the steel to prevent corrosion. This protection is both physical and chemical in nature. The high pH of the concrete pore solution leads to spontaneous formation of a protective iron oxide film around the

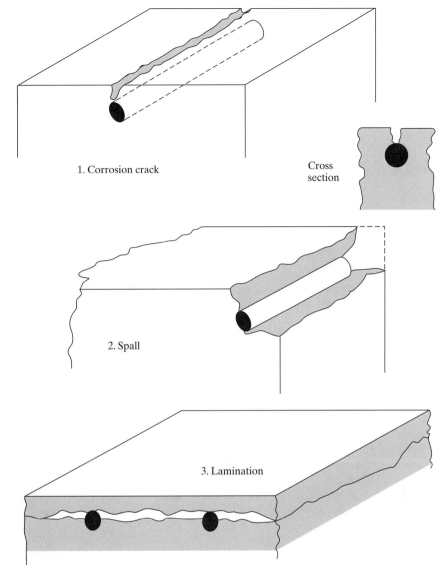

1. Corrosion crack

Cross section

2. Spall

3. Lamination

Figure 11.43
Spalling of concrete cover due to rust formation during the corrosion of steel in concrete (after R. D. Browne and M. P. Geoghegan, "The Corrosion of Concrete Marine Structure: The Present Situation," pp. 79–98 in *Proc. Symp. Corrosion of Steel Reinforcement in Concrete Construction,* Soc. Chem. Ind., UK, 1979).

steel bar, known as the passivation film. As long as this film is stable, the steel is immune to corrosion. Two main causes can lead to disruption of the passivation film:

1. Penetration of chlorides into the concrete; depassivation occurs when the chloride ion content reaches a value in the range of 0.2% to 0.4% by weight of the portland cement in the concrete adjacent to the steel. Conditions for depassivation of this kind exist in the marine environment and in structures sprayed with salt to prevent freezing (bridge decks).

2. Carbonation of the concrete leading to reduction in pH. Depassivation occurs as pH approaches 11. This process can be generated in any climate which promotes carbonation, but it is particularly accelerated in contaminated environments such as industrial zones, or in concrete which is cracked.

In the United States, chloride-induced corrosion is of primary concern while that due to carbonation is rare, although in other climates carbonation may dominate. Once depassivation has occurred, an electrochemical corrosion process can be initiated. However, if the rate of corrosion is sufficiently low that damage does not occur during the designed life, corrosion may not be of practical concern. Several means should be taken to prevent depassivation, or at least ensure that it is sufficiently delayed, and to slow down the rate of corrosion if depassivation occurred. They include good-quality concrete with sufficient depth of cover (lower permeability and diffusion; see Sec. 11.3.4), and design of the structure so that the width of loading and drying cracks will be limited. These means will slow down the ingress of chlorides, CO_2, and O_2, into the concrete; the first two can lead to depassivation and the latter is required for corrosion to proceed. The dense concrete is also of a higher electrical resistivity, which will slow down the flow of ions in the pores solution in the concrete between the anode and the cathode (Fig. 11.42). The various specifications and standards present requirements of this kind in terms of the depth of cover of the concrete, its water/cement ratio or strength, and the maximum crack width, as a function of the environmental conditions and the nature of the structure. For example, the requirements are more stringent for prestressed concrete components, where the damage due to corrosion is higher and the steel is more sensitive to corrosion because of its higher stress.

11.6   CONCRETE MIX DESIGN

The requirements for concrete performance can be different depending on the end use and the production process (Table 11.14). The proportioning and choice of concrete ingredients should be such that the appropriate requirements are met satisfactorily, taking into account the properties of the fresh and hardened material and economy (portland cement is the most expensive ingredient). First, the choice of ingredients (type of cement, aggregates, admixtures) should be made and guidelines for these are provided in Table 11.14. The second stage should be proportioning of the ingredients, which include the amount of cement, water, and admixtures and the grading and content of the aggregates.

The guiding principle is to meet the criteria for design strength and durability. In both cases, the w/c ratio is the primary parameter. An example of strength-w/c ratio relationships is given in Fig. 11.26. Allowance must be made for the variability in strength, which can be described by Gaussian statistics. Thus, the average concrete strength taken for the mix design should be higher than the specified strength, so that the probability of concrete strength falling below the specified value will be small, usually less than 1%. Relations between average and specified strength are usually included in national and international codes, such as the ACI Building Code. The difference between the average and specified strength values is in the range of 3 to 10 MPa, depending on the strength level and the variability in the production process.

Guidance for durability criteria is based on exposure conditions expected; then requirements may be set both for maximum w/c ratio and minimum air content when freezing is expected. An example is given in Table 11.15. Such requirements are usually specified by local standards. Of the two w/c ratios determined, on the basis of strength and exposure conditions, the lower one will be taken because it would fulfil both set of conditions.

The required consistency of the fresh concrete is controlled by the water content. An example is given in Fig. 11.20 and must be determined for the specific aggregate used. The choice of consistency is specified in terms of the slump of the

TABLE 11.14 Common Technical Requirements for Concrete

Desired Property	Typical application(s)	Materials Selction
High workability (without segregation of components)	1. Underwater placements 2. Difficult placement, deep forms, congested reinforcemnt 3. Self-leveling floors	Water-reducing admixtures Mineral admixtures Small-sized aggregates
Rapid setting	1. Shotcreting	Accelerating admixtures
Slow setting	1. Elimination of cold joints 2. Avoidance of form deflection cracking 3. Offset high temperatures for easier placement	Retarding admixtures
High early strength	1. Early form removal (slip casting, precasting) 2. Winter concreting 3. Prestressing	High early strength cement (Type III) Water-reducing admixtures (plasticizers, superplasticizes) Acceleration admixtures

Low heat of hydration	1. Mass construction:
	    (a) thick slabs, walls
	
	    (b) dams
		
		Moderate heat of hydration cement (Type II). Blended cement (Type IPM) or mineral admixtures
		Low heat of hydration cement (Type IV)
		Blended cement (Type IP) or mineral admixtures

High durability	1. Frost resistance 2. Sulfate resistance (sea water, groundwater) 3. Alkai-aggregate resistance	Air-entraining admixtures Sulfate-resistant cement (Type II or V) Blended cements or mineral admixtures Water-reducing agents Mineral admixtures Water-reducing agents Low alkali cements Unreactive aggregates
Low permeability	1. Water-retaining structures 2. Gas-retaining structures 3. General durability	Water-reducing admixtures Mineral admixtures
Diminsional stability	1. Water-and gas-retaining structures 2. High-rise construction 3. Critical dimensioning	Water-reducing admixtures High modulus aggregates
Low thermal expansion	1. Refractory applications 2. Wide variations in ambient temperatures	Aggregates with low thermal expansion
Low density	1. High rise construction 2. Insulating concretes	Lightweight aggregates
High density	1. Radiation shielding 2. Counterweighting	Heavyweight aggregates
Low thermal conductivity	1. Insulating conceretes	Lightweight aggregate

concrete, which is a function of the production process of the concrete and the equipment to be used for its transportation and consolidation.

Having established the w/c ratio and water content, the cement content (in units of mass, kilograms, per cubic meter of concrete) can be calculated. At this stage the volume content of the aggregates can be determined as the difference between

TABLE 11.15   Requirements for Concrete Subjected to Harsh Exposure Conditions (after CSA Standard A23.1)

Exposure Classifi cation	Exposure Condition	Typical Examples	Min. Specified 28-day Compressive Strength (MPa)	Max.Water: Cementing Materials Ratio	Air Content for 20 mm Aggregate Concrete (% Vol.)
C-1	Concrete for which protection against corrosion of rein- forcement is deemed critical	Bridge decks, sus- pended parking floors and ramps; portions of marine structures located within the tidal zone	35	0.40	5–8
C-2	Concrete not falling under C-1 but subjected to cycles of freezing and thawing	Pavements, side- walks, curbs, and gutters	32	0.45	5–8
C-3	Concrete in a satura- ted condition, not falling under C-1 and not subjected to cycles of freezing and thawing	Portions of marine structures perma- nently submerged	30	0.50	4–7
C-4	Concrete in a rela- tively dry condition, not falling under C-1 and not subjected to cycles of freezing and thawing	Slabs on grade in heated buildings	25	0.55	4–7
F-1	Concrete subjected to freezing and thawing in a saturated condition		30	0.50	5–8
F-2	Concrete subjected to freezing and thawing in an unsaturated condition		25	0.55	4–7

one cubic meter and the sum of the volumes of the cement and water calculated from their weight and density and the estimated content of air (Table 11.15). The last step will determine the distribution of the various fractions of the aggregate and the content of each fraction per unit volume of concrete (1 m^3). The grading should pro- vide densest packing, but with sufficient content of fine materials to assure cohe- siveness of the fresh mix. This will minimize the paste requirements and make more economical concrete. This is the stage at which different mix design procedures be- come available. One of them, developed by the American Concrete Institute, is based on determining the volume of the coarse aggregate per unit volume of con- crete as a function of the maximum size of coarse aggregate and the fineness mod- ulus of the fine aggregate. Full details of the calculation are given in the ACI document entitled *Standard Practice for Selecting Proportions for Normal, Heavy- weight and Mass Concrete.*

BIBLIOGRAPHY

S. MINDESS and J. F. YOUNG, *Concrete*, Prentice-Hall, Englewood Cliffs, New Jersey, 1981.

P. K. MEHTA and P. J. M. MONTEIRO, *Concrete: Structure, Properties and Materials,* 2nd ed., Macmillan, 1993.

J. M. ILLSTON, ed. *Construction Materials,* 2nd ed., E & FN Spoon., 1994.

M. GANNI, *Cement and Concrete*, E & FN Spoon, 1997.

PROBLEMS

11.1. Why is gypsum always interground with cement clinker during the final stages of Portland cement manufacture?

11.2. Classify the following cements according to the most appropriate ASTM classification.
   (a) $C_3S = 55\%$, $C_2S = 23\%$, $C_3A = 11\%$, $C_4AF = 9\%$
   (b) $C_3S = 43\%$, $C_2S = 40\%$, $C_3A = 1\%$, $C_4AF = 10\%$
   (c) $C_3S = 50\%$, $C_2S = 28\%$, $C_3A = 7\%$, $C_4AF = 11\%$
   (d) $C_3S = 60\%$, $C_2S = 15\%$, $C_3A = 13\%$, $C_4AF = 8\%$

11.3. Which ASTM cement type is most appropriate for the following applications? What blended cement could be substituted for the original choice? Give reasons for your choices.
   (a) Two foot thick foundation mat for a high-rise building.
   (b) Construction of a concrete foundation in late Fall.
   (c) Slip-forming of a highway pavement.
   (d) Construction of a sea-wall on the Gulf Coast.

11.4. The cement paste binder in a concrete made with a w/c ratio of 0.52 is determined to be 70% hydrated.
   (a) Calculate (i) the capillary porosity and (ii) total porosity of the paste.
   (b) Determine these porosities if the w/c ratio is reduced to 0.42.
   (c) Determine these porosities when the paste is cured for a longer time so that the cement is 85% hydrated.

11.5. Estimate the strength and elastic modulus for each of the three concretes in Q.4.

11.6. Describe the origins of the following types of cracks, and their implications for concrete performance.
   (a) Bond cracks
   (b) Plastic shrinkage cracks
   (c) Drying shrinkage cracks

11.7. **(i)** What is the most critical microstructural component that influences concrete durability in the following situations?
   (a) Exposure to high sulfate levels
   (b) Exposure to deicing salts
   (c) Subjected to frequent freezing and thawing.
   (d) Use of aggregates containing reactive silica.
   (ii) What admixture can be added to the concrete to provide protection from deterioration in each situation?
   (iii) How does the admixture you have chosen protect the concrete?

12

Asphalt Cements and Asphalt Concretes

12.1 INTRODUCTION

Asphalt concrete, which is a mixture of asphalt cement and mineral aggregates, is used primarily as the surface layer (course) in flexible pavement structures (Fig. 12.1). The asphalt cement acts as the binder phase on the mixture and controls most of its physical and mechanical properties, as does the portland cement in a portland cement concrete mixture.

In general practice, the terminology associated with asphalt materials is not well defined. Both ASTM and the Asphalt Institute (AI) define asphalt as "a dark brown to black cementitious material in which the predominating constituents are

Figure 12.1
Conventional flexible pavement structure with asphalt concrete surface layer (after *The Asphalt Handbook,* Manual Series No. 4 [MS-4], The Asphalt Institute, 1989).

- Asphalt concrete surface
- Granular base
- Granular subbase
- Prepared subgrade

bitumens which occur in nature or are obtained in petroleum processing."[1] This definition places no restriction on the inorganic content (natural or added as aggregate) of the asphalt, and hence there will be confusion if this term is applied to either the cement (i.e., the binder) or the cement-aggregate mixture (i.e., the concrete) without qualification. In an attempt to eliminate this confusion, the following terms will be used throughout this chapter:

Asphalt cement: a viscous, cementitious material composed principally of high-molecular weight hydrocarbons, and

Asphalt concrete: a complex material consisting of a mixture of asphalt cement and (mineral) aggregates.

This chapter deals first with asphalt cement, which is the binder, and thereafter with asphalt concrete, which is the particulate composite consisting of the binder and the aggregates.

12.2 ASPHALT CEMENTS

12.2.1 Introduction

Bituminous materials, including asphalt cements, occur naturally in several forms and have been used for bonding, coating, and waterproofing purposes since the dawn of civilization. Natural materials were used to construct the first modern asphalt pavements in the mid-1800s in Europe and, later, North America. Petroleum asphalts, refined from crude oil using a variety of techniques, were developed in the early 1900s, and very quickly replaced natural materials for most purposes.

Most petroleum asphalt (75–80%) is asphalt cement used for road construction or paving purposes, with the remainder being used by the roofing industry or in the manufacture of numerous miscellaneous products, such as linings, sealants, and insulation. In this chapter we are interested in the use of asphalt cements for the construction of various types of pavement structures, and the discussion in this chapter will be restricted to this topic.

Asphalt cements are essentially thermoplastic materials (Chapter 2) and their properties are therefore very sensitive to temperature. Above their glass transition temperature, T_g, which is approximately 140°C, they behave like Newtonian fluids. When cooled below T_g, the inter-molecular secondary bonds reform and are strong enough that the material acts as a viscoelastic solid (Chapter 7). If the temperature is very low, it will behave as a brittle elastic solid. In view of these characteristics, the production of asphalt concrete usually consists of mixing and placing the material after the constituents (asphalt cement and aggregates) have been heated to a temperature of about 140°C. Hardening of this 'hot-mixed, hot-laid' asphalt then occurs as it cools in place. Because the viscosity of the asphalt cement is high even above the glass transition temperature, the asphalt concrete mix does not have the flow properties needed for casting, and it is placed and consolidated by a process involving static or vibratory compaction.

[1]The Asphalt Institute defines a bitumen as "a mixture of hydrocarbons of natural or pyrogeneous origin, or a combination of both; frequently accompanied by nonmetallic derivatives which may be gaseous, liquid, semisolid, or solid; and which are completely soluble in carbon disulfide."

Asphalt concrete can also be produced by a process which does not require heating of the constituent materials. An asphalt cement is first combined with a solvent to form a so-called 'liquid' asphalt which is fluid at ambient temperatures (section 12.3). This material is then mixed with aggregates to form a 'cold-mixed' asphalt concrete, which is placed and consolidated similarly to hot-mixed, hot-laid asphalt concrete mixtures. Hardening occurs on the solvent evaporates and secondary bonds are reformed in the asphalt cement.

12.2.2 Composition and Structure

Chemical Composition

Asphalt cements have a very complex and variable chemical composition, which is dependent on the origin of the crude oil or petroleum from which they are derived and the processing to which they have been subjected. They are composed of high-molecular-weight hydrocarbons having the general formula $C_nH_{2n+b}X_d$, where X represents such elements as sulfur, nitrogen, oxygen, or trace metals; d is usually small; and b may be negative. The elemental composition generally lies within the following limits:

Carbon	80–87%	Nitrogen	0–1%
Hydrogen	9–11%	Sulfur	0.5–7%
Oxygen	2–8%	Trace metals[a]	0–0.5%

[a] Iron, nickel, vanadium, and calcium.

The number of carbon atoms (n) in asphalt cement molecules ranges from about 25 to 150, giving molecular weights from about 300 to 2000. These molecules are of three main types: (1) aliphatic or paraffinic, in which the carbon atoms are linked in straight or branched chains; (2) napthenic, in which the carbon atoms are linked in simple or complex (condensed) saturated rings (*saturated* means that the highest possible hydrogen to carbon ratio is present); and (3) aromatic, in which the carbon atoms are linked in especially stable benzene rings. Asphalt cements consist of a complex combination of all three types of molecules.

Although the content of oxygen, nitrogen, sulfur, and/or trace metals in asphalt cement is relatively small, these elements have a substantial effect on the physical and rheological properties of the material. The heteroatoms nitrogen, oxygen and sulfer form functional or polar groups and hydrocarbon molecules to which such groups are attached are capable of strong intermolecular associations through secondary bonding mechanisms. As a result, the material behaves as if it has a much higher molecular weight, which affects properties such as boiling point, solubility, and viscosity.

Because asphalt cements contain such a large number of molecules with differing chemical composition and structure, complete chemical analysis is generally impractical, if not impossible. Commonly, therefore, an asphalt cement is analyzed by separation into a small number of component fractions on the basis of the size and the reactivity and/or polarity of the various types of molecules present. However, since the separated fractions are still complex mixtures and not distinct chemical species, they can only be defined in terms of the particular procedure used, since the fractionation is not necessarily dependent on chemical composition. The same generic fraction can vary considerably in composition and properties from one asphalt cement source to another. In addition, the correlation between compositional characteristics based on such separation techniques and the properties or in-service performance is uncertain at this time. Nevertheless, component fractionation can be used to evaluate changes which occur in a particular asphalt cement during its manufacture and use.

Physical Structure

To understand the behavior and properties of asphalt cements, it is necessary to have some picture of their structure (i.e., the physical arrangement of their components). As a first approach, an asphalt cement can be viewed as a colloidal suspension consisting of two phases, a dispersion (continuous) phase and a dispersed (discontinuous) phase, with an interfacial material which acts to prevent coalescence of the dispersed phase. The constituents of these phases, their approximate chemical composition, and their general contribution to the properties of an asphalt cement are summarized in Table 12.1. Their general molecular structure is shown schematically on the three component (ternary) phase diagram (chapter 3) in Fig. 12.2, which represents possible asphalt cement compositions.

TABLE 12.1 Constituents of Asphalt Cement

Phase	Component	C/H Ratio[a]	Contribution
Dispersion	Oils	< 0.4	Viscosity and fluidity
Dispersed	Asphaltenes[b]	> 0.8	Strength and stiffness
Interfacial	Resins	~0.6	Adhesion and ductility

[a] Number of carbon atoms/number of hydrogen atoms
[b] Typical formula: $C_{84}H_{97}S_{3.2}O_{2.5} = C_nH_{2n-71} X_{5.7}$, C/H = 0.87

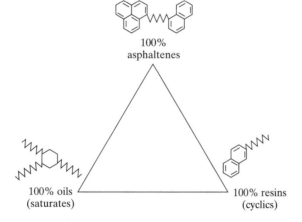

Figure 12.2
Ternary phase diagram representing composition of asphalt cement (after R. Nicholls, *Composite Construction Materials,* Prentice Hall, 1976, p. 194)

100%
asphaltenes

100% oils
(saturates)

100% resins
(cyclics)

This phase separation is essentially based on molecular size, and hence on the solubilities of the different components. Yet there are differences which are due to molecular structure and polarity. The oil phase consists primarily of nonpolar, uncondensed hydrocarbon ring molecules saturated with long-chain molecules, whereas the asphaltene phase consists of polar, condensed aromatic ring molecules joined with chain molecules. The resins have an in-between composition, with partially condensed ring molecules and some sidechains; without their presence, the suspension would break down since the other two components are not mutually soluble.

It should be emphasized that the phases exist as a continuum, with no distinct boundaries between them. The most polar and largest molecules, making up the asphaltene component, tend to interact and associate and hence are concentrated in colloidal droplets or micelles. The degree of association, and therefore the size of the micelles, depends on the asphaltene concentration, the dispersing ability of the other components, and the temperature. These micelles, each surrounded by a layer of resins, are distributed randomly throughout the oil phase, with the polarity and size of the molecules in each of the phases generally decreasing as the distance from the central core increases. In effect, the resins act as a solvent for the asphaltenes, and the oils as a solvent for the resins, resulting in a chemically stable structure.

On the basis of their gross structural makeup, asphalt cements can be classified into three different types: sol, gel, and sol-gel (Fig. 12.3). In the sol type, the asphaltene micelles are freely separated and widely dispersed in the oil phase, and the asphalt cement exhibits essentially viscous (Newtonian) fluid behavior with little or

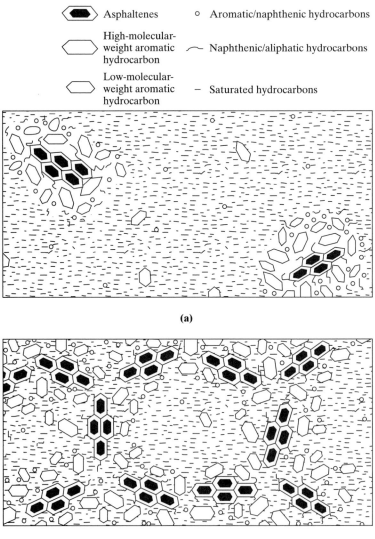

Figure 12.3
The internal structure of asphalt cement (a) sol, (b) gel (after the *Shell Bitumen Handbook*, 1990, taken from J. M. Illston, ed. *Construction Materials*, E&FN SPON, 1994).

no elasticity. In the gel type, the micelles are still discrete, but, through intermolecular attraction, are strongly bound in a complex, three-dimensional network, so the asphalt cement exhibits elastic, inelastic, and permanent deformational (non-Newtonian) behavior. The structure of the sol-gel type is intermediate between that of the other two, with the micelles bound, but not as closely as in the gel type. When load is applied, sol-gel asphalt cements initially exhibit elastic behavior, followed by viscous behavior, and hence they are viscoelastic in nature. This type constitutes the greater proportion of asphalt cements used for road construction purposes.

The structure of an asphalt cement and therefore its properties, as noted previously, are a function of temperature as well as its chemical nature and the relative volumes of its constituents. With increasing temperature, the asphaltenes become more dissolved in the resins, which in turn become more dissolved in the oils, and the material becomes less viscous. With decreasing temperature, the asphaltenes become less soluble and the micelles increasingly bound in an ordered structure, so the material becomes more viscous. Eventually, as the temperature is decreased below the glass transition temperature (T_g), the structure is effectively frozen and the material becomes rigid and brittle and behaves as a viscoelastic solid. T_g is dependent to some extent on the composition of the asphalt cement, and in particular the asphaltene content.

12.2.3 Properties

Because of the difficulty of complete chemical characterization, specifications for asphalt cements are currently based only on the physical and rheological properties of these materials. However, the correlation between physical and rheological properties and engineering performance has been shown to be reliable in most cases, although all of the tests employed to measure these properties are limited in some way. Many of these tests are empirical, and most do not provide information for the entire range of typical in-service operating temperatures. Nevertheless, physical and rheological properties will continue to provide a reasonable means of evaluating asphalt cement quality without necessitating the difficult task of analyzing these materials chemically.

Aging

It must be recognized that the properties of an asphalt cement in an asphalt concrete mixture will differ significantly from its properties prior to the production of the mixture. Asphalt cements are subjected to heating for variable time periods and at a wide range of temperatures during the mixing, curing, and service life of an asphaltic concrete. Substantial changes occur in the structure and composition of the asphalt molecules during these processes, due primarily to volatilization of light hydrocarbon fractions, and oxidation by reacting with oxygen from the environment. These changes, in turn, cause the asphalt cement to harden or become less ductile, a phenomenon known as age hardening or aging. Furthermore, asphalt cements exist in the form of thin films in asphaltic concretes, as shown schematically in Fig. 12.4. Their behavior and properties in this form are substantially different than in the 'bulk' material, and the oxidation or age-hardening reaction occurs at a much faster rate.

Standardized equipment and procedures have been developed to simulate the mix production, construction, and in-service aging processes and thereby allow

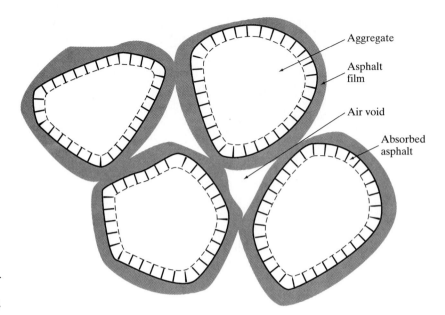

Figure 12.4
Structure of asphalt concrete paving mixture showing aggregates, air voids, and asphalt cement binder.

evaluation of the resulting changes in asphalt cement properties. Two commonly used artificial aging methods are the rolling thin film oven (RTFO) procedure (ASTM D-2872 or AASHTO[2] T-240) and the pressure aging vessel (PAV) procedure (AASHTO Provisional Standard PP1). The RTFO procedure, in which a thin film of asphalt cement is continuously exposed to heat and airflow in an oven, is supposed to simulate the effects of asphalt concrete mixing and construction procedures on the properties of the cement. It serves two purposes, to provide an aged asphalt cement sample for additional testing, and to provide an indication of the mass quantity of the volatiles lost during the heating and mixing process.

The PAV procedure, in which a thin film sample of asphalt cement previously aged in the RTFO is exposed to high pressure and temperature, is supposed to simulate the effects of long-term in-service conditions. Hence, the use of these procedures allows samples to be prepared and tested under conditions which represent an asphalt cement at various critical stages in its life.

Viscosity and Consistency

RHEOLOGICAL BEHAVIOR

Viscosity (Chapter 7) is the fundamental material property relating the rate of shear strain in a fluid to the applied shear stress; in practical terms, it is the resistance to flow. In Newtonian fluids, it is described mathematically by the coefficient of viscosity, η, which is expressed in units called poise (P), where 1 poise is 1×10^{-1} Pascal seconds (Pa s). The engineering term *consistency* is an empirical measure of the resistance offered by a fluid to continuous deformation when it is subjected to a shearing stress. In general, at typical asphalt concrete mixing temperatures (~140°C), sol-type asphalt cements are Newtonian fluids, whereas gel-type and sol-gel-type are non-Newtonian. Since the flow curve of non-Newtonian fluids can be represented by

---

[2]American Association of State Highway and Transportation Officials.

a power relationship (Eq. 7.7) their $\log(\sigma)$-$\log(\dot{\gamma})$ plots are linear and, therefore, often used by paving engineers (Fig. 12.5).

The rheological behavior of asphalt cements is very temperature dependent, and therefore the temperature of a material must be stated when reporting its rheological properties. As shown in Fig. 12.6, the viscosity-temperature relationship is inverse: Viscosity decreases as temperature increases. This relation will be different for different asphalt compositions. Viscosity-temperature plots for all asphalt cements appear to approach a limiting value of 10^9 poise as temperature is decreased. Prolonged aging at high temperatures, as well as the film thickness, would influence such relations.

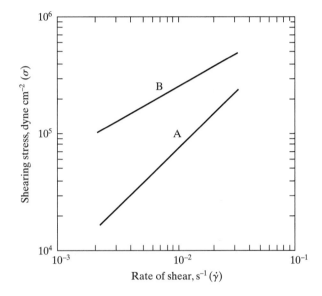

Figure 12.5
Flow curves representing Newtonian(A) and Non-Newtonian (B) rheological behavior of asphalt cements.

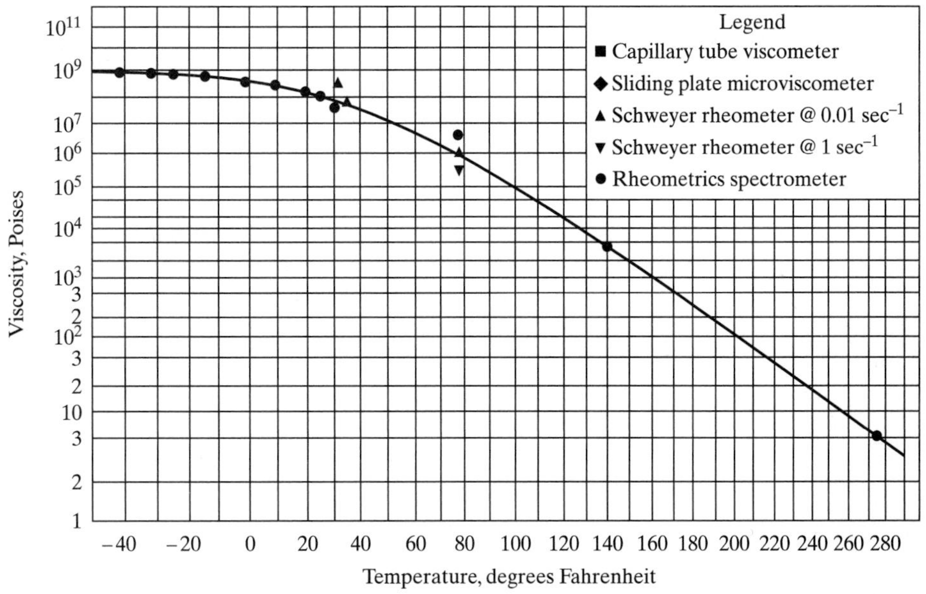

Figure 12.6
Effect of temperature on the viscosity of asphalt cement (after J. W. Button, J. A. Epps, D. N. Little, and B. M. Gallway, "Asphalt Temperature Susceptability and Its Effect on Pavements," Transportation Research Record 843: Asphalts, Asphalt Mixtures and Additives, Transportation Research Board, National Research Council, Washington, D.C., 1982, p. 123).

As noted previously, the viscosity/consistency of asphalt cements increases with time, due to aging or age hardening. Viscosity/consistency versus time of aging relationships are nonlinear and of the form

$$\eta = bt^m \tag{12.1}$$

or

$$\log \eta = \log b + m \log t \tag{12.2}$$

where η = viscosity (Poise)
 t = time of aging (hours)
 b = constant = intercept of log-log plot, and
 m = constant = slope of log-log plot.

The slope, m, of the log n-log t plot is called the degree of aging, or asphalt aging index; there is reasonable correlation between this parameter and the degree of complex or non-Newtonian flow of the material.

The greatest extent of asphalt cement aging typically occurs in the surface of an asphalt concrete pavement, where the binder is most exposed to oxygen and ultraviolet light.

RHEOLOGICAL TESTS

The consistency of asphalt cements is most commonly measured using the penetration test (ASTM D-5), shown schematically in Fig. 12.7. Consistency is measured in terms of the depth (in units of 0.1 mm) that a standard needle penetrates a sample of the material under standard conditions of loading, time, and temperature. Penetration does not represent a fundamental material property and is suitable for comparative purposes only, but it was the principal test method for grading asphalt cements for paving purposes, and still continues to be used in some parts of North America. There is no unique relationship between viscosity and penetration that is applicable to all asphalt cements.

There are several methods for measuring the viscosity of asphalt cements. Capillary flow under either gravity (ASTM D-2170) or vacuum (ASTM D-2171) is the most common. However, with the advent of more modern classification systems of asphalt cements, there is a need to determine the flow properties by tests which provide data of greater physical significance. Two such tests, the rotational coaxial cylinder viscometer (Chapter 7), and the dynamic shear rheometer are required for the Superpave™ asphalt cement classification system (Section 12.2.4). The rotational

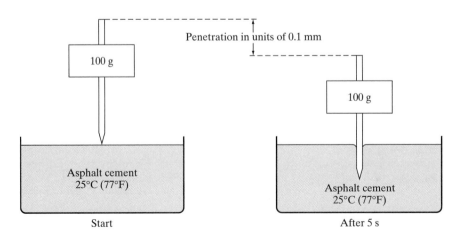

Figure 12.7
Penetration test (ASTM D-5) (*The Asphalt Handbook,* Manual Series No. 4 [MS-4], The Asphalt Institute, 1989, p. 39).

viscometer (RV) test (ASTM D-4402) is used to ensure that original or 'tank' asphalt cement is fluid enough to mix easily with an aggregate to produce asphalt concrete. The more complex dynamic shear rheometer (DSR) test (AASHTO Provisional Standard TP5) measures the rheological properties of both original and aged (RTFO and/or PAV) materials at intermediate and high temperatures. With this apparatus, the elastic as well as the vicous components of the behavior of asphalt cements can be determined, and a more complete picture obtained of their behavior under in-service conditions.

Stiffness

Like any other viscoelastic material, the deformational behavior of asphalt cements is also dependent on the rate or duration of loading or stress application. At low temperatures and/or short durations of loading, the behavior is mainly elastic; at high temperatures and/or long durations of loading, the viscous component becomes the dominant one. The behavior of an asphalt cement over the practical range of conditions encountered in a pavement structure lies between these extremes and can be described in terms of a stiffness or stiffness modulus (which is the creep modulus defined in Chapter 7), defined as

$$S_{t,T} = (\sigma / \varepsilon)_{t,T} \tag{12.3}$$

where $S_{t,T}$ = the stiffness or stiffness modulus of the material for a particular time of loading (t) and temperature (T),
σ = stress at time t and temperature T, and
ε = strain at time t and temperature T.

The time-of-loading and temperature dependence of the stiffness of an asphalt material is shown in Fig. 12.8. At very short loading times, the curves become horizontal, indicating essentially elastic behavior with the stiffness approximately equal to the elastic modulus value. At intermediate loading times, the material shows viscoelastic behavior, with both elastic and viscous deformations contributing significantly; the stiffness decreases nonlinearly with an increase in the loading time. At very long loading times, the behavior approaches that of a purely viscous Newtonian fluid, with the stiffness continuing to decrease, but approximately linearly, and approaching $3\eta/t$, where η is the viscosity indicating purely viscous behavior.

The similarity of the shape of the curves in Fig. 12.8 indicates that they could be made to coincide by shifting them horizontally. That is, there is a value of temperature which has the same effect on the response of the material as a certain loading time. This interchangeability of time and temperature is an important aspect of

Figure 12.8
Variation of asphalt cement (bitumen) stiffness (S_b) with time and temperature (after S. F. Brown, "Material Characteristics for Analytical Pavement Design, in *Development in Highway Pavement Engineering*, Vol. I, P. S. Pell, ed., Applied Science Publishers, London, 1978, p. 53).

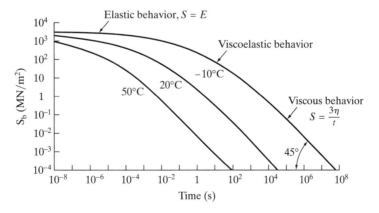

viscoelatic behavior (see the discussion in Chapter 7). As a consequence, laboratory test results determined over a limited time scale can be extended to very large times by carrying out such tests at different temperatures.

At low temperatures, the stiffness of asphalt cement is too high to permit determination of rheological properties using the rheological methods described previously. Hence a method based on a bending beam rheometer (BBR) is used in the Superpave™ classification system. In this method (AASHTO Provisional Standard TP1), both RTFO and PAV aged samples are subjected to a constant bending load at constant temperature (Fig. 12.9). The variation of the resulting deflection with time (creep) is used to determine the stiffness characteristics of the material.

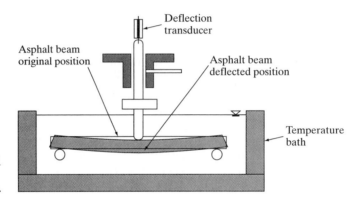

Figure 12.9
Bending beam rheometer test (after Superpave—Performance Graded Asphalt Binder Specifications and Testing, Asphalt Institute Superpave Series No. 1 [SP-1], Asphalt Institute, Lexington, Ky., 1994, p. 33).

Temperature Susceptibility

The change in stiffness of an asphalt cement with temperature is fundamentally associated with a change in viscosity and hence in the proportional importance of the elastic and viscous components of the deformational behavior of the viscoelastic material. The greater the *rate* of change of viscosity/consistency with temperature, the more temperature susceptible the asphalt cement, (i.e., the more liable it is to fracture, or crack, at low temperatures or to deform excessively, rut, at high temperatures; see Fig. 12.10). There does not appear to be any correlation between the parameters used to describe temperature susceptibility and the structure and/or composition of an asphalt cement.

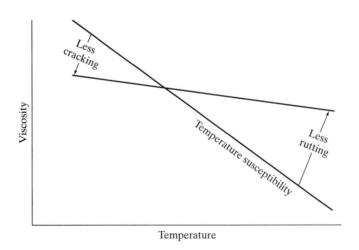

Figure 12.10
Temperature susceptibility of asphalt cements.

Temperature susceptibility can be evaluated using the viscosity-temperature susceptibility (VTS), defined as

$$\text{VTS} = \frac{\log \eta_2 - \log \eta_1}{\log T_2 - \log T_1} \qquad (12.4)$$

where η_1 is the viscosity at temperature T_1 and η_2 is the viscosity at temperature T_2. This relationship has been used to develop a standard viscosity-temperature chart for asphalt cements (ASTM D-2493), similar to Fig. 12.6. In most VTS calculations, viscosities measured at 60°C (140°F) and 135°C (275°F) are used, but other temperature ranges can be selected.

Tensile Properties

The tensile strength of an asphalt cement is important under conditions which promote elastic behavior and brittle fracture. The strength of the material is dependent on its viscoelastic and physical characteristics and the rate of loading and temperature conditions to which it is subjected. It has been generally observed that tensile strength reaches a maximum of 2 MPa to 4 MPa at low temperatures and decreases as load duration and temperature increase, as shown in Fig. 12.11. An increase in the strength of asphalt cements with decreasing film thickness has been observed, and is generally attributed to (1) a decrease in the ability of the material to flow, (2) a decrease in the probability of occurrence of flaws and/or cavities in the material (size effect, Chapter 5), and (3) an increase in the effect of molecular orientation at the surface of the material. In addition to strength, there is a need to evaluate the performance of asphalt cement with respect to ductility or elongation before rupture, and, in particular its sensitivity to service temperature and loading conditions. Therefore a direct tension test (AASHTO Provisional Standard TP3) is also used in the Superpave™ classification system.

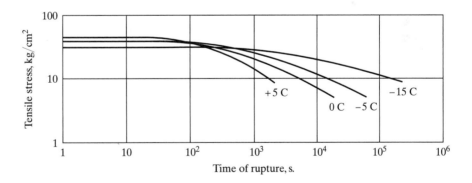

Figure 12.11
Effect of duration of load and temperature on the tensile strength of a sol-type asphalt cement.

12.2.4 Grading of Asphalt Cements

Asphalt cements are graded or classified according to their rheological and physical properties at temperatures representative of the preparation and in-service conditions of asphalt concrete paving mixtures.

Traditionally, asphalt cements have been graded on the basis of their penetration and/or viscosity. Viscosity grading systems are the most widely used, and are normally based on the viscosity of the original (as supplied, or unaged) asphalt cement at 60°C and 135°C (ASTM D-3381). In these systems, different grades have different physical property requirements, but these are measured at the same temperature for all grades. It has been found, however, that there is little correlation between material properties at the high temperatures on which viscosity grading is based and in-service performance, particularly at low ambient temperatures.

A new performance-based grading system (AASHTO Provisional Standard MP1) has been developed as part of the Superpave™ program.[3] In this system, all grades of asphalt binders, which includes modified as well as unmodified asphalt cements, must meet the same requirements, but at different temperatures for different grades, depending on the climate in which the binder will be used. Tests on the binders therefore are conducted at both high and low temperatures representative of the in-service environmental conditions.

12.3 LIQUID ASPHALTS

The viscosity of an asphalt cement can be reduced by combining it with a liquid of lower viscosity (i.e., a solvent), resulting in the formation of a liquid asphalt. After placement of this material, the viscosity and hence the viscoelastic nature of the material gradually increase as the solvent evaporates, leaving the asphalt cement. There are two basic types of liquid asphalts: cutbacks and emulsions.

Cutback asphalts are produced by blending an asphalt cement with a hydrocarbon solvent. There are three common types of cutbacks, which differ on the basis of the rate at which the material cures or hardens: slow-curing, medium-curing, and rapid-curing. The curing rate depends on the rate at which the solvent evaporates and hence on the type of solvent used. The composition of typical cutbacks is shown in Table 12.2.

TABLE 12.2   Composition of Cutback Asphalts

Type	Base Asphalt Cement	Solvent	Solvent Concentration (% volume)
Slow-curing (SC)	Low viscosity/ high penetration	Diesel fuel	0–50
Medium-curing (MC)	Medium viscosity/ medium penetration	Kerosene	15–45
Rapid-curing (RC)	High viscosity/ low penetration	Naptha/gasoline	15–45

Emulsified asphalts are produced by breaking asphalt cement into very fine droplets or particles and dispersing these in a mixture of water and a surface-active emulsifying agent, using the principles described in Chapter 4 (see Fig. 4.7). These liquid asphalts are classified as anionic, cationic, or nonionic, according to the type of ionic charge which is induced on the dispersed asphalt cement droplets. The charged asphalt droplets are attracted to oppositely charged aggregate particles, as shown in Fig. 12.12. Cationic emulsions are most commonly used.

A normal emulsion generally contains 55% to 70% asphalt cement and 0.5% to 3.0% emulsifying agent. It becomes unstable as the asphalt cement content increases, since the dispersed asphalt cement droplets begin to coalesce at a content of about 80%; the emulsion is then said to be broken. The breaking process is brought

[3]The Superpave™ (**Su**perior **per**forming asphalt **pave**ments) program is one of the products of the Strategic Highway Research Program (SHRP) carried out in the United States between 1987 and 1993. This $150 million research program was established by the United States congress to improve the performance and durability of highways and to make them safer for motorists and highway workers.

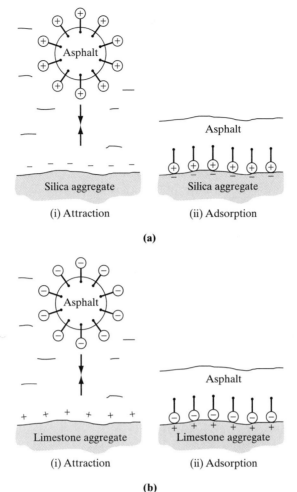

Figure 12.12
The mechanism of asphalt emulsion attraction to aggregate surface: (a) cationic emulsion attracted to negatively charged silica aggregate, (b) anionic emulsion attracted to positively charged limestone aggregate.

about by the evaporation of the water phase from the emulsion, and is accompanied by an increase in viscosity.

12.4 BINDER-AGGREGATE BONDING

Interactions between the asphalt cement and the aggregate at their interface, or plane of contact, are responsible for the formation of an interfacial region in the asphalt cement, as well as the bonding between them. These interactions take place during all three stages in the life of the resulting asphalt concrete mixture: mixing (when the asphalt cement is in a fluid state at high temperature), curing (when it cools down and changes into a viscoelastic solid), and aging (when it is exposed to environmental effects and loading).

Bond formation between an asphalt cement and an aggregate is a complex phenomenon that depends on the physical and chemical properties of both components. The strength of the bond formed is dependent on the wettability, viscosity (and hence temperature), composition (particularly as regarding oxygen-containing functional groups), and durability of the asphalt cement, and the surface chemistry (and hence mineral composition), surface texture, porosity, and surface condition (cleanliness and dryness) of the aggregate.

The ability of an asphalt cement to form a good bond with an aggregate particle depends strongly on how well it wets the aggregate surface (Chapter 4). This depends on the values of the surface or interfacial energies of the various phases in the system under consideration. In the absence of water, the surface energy of asphalt cements is sufficiently low that they will wet the surface of most common aggregates. However, as water is more polar, its surface energy is much lower than that of asphalt cements, and therefore when it is present it will preferentially wet the aggregate surface and prevent or disrupt an asphalt cement-aggregate bond. Thus, aggregates must be thoroughly dry when mixed with an asphalt cement to produce an asphalt concrete.

The mechanism of wetting is essentially that of adsorption of asphalt molecules on the surface of the aggregate (Fig. 12.13), which is controlled by electrostatic interactions due to the polar nature of the asphalt cement molecules. In addition, there will be some mechanical component to bonding if the asphalt cement is able to flow into pores and irregularities on the surface of the aggregate particles.

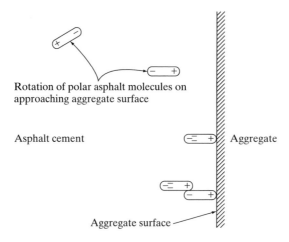

Figure 12.13
Adsorption mechanism of polar asphalt molecules on aggregate surface (after E. K. Ensley, "Multilayer Adsorption with Molecular Orientation of Asphalt on Mineral Aggregate and Other Substrates," *J. Applied Chemi. & Biotechnol.*, 25, 1975, pp. 671–682).

Failure of the asphalt cement-aggregate bond can occur as a result of (1) abnormal loading conditions and/or stress concentrations, such as flaws and irregularities at the interface, or (2) displacement or detachment of the binder from the aggregate surface by water. This latter mechanism, called stripping in practice, is a well-known phenomenon and is discussed in some detail later in this chapter.

12.5   ASPHALT CONCRETE

12.5.1 Introduction

Asphalt concretes can be classified according to the following characteristics:

1. *Type of binder.* Asphalt cements are the primary type of binder used, but liquid asphalts (cutbacks and emulsions) are commonly used in certain applications.

2. *Aggregate gradation.* Mixtures can be classified as dense-graded (well-graded), where the aggregate is uniformly graded from the maximum size down to the filler (Section 12.5.7) to obtain a minimum void content and a high stability, or open-graded, where there is little or no fine material so that the void spaces in the compacted aggregate are relatively large.

3. *Production methods.* There are two common types of asphalt concrete: hot-mixed, hot-laid mixtures, which are mixed, placed, and compacted at various elevated temperatures determined by the viscosity-temperature relationship of the binder; and cold-mixed, cold-laid mixtures, which are mixed, placed, and compacted at, or slightly above, ambient temperatures, using liquid asphalt as the binder.

Unless noted otherwise, the discussion in the remainder of this chapter will apply to well-compacted, hot-mixed and hot-laid asphalt concrete mixtures containing asphalt cement and a dense-graded aggregate.

12.5.2 Composition and Structure

An asphalt concrete is essentially a complex or composite material, consisting of two components or phases, an asphalt cement and an aggregate, which are physically combined. To characterize the structure and, hence, the behavior of such a material fully, it is necessary to identify some of its geometrical and compositional variables. As shown in Fig. 12.14, an asphalt concrete mixture can be proportioned so that either the asphalt cement or the aggregate acts as the continuous phase. In the first case, Fig. 12.14(a), the deformational behavior is more viscous or viscoelastic, since the asphalt cement is the matrix and the aggregate is the filler. In the second case, Fig. 12.14(b), the behavior is essentially that of a solid, since the aggregate particles are in intimate point-to-point contact acting as a continuous phase, with the asphalt cement filling the void spaces between the particles and helping to restrain their relative motion. The optimum structure is, to a considerable extent, dependent on the desired properties of the mixture. In general, however, it more closely approximates the second case, consisting of a dense, interlocked structure of aggregate particles with sufficient asphalt cement between the particles and in the interparticle void spaces to bind them together and protect them from detrimental environmental effects.

As suggested by the preceding discussion, aggregates constitute 90% or more, by weight, of asphalt concretes, with asphalt cement (and, in practice, air voids) making up the remainder of the mixture. The important characteristics of aggregates, from the point of view of their use in these mixtures, were described in detail in Chapter 10.

Figure 12.14
Asphalt concrete mixture with (a) open-graded aggregate and excess asphalt cement, (b) dense-graded aggregate and sufficient asphalt cement.

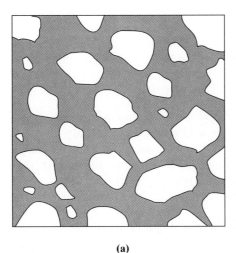

(a)

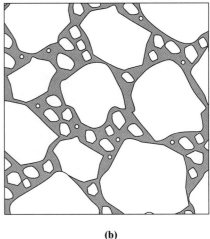

(b)

Densification of asphalt concrete mixtures by compaction is of paramount importance in achieving the desired behavior and properties. In practice, asphalt concretes are compacted to a certain percentage (typically greater than 90%) of their theoretical maximum density, (i.e., the density of a voidless mixture of the asphalt cement and aggregate). In the field, this is normally achieved through the use of self-propelled steel-wheeled or pneumatic-tired rollers, and must be carried out while the asphalt cement is sufficiently fluid that it acts as a lubricant and assists the repositioning of the aggregate particles, if necessary, into a denser and more stable configuration. As noted previously, the discussion in this chapter will apply to sufficiently well-compacted asphalt concrete mixtures.

12.5.3 Response to Applied Loads

The response of an asphalt concrete to applied loads is a function of several characteristics of the material, which are discussed in this section.

Stiffness

Stiffness, as defined previously in this chapter (Eq. 12.3), is expected to be dependent on both the asphalt cement and the concentration and geometric properties of the aggregate. From an engineering standpoint, simple axial loading conditions can be used to determine the stiffness characteristics of these mixtures.

Data obtained under both creep (long time of loading) and dynamic (short time of loading) tests indicate that the stiffness of an asphalt concrete is dependent on the stiffness of the asphalt cement and the volume concentration of the aggregate in the mixture (Fig. 12.15):

$$S_{con} = S_{ac}\left(1 + \frac{2.5}{n}\frac{C_v}{1 - C_v}\right)^n \tag{12.5}$$

where S_{con} = the stiffness of the mixture,
S_{ac} = the stiffness of the asphalt cement,
$n = 0.83 \log[(4 \times 10^5)/S_{ac,}]$ and
$C_v = \dfrac{\text{volume of compacted aggregate}}{\text{volume of compacted mixture}}.$

This equation is applicable to well-compacted mixtures with about 3% air voids and C_v values ranging from 0.7 to 0.9. Although not implicit in Eq. 12.5, mixture stiffness is also influenced by aggregate characteristics, such as surface texture, particle shape, and gradation: Rougher, more angular, dense-graded aggregates generally produce mixtures with greater stiffness than do smoother, rounded, open-graded aggregates.

Stability

Stability is the property of the asphalt concrete which allows it to resist permanent deformation under traffic loading. Stability can be determined by the Marshall test, (ASTM D-1559) which is an unconfined compression test using the special configuration shown in Fig. 12.16(a). In this test, the maximum load is defined as stability, in load units (Fig. 12.16(b)).

Since stability is mainly due to mechanical or frictional interlock between the aggregate particles, it is a maximum in a mix with a dense-graded aggregate and an asphalt cement content which is just sufficient to fill the interparticle void space (Fig. 12.14(b)). If the asphalt cement content is too high, the stability of the mixture

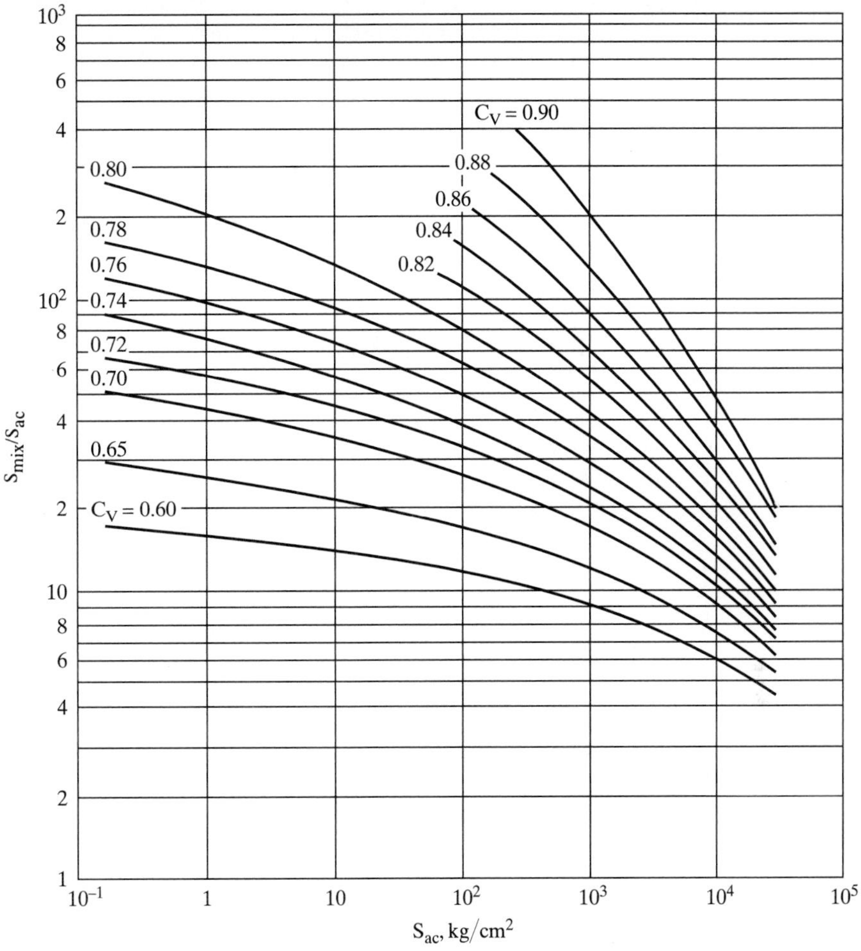

$C_V = 0.90$
0.88
0.86
0.84
0.82
0.80
0.78
0.76
0.74
0.72
0.70
0.65
$C_V = 0.60$
S_{mix}/S_{ac}
S_{ac}, kg/cm²

Figure 12.15
Effect of asphalt cement stiffness and aggregate content on the stiffness of asphalt concrete (after W. Heukelom and A. J. G. Klomp, "Road Design and Dynamic Loading," *Proc. Association of Asphalt Paving Technologies*, 33, 1964, pp. 92–125).

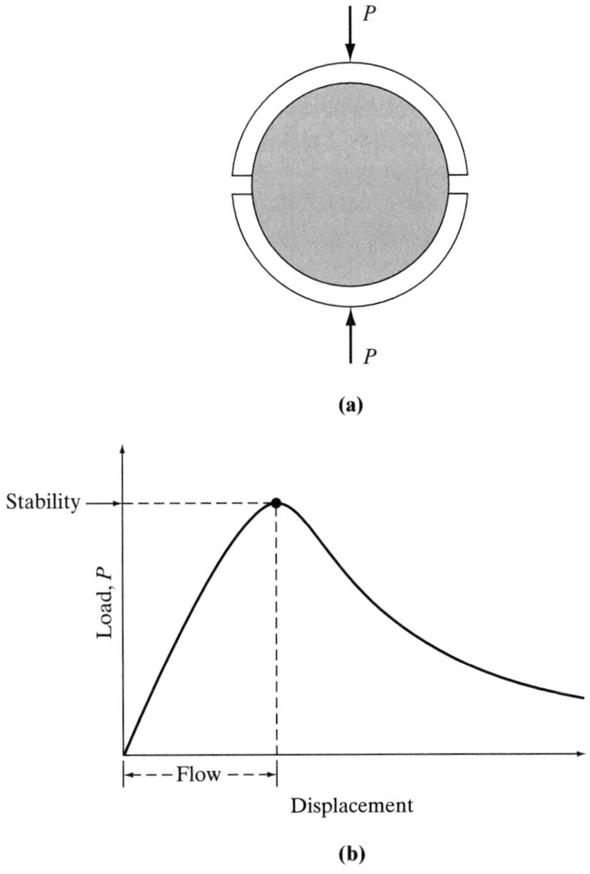

P
P
(a)
Stability
Load, P
Flow
Displacement
(b)

Figure 12.16
Marshall stability test configuration (a) and resulting load-displacement curve (b) showing stability and flow values.

will be reduced and the asphalt cement will act as a lubricant and reduce the degree of aggregate particle contact (Fig. 12.14(a)). If, on the other hand, the asphalt cement content is less than that required to fill the void space, the stability of the mixture is low because aggregate particles are able to move relative to each other under the applied load. There is, therefore, an optimum asphalt cement content at which the stability of a mixture containing a given aggregate and asphalt cement, at a given level of compaction, is a maximum, as shown in Fig. 12.17(c).

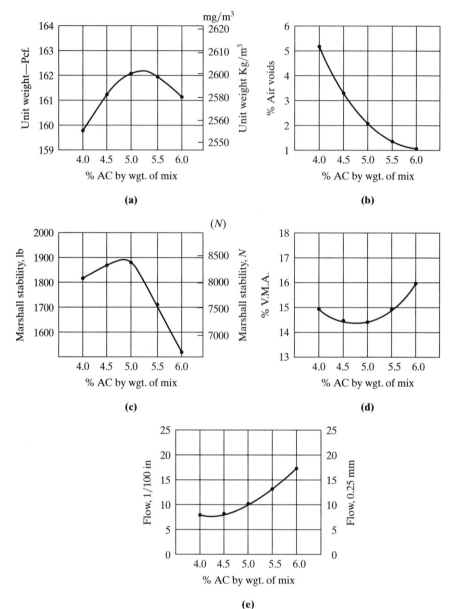

Figure 12.17
Typical Marshall test property vs. asphalt cement (AC) content curves for asphalt concrete unit weight (a), air voids content (b), stability (c), Voids-in-Mineral-Aggregate (V.M.A.) (d), and flow (e) (after *The Asphalt Handbook,* Manual Series No. 4 [MS-4], The Asphalt Institute, 1989).

Flexibility

Flexibility is the ability of an asphalt concrete to conform, without fracturing, to minor, long-term settlement of the base or subgrade layers beneath it in a pavement structure. Generally, flexibility is enhanced by relatively open-graded aggregate mixtures and high asphalt cement contents, factors which contribute to a lower mixture stability. A compromise therefore has to be reached between good stability and sufficient flexibility.

Chap. 12   Asphalt Cements and Asphalt Concretes

Fatigue Resistance

Fatigue resistance refers to the ability of the mixture to resist failure, by fracture or cracking, under the repeated loading conditions associated with vehicular traffic. Mixtures with higher asphalt cement contents and densely-graded aggregates generally have greater fatigue resistance than those with low cement contents or open-graded aggregates.

Tensile (Fracture) Strength

The maximum tensile strength or fracture strength of the mixture is dependent on the tensile strength of the binder, the adhesion between the binder and the aggregate, the amount of binder in the mixture, and the voids content of the mixture. As we have seen, the tensile strength of the binder itself is directly related to its viscoelastic and physical characteristics; it is approximately constant at low temperatures and high loading rates, and it decreases as temperature increases or loading rate decreases.

Under rapid loading rates and/or low temperatures, the maximum tensile strength of asphalt concretes is 4 MPa to 10 MPa, and the strain at failure is 1000×10^{-6} to 1200×10^{-6}. Furthermore, as the stress-strain behavior under these conditions approaches linearity (i.e., elastic behavior), the limiting values of tensile strength and strain at failure can be associated with a corresponding limiting value of stiffness modulus (Fig. 12.8).

The tensile strength of the mixture is important under the following conditions: (1) when an asphalt concrete pavement layer is subjected to heavy loads at low temperatures, (2) when the subgrade beneath the layer is relatively weak, and (3) when high tensile stresses are induced in the layer by temperature changes, volume changes in the subgrade, or volume changes in the mixture itself.

Permanent Deformation

Excessive permanent or plastic deformation in an asphalt concrete occurs as a result of normal and/or shear stresses caused by the applied load. Under traffic loading, this deformation can take several forms, known in practice as rutting, shoving (or pushing), slippage, or corrugation. Rutting, the most common form, consists of channelized depressions in the wheelpath which develop from densification (consolidation) and/or lateral movement of the mixture.

Several factors influence the tendency of an asphalt concrete to rut; those related to the mixture itself include asphalt cement content, filler content, and shape of the aggregate particles, with higher cement and/or filler contents and rounded coarse and/or fine aggregate particles resulting in an increased rutting tendency. Laboratory studies have shown that these factors also contribute to an increase in the tendency of a mixture to creep, and that there is a correlation between creep and rutting test data. The mechanisms of deformation are thought to be similar, and hence it may be possible to assess the rutting potential of a mixture using a relatively simple laboratory creep test.

12.5.4 Response to Moisture

Permeability

The ease with which air, water, and/or water vapor will pass into or through an asphalt concrete is dependent on its permeability, or the content and degree of interconnection of the voids in the mixture. High asphalt cement contents, dense aggregate gradations, and good compaction result in low permeability and hence good durability.

Durability

The ability of an asphalt concrete to resist disintegration due to weathering and the abrasive action of traffic is its durability. Durability is maximized when the aggregate particles are completely covered by asphalt cement and the mixture contains no air voids to facilitate the entry of water (and the resulting possibility of stripping and/or freezing) or light and air (which increase the rate of age hardening of the binder). Durability considerations therefore limit the allowable air content in the placed and compacted paving mixture to less than approximately 5%.

The durability characteristics of the aggregate are also important: It must be sufficiently strong and tough to resist the forces associated with the traffic loading and, where necessary, to resist breakdown due to freeze-thaw action.

Stripping (Moisture-induced Damage)

Stripping refers to the breaking of the adhesive bond between the aggregate particles and the asphalt cement binder through the action of water. Several mechanisms contribute to stripping, the most impotant being (1) spontaneous emulsion formation as water droplets migrate through the asphalt film to the asphalt-aggregate interface, and (2) rupture of the asphalt film by interfacial tension at the air-water-asphalt interface. There are two fundamental classes of tests for adhesion of asphalt to aggregate, stripping tests (ASTM D-3625) and loss of strength tests (ASTM D-1057).

In practice, many of the factors contributing to the potential for stripping in asphalt concrete pavement mixtures can be avoided by paying particular attention to construction details. Examples of such details include the following:

1. Reducing the void content of the asphalt concrete;
2. Not using known hydrophilic aggregates;
3. Washing aggregates to remove any coatings;
4. Precoating aggregates with bitumen or diluents prior to mixing with asphalt cement; and,
5. Using higher temperatures in the mixing phase to drive off water and to reduce asphalt cement viscosity, thereby facilitating aggregate coating.

12.5.5 Response to Temperature

Thermal Expansion and Contraction

The coefficient of thermal expansion and contraction of the mixture reflects the proportional influence of the coefficient for each of its components. For asphalt cements, the coefficient is essentially constant at $6 \times 10^{-4}/°C$ for temperatures above the glass transition temperature, and at $2 \times 10^{-4}/°C$ to $4 \times 10^{-4}/°C$ for temperatures below. For aggregates, the coefficient is much smaller, typically between approximately $5 \times 10^{-6}/°C$ and $13 \times 10^{-6}/°C$. Expansion or, more importantly, contraction of an asphalt concrete mixture with temperature is therefore largely a reflection of the response of the asphalt cement binder.

Thermal cracking (or low-temperature cracking, as it is more commonly called) is a consequence of the thermal contraction or shrinkage of an asphalt concrete under freezing conditions. It occurs in practice when this thermal shrinkage is resisted (for example, by friction with an underlying layer in a pavement structure), and the tensile stresses thus developed exceed the tensile (fracture) strength of the mixture. It generally starts at the exposed surface of an asphalt concrete layer and progresses down through the layer with time. Asphalt concrete mixtures which have a high stiffness modulus at low temperatures are most prone to low-temperature shrinkage cracking.

12.5.6 Response to Chemicals

Asphalt cements are highly resistant to the action of most acids, alkalis, and salts, as are most aggregates used to produce asphalt concrete mixtures. Therefore, we would expect these mixtures to show little or no response when exposed to such chemicals. However, asphalt cements readily dissolve in petroleum solvents of varying volatility, which may present difficulties with asphalt concrete pavements in parking areas for vehicles using petroleum-based fuels or lubricants (spills of such fluids can result in disintegration and failure of the asphalt concrete).

12.5.7 Additives and Fillers

It is common to add additives and fillers to the asphalt cement or asphalt concrete to improve their performance.

Antistripping Agents

Stripping is recognized as a major durability problem in asphalt concrete paving mixtures throughout much of North America, and antistripping agents are widely used. Many of these agents are organic compounds (either cationic or anionic in nature or a combination of both), or inorganic chemicals used alone or in combination with organic acids. They migrate to the aggregate-asphalt interface and enhance the bonding and bond stability between these two components by forming water resistant complexes.

Hydrated lime, or calcium hydroxide, is the most commonly used adhesion promoter. It acts to stabilize any fine material or clays which may have contaminated the aggregate surface, to provide calcium binding sites for asphalt cement, and as a filler.

The addition of antistripping agents to the asphalt cement prior to mixing with the aggregate is a common but inefficient means of applying these agents to the binder-aggregate interface, where they function in the mixture. Migration of the agents to the interface can take place only while the asphalt cement viscosity is low enough to facilitate this movement. Application of the agent directly to the aggregate surface, through pretreatment of the aggregate, while less convenient and initially more costly, may be a more effective means of introducing these agents into the mix.

Asphalt Cement Modifiers

Antioxidants, such as lead diethyldithiocarbamate, are added to asphalt cements to retard the oxidative aging or hardening process. These materials must be stable and nonvolatile at the temperatures encountered in the mixing process.

Natural and synthetic rubbers have been used extensively in asphalt concrete mixtures; styrene-butadiene rubber (SBR) is the material most widely used. These materials can be added to asphalt cement in the form of an emulsion or latex or as finely divided solids. In general, small quantities of rubber (i.e., less than 5% by weight), will increase the viscosity of an asphalt cement, improve its adhesion to the aggregate particles, retard its rate of oxidation, and reduce its temperature susceptibility.

Recycling Agents

Recycling of asphalt materials is becoming a common practice due to the increasing cost of asphalt cements, the decreasing availability and increasing cost of new aggregates, and the reduction of available funds for transportation facilities. Recycling agents are hydrocarbon products with physical and chemical characteristics selected to restore aged asphalt cements to meet desired specification requirements. There are essentially two common types: softening agents and rejuvenating agents. Softening agents are normally crude oil fractions of suitable viscosity which act simply to decrease the viscosity of the aged asphalt cement. Rejuvenating agents have a composition similar to that part of the asphalt cement which is changed during the aging process and act to recover the chemical and physical properties of the aged material so that it is effectively a new asphalt cement.

Extenders

An extender is an additive which replaces a part of the asphalt cement that would normally be used in an asphalt concrete mixture and hence allows a reduction in its content. It may additionally result in performance improvements or better economy. Elemental sulfur is by far the most popular and most commonly used extender. There are two methods for dispersing sulfur in an asphalt concrete mixture: (1) pre-blending, in which the liquid sulfur and hot asphalt cement are mixed before being added to the aggregate, or (2) mixer blending, in which the liquid sulfur and hot asphalt cement are separately added to the aggregate and blending takes place during mixing. Obviously, both processes require certain additions or modifications to the asphalt concrete mixing plant.

The principal effect of the addition of sulfur is to reduce the viscosity of an asphalt cement over the normal temperature range used in mixing and placing asphalt concrete mixtures. In addition, sulfur acts to increase the stiffness of an asphalt cement, and hence that of an asphalt concrete mixture, at higher service temperatures while having little or no effect at lower service temperatures. This results in an improvement in the temperature susceptibility of the binder and the concrete mixture.

Fillers

Fillers are very fine materials sometimes added to the aggregate in an asphalt concrete mixture to achieve a gradation which more closely approximates the maximum-density, minimum-voids content condition. Pulverized limestone is the most common manufactured material used as a filler, but stone dust, hydrated lime, portland cement, and other types of finely divided mineral matter are also used.

Because of their size, filler particles play a dual role in asphalt concrete mixtures. They fill the void spaces and provide contact points between the larger particles of aggregate. Also, the addition of the filler generally results in a 'hardening' of the asphalt cement (i.e., an increase in its viscosity, or consistency, and tensile strength, and a decrease in its ductility). The magnitude of these effects is dependent on the type of filler used as well as its concentration.

12.5.8 Mix Design Methods

The design of an asphalt concrete mixture is basically a matter of selecting and proportioning the component materials to obtain the desired qualities and properties in the finished product. In addition to the properties of the component materials themselves, economic factors, such as the relative costs of the different aggregates in the blend (coarse, fine, and mineral filler, if used), the asphalt cement content, and the proposed construction method, should also be considered.

The mix design process consists of two stages. In the first stage, an aggregate blend is selected to satisfy the gradation and other requirements of the appropriate specifications or standards for the type of asphalt concrete required (e.g., ASTM D-3515). In the second, an asphalt cement content which will achieve a balance among the following asphalt concrete properties is determined:

Stability: Stability is a maximum at low asphalt cement contents where the mechanical and frictional interlock between aggregate particles is high.

Durability: Durability is a maximum at high asphalt cement contents where the surface of the aggregate particles is completely covered with asphalt cement, and the voids between the aggregate particles (VMA) are completely filled.

Flexibility: Flexibility is also a maximum at high asphalt cement contents.

Skid resistance: The skid resistance of the surface of the asphalt concrete mixture is a maximum at low asphalt cement contents.

Workability: Workability is also a maximum at high asphalt cement contents.

The Marshall test property curves shown in Figure 12.17 illustrate that there is an inherent contradiction in the asphalt cement content required to optimize different asphalt concrete mixture properties. Therefore, the various mix design methods that have been developed attempt to provide an asphalt cement content which achieves a sound compromise between conflicting requirements. These methods typically involve the use of some combination of total surface area of aggregates, void space between aggregates, and physical and mechanical properties of the mixture to determine an appropriate asphalt cement content.

BIBLIOGRAPHY

F. L. ROBERTS, P. S. KANDAHL, E. R. BROWN, D.-Y. LEE, and T. W. KENNEDY, *Hot Mix Asphalt Materials, Mixture Design, and Construction* (First Edition) NAPA Education Foundation, 1991

Mix Design Methods for Asphalt Concrete and Other Hot-Mix Types (Sixth Edition). Manual Series No. 2(MS-2). The Asphalt Institute

The Asphalt Handbook (1989 Edition). Manual Series No. 4(MS-4). The Asphalt Institute

Performance Graded Asphalt Binder Specification and Testing. Superpave Series No.1 (SP-1). The Asphalt Institute

PROBLEMS

1. How does the behavior of an asphalt cement change as its temperature is lowered from above its glass temperature (T_g) to approximately 0°C? What are the practical consequences of this change in behavior on the mixing and placing of an asphalt concrete mixture?
2. How do heteroatoms (oxygen, nitrogen and sulfur) in asphalt cement molecules affect the behavior and properties of the material?
3. Briefly describe the differences in the structure and properties of the sol, gel, and sol-gel asphalt cements.
4. What is the effect of temperature on the structure and rheological properties (e.g., viscosity) of an asphalt cement?

5. What is 'aging' (age-hardening) of an asphalt cement? Briefly describe two methods of artificially aging an asphalt cement.

6. What is the relationship between the viscosity and the 'time of aging' of an asphalt cement? What is the 'asphalt aging index'?

7. Describe the variation in the stiffness or stiffness modulus of an asphalt cement with time-of-loading.

8. What is the 'temperature susceptibility' of an asphalt cement? What are the practical consequences of using an asphalt cement with a high temperature susceptibility?

9. What is the effect of a decrease in the thickness of an asphalt cement film on the apparent tensile strength of the material? What are the probable causes of this effect?

10. What are the differences between traditional viscosity grading systems and the Superpave™ grading system for asphalt cements, in terms of the relationship between properties and the temperatures at which they are measured?

11. What is a 'liquid' asphalt? How is an emulsified liquid asphalt produced? What is 'breaking' of an emulsified liquid asphalt?

12. What are the principal factors affecting the bonding of an asphalt cement to an aggregate particle surface?

13. Identify and describe three characteristics that can be used to classify asphalt concrete mixtures.

14. In terms of aggregate gradation and asphalt cement content, what is the optimum structure for an asphalt concrete mixture?

15. In practice, how is adequate densification of asphalt concrete mixtures achieved?

16. What properties of the aggregate used in an asphalt concrete mixture will affect the stiffness of the mixture?

17. What is the principal factor affecting the stability of an asphalt concrete mixture?

18. Under what practical circumstances is the tensile strength of an asphalt concrete mixture important?

19. What is 'rutting' of an asphalt concrete mixture? What factors affect the tendency of an asphalt concrete to rut?

20. What is 'stripping' of an asphalt concrete mixture? Identify three measures that can be taken to reduce the potential for stripping to occur.

21. What is the cause of thermal or low-temperature cracking of asphalt concrete mixtures, and why does it occur?

22. What are the effects of the addition of small quantities (less than 5%) of natural or synthetic rubbers on the properties of asphalt cements?

23. What is an 'extender'? What is the most popular and commonly used extender? How does it affect the properties of an asphalt cement?

24. What is a 'filler'? What is the most popular and commonly used filler? How does it affect the properties of an asphalt concrete mixture?

25. Describe the effect of increasing asphalt cement content on the (a) stability, (b) flexibility, (c) durability, and (d) skid resistance of an asphalt concrete mixture?

IV

STEEL, WOOD, POLYMERS, AND COMPOSITES

13

Structural Steels

13.1 INTRODUCTION

Steel is the most useful of the industrial metals. Thousands of varieties have been developed to meet specific needs. As civil engineers, our interest is primarily in structural quality steels, which are steels used to produce the various sections (shapes, plates, and bars) from which structural members (beams, girders, columns, struts, ties, and hangers) are fabricated. The primary purpose of such members is to support loads or resist forces acting on structures such as bridges; industrial, institutional, commercial, and residential buildings; and cranes and other construction equipment.

Steel is also used as a reinforcement material (bars, wires, and strands) in portland cement concrete structural members, to resist tensile stresses which could lead to the failure of the more brittle concrete. In addition, it is used in many other construction applications which are nonstructural in nature, including roofing, siding, and decking; joists and studs; flooring; and interior and exterior partition walls.

Most structural steel sections are formed from large castings by a series of rolling operations, which shape the steel, reduce it in cross section, and elongate it. The final mechanical properties of the 'as-rolled' steel are strongly influenced by the temperatures at which these operations are carried out, the extent to which the material is deformed (strained), and the rate at which it is cooled after rolling is completed, as well as the chemical composition of the steel. In addition, the properties can be changed by subsequent heat treatment of the steel section.

13.2.1 Composition

The chemical composition of steel is very important since it has a significant effect on the microstructure of the material and hence on its mechanical behavior and properties.

Steel is basically an alloy of iron and carbon, but several elements are used in various proportions and combinations to produce different types. By definition, a steel has a maximum carbon content of 2.0%. Structural steels normally contain less than 0.30% carbon, however, and in terms of chemical composition can be classified as either plain carbon or low-alloy steels. In plain carbon steels, the amounts of carbon and manganese, the principal strengthening elements, are restricted, and other alloying elements are not normally included. In low-alloy steels, the carbon content is also restricted, and increased strength is achieved through the use of alloying elements such as nickel, chromium, and molybdenum. The alloying additions usually do not exceed a total of 8%.

Carbon is by far the most important element in steel: The changes in composition (cementite or iron carbide [Fe_3C] content), microstructure (pearlite content), and mechanical properties (strength and ductility, in particular) resulting from changes in the carbon content in annealed plain carbon steels are shown very clearly in Fig. 13.1. In the range of carbon contents shown in Fig. 13.1, a significant increase in strength and decrease in ductility are produced by an increase in the carbon content in the steel.

Steels can be classified on the basis of composition. The most widely used system for identifying or designating carbon and alloy steels is that developed by the American Iron and Steel Institute (AISI) and the Society of Automotive Engineers (SAE). The AISI-SAE system employs a four- or five-digit designation, where the major alloying elements in a steel are indicated by the first two digits, and the amount of carbon, in hundredths of a percent, by the last two or three. Examples of several steels with their AISI-SAE designation and alloy composition are shown in Table 13.1. Structural quality steels are further classified on the basis of strength, using a system developed by the American Society for Testing and Materials (ASTM), as discussed in Section 13.6.

13.2.2 Microstructure

The microstructure of metals and alloys—that is, the geometric arrangements, volume fractions, sizes, and morphologies of the constituent phases and/or grains, as observed under the microscope—plays an important part in determining the mechanical properties of the material. In general, strength depends on the nature, distribution, and size of the phases and/or grains present: Fine-grained structures are stronger than coarse-grained structures. In two-phase alloys, where one phase is hard and brittle relative to the other, optimum properties are obtained where the hard and brittle phase is uniformly distributed as isolated particles. If the brittle phase occurs as a continuous network, then fracture will follow this network, making the alloy as a whole brittle. The distribution of phases depends initially on the conditions under which the alloy solidified, but it may be varied by mechanical working and heat treatment.

In a binary, metallic alloy system, the relationship between the composition of the phases present and temperature may conveniently be represented by an

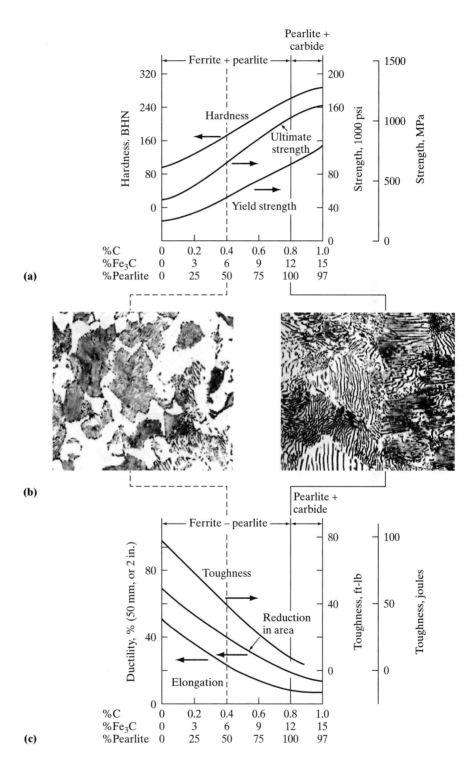

Figure 13.1
Properties versus carbon content (annealed, plain carbon steels): (a) hardness and strength versus the amount of carbon, Fe₃C, and pearlite; (b) microstructure of 0.40% C (left) and 0.80% C (right) steels (×500); and (c) ductility and toughness versus the amount of carbon, Fe₃C, and pearlite (from L. H. Van Vlack, *Elements of Materials Science and Engineering,* 5th ed., Addison-Wesley, 1985).

TABLE 13.1 Examples of AISI-SAE System of Steel Designations

		Example	
AISI-SAE	Type of Steel	Designation	Composition (weight %)
10XX	Plain carbon	1020	0.18–0.23 C
			0.30–0.60 Mn
		1040	0.37–0.44 C
			0.60–0.90 Mn
13XX	Manganese	1340	0.38–0.43 C
			1.60–1.90 Mn
			0.15–0.30 Si
41XX	Chromium-molybdenum	4140	0.30–0.43 C
			0.75–1.00 Mn
			0.15–0.30 Si
			0.80–1.10 Cr
			0.15–0.25 Mo

equilibrium or phase diagram (see Chapter 3). The following features should be noted in the iron-cementite, Fe—Fe_3C, phase diagram (Fig. 13.2):

1. Iron goes through two allotropic transformations during heating or cooling: On continued cooling from a liquid melt, it first forms a body-centered cubic (BCC) structure, then a face-centered cubic (FCC) structure, and finally another BCC structure. All three allotropes will form interstitial solid solutions with carbon, identified as delta iron (δ), austenite (γ), and ferrite (α), respectively. A greater number of carbon atoms can be accommodated in the austenite than in the other two phases, since the interstitial holes in the FCC lattice are somewhat larger than those in the BCC lattices. Hence the maximum solubility of carbon in austenite is 2.0%, while it is much lower in delta iron (0.10%) and ferrite (0.025%). These solid solutions are relatively soft and ductile, but stronger than pure iron due to solid solution strengthening by the interstitial carbon atoms.

2. Cementite, or iron carbide, Fe_3C, is a stoichiometric intermetallic compound formed when the solubility of carbon in solid iron is exceeded. This compound contains 6.67% C, is extremely hard and brittle, and is present in all commercial steels. The degree of dispersion strengthening and hence the properties of the steel are controlled by properly regulating the amount, size, and shape of the Fe_3C phase.

3. A eutectoid reaction occurs as the austenite (γ) cools below 727°C. That is,

$$\gamma \rightarrow \alpha + Fe_3C.$$

Since the two phases that form have different compositions, atoms must diffuse during the reaction: Most of the carbon in the austenite diffuses to the Fe_3C, and most of the iron to the ferrite (α). Since this redistribution of atoms is easiest if the diffusion distances are short, the α and Fe_3C grow as thin lamellae, or plates, forming a structure called pearlite (Fig. 13.3).

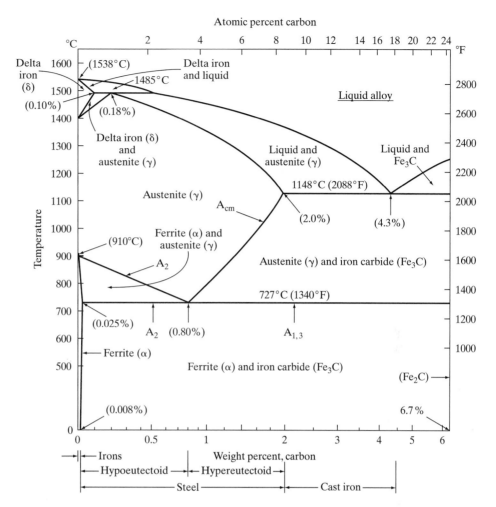

Atomic percent carbon

Figure 13.2
Iron-iron carbide equilibrium phase diagram (iron rich portion) (P. A. Thornton and V. J. Colangelo, Fundamentals of Engineering Materials, Prentice-Hall, Inc., 1985).

Figure 13.3
Photomicrograph illustrating the lamellar nature of pearlite (×1125) (Courtesy of Theresa Brassard).

Since the structural steels are hypoeutectoid—that is, they contain less carbon than the eutectoid composition (0.80%)—the primary microconstituent is ferrite (α). When a hypoeutectoid alloy cools under equilibrium conditions from some temperature above, say, 900°C, the following occurs (Fig. 13.4):

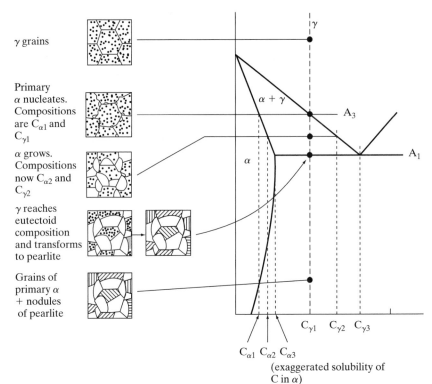

Figure 13.4
Microstructural formation during the slow cooling of a hypoeutectoid steel from the melt (M. F. Ashby and D. R. H. Jones, Engineering Materials 2, An Introduction to Microstructures, Processing and Design, Pergaman Press, 1986).

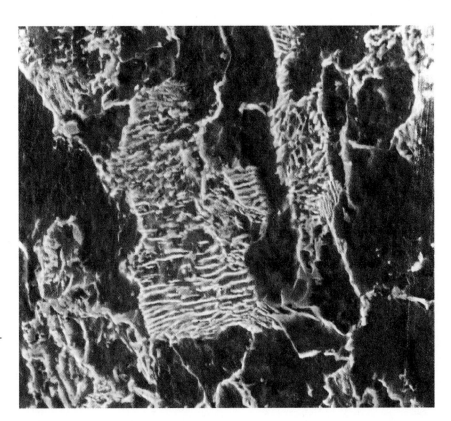

Figure 13.5
Photomicrograph showing pearlite "islands" (striped regions) surrounded by primary ferrite (×8000). Courtesy of Mary Mager, Dept. of Metals and Materials Engineering, University of British Columbia.

1. Just below the A_3 temperature, ferrite precipitates and grows, usually at the austenite grain boundaries; primary ferrite continues to grow until the temperature falls to A_1.

2. At the A_1 temperature, the remaining austenite is surrounded by ferrite and its composition has changed to the eutectoid composition (0.80% C); subsequent cooling causes all of this austenite to transform to pearlite by the eutectoid reaction.

The final structure contains two phases—ferrite and cementite—arranged as two microconstituents: primary ferrite and pearlite (Fig. 13.5). The pearlite exists as "islands" surrounded by ferrite; this permits the alloy to be ductile, due to the continuous primary ferrite, yet strong, due to the discontinuous, dispersion-strengthened pearlite.

13.3 STRENGTHENING MECHANISMS

The use of steel as a structural material is based, in large part, on its strength and ductility. Both of these properties are related to the movement of dislocations through the crystal lattice of the material (see Chapters 2 and 5). Measurable plastic deformation, or yielding, requires that there be an applied stress and that two other processes occur consecutively: (1) Dislocations must be generated at some type of dislocation source, and (2) these dislocations must move appreciable distances through the crystal. The mechanism, either dislocation generation or dislocation movement, that requires the higher stress controls the yielding behavior, and thus the mechanical properties, of the steel.

Strengthening of a metal to increase its resistance to yielding or plastic deformation can be obtained by changes in microstructure that impede the motion of dislocations. Structural steels can be strengthened through mechanisms such as (1) the introduction of interstitial and substitutional atoms (alloying), (2) the generation and concentration of dislocations (work or strain hardening), and (3) the formation of additional grain boundaries (heat treatment). Alloying and heat treatment processes combine synergistically to produce a tremendous variety of microstructures, and hence properties, in structural steels.

13.3.1 Alloying

Alloying elements affect the microstructure and, therefore, the properties of steel through the following mechanisms:

1. Formation of a solid solution with iron, resulting in solid-solution strengthening and increased corrosion resistance (carbon, chromium, manganese, nickel, copper, and silicon).

2. Formation of a carbide (i.e., a binary compound of carbon and a more electropositive element); such carbides impart additional hardness and elevated temperature strength (titanium, vanadium, and molybdenum).

3. Formation of an undissolved, second phase which promotes machineability (lead, sulfur, and phosphorus).

In addition, all alloying elements promote hardenability (a measure of the capability of a steel to harden throughout the cross section of a bar or component during heat treatment) and facilitate the nonequilibrium transformation of austenite to microstructural products other than pearlite. This may be beneficial in high strength steels, but it can cause problems in welding.

Elements which readily form oxides (aluminum, silicon, calcium, and manganese) are also added to steel to remove dissolved gaseous oxygen. Otherwise, this dissolved oxygen will come out of solution on cooling to form small bubbles in the solid steel. The chemical composition and mechanical properties of deoxidized or 'killed' steels are more uniform than those of other types.

The raw materials used to make steel contain other elements which are undesirable but appear in the resulting products. The amounts of these residual elements are usually held to acceptable limits by careful steel-making practice, and they are not generally reported. However, certain of these elements (sulfur and phosphorus, for example) negatively affect the mechanical properties of steel. Hence restrictions are placed on the amounts of such elements allowed in most grades of steel, and these amounts are normally reported in the analysis of both plain carbon and low-alloy steels.

The strengthening and hardening that result from a given alloying element in solid solution depend on the difference between its atom size and electron structure and those of the solvent metal. In dilute solid solutions, the amount of strengthening is roughly proportional to the concentration of the alloying element. If more than one solute element is present, the total strengthening is approximately the sum of the characteristic effects of each element taken alone.

Solute atoms that are simply distributed at random in the solvent impede the motion of dislocations to a limited extent, and therefore impart only modest strengthening. However, if the solute atoms collect preferentially around existing dislocations, these dislocations are effectively locked and the force required to move them may be increased greatly. Both substitutional solid-solution atoms, which are either appreciably larger or smaller than the matrix atoms, and interstitial atoms can lock dislocations. Carbon and nitrogen are interstitial alloys that exert a pronounced strengthening effect on iron in this way.

The properties of the steel for a given alloying composition can be further modified by changing the microstructure by a variety of mechanical treatments and heat treatments which involve solid-state transformations. These are discussed in the following sections.

13.3.2 Work (Strain) Hardening

All pure metals and alloys can be strengthened by cold working, or plastic deformation below the recrystallization temperature. Recrystallization is a process in which new strain-free grains are nucleated and grow. It occurs when the metal or alloy is heated above a certain temperature—the recrystallization temperature. Because the recrystallization temperature is several hundred degrees above room temperature for steels, shaping or working at and near ambient temperatures is cold-working. In this process, excessive concentrations of dislocations are generated during initial straining (working), and a complex network of dislocations is produced. This network makes further dislocation motion extremely difficult because of mutual interference. The resistance to further deformation increases with increasing amounts of deformation, a phenomenon called strain-hardening, or work-hardening. Thus, easy slip in the crystal lattice is reduced and the alloy is strengthened.

However, the ductility and toughness of the metal or alloy decrease when it is work- or strain-hardened (Fig. 13.6). Materials strengthened appreciably in this way, therefore, tend to have low ductility. Furthermore, materials strengthened by work-

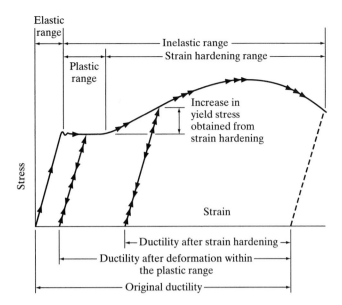

Figure 13.6
Change in mechanical
properties associated with
work- or strain-hardening.

hardening cannot be joined by welding without softening (annealing) the material in the vicinity of the weld. For these reasons, this method of strengthening finds little application in structural steels.

13.3.3 Heat Treatment

Structural or mild steels are usually employed in the 'as-rolled' or non-heat-treated condition. However, to obtain certain properties, it may be necessary to alter the microstructure of the material from that found in this condition by subjecting the steel to various heat treatments. All of these processes are directed toward producing the mixture of ferrite (α) and cementite (Fe_3C) that gives the proper combination of properties.

In general, a heat treatment may be defined as an operation or series of operations involving the heating and cooling of a metal or alloy in the solid state to produce desirable conditions or properties. Of primary interest, insofar as structural steels are concerned, are changes which occur on cooling of the material from some point above one of the critical temperatures (i.e., A_1 or A_3; Fig. 13.4). The temperature to which the material is heated before cooling is begun determines the phases which are present at the end of the cooling process, as well as at the beginning, whereas the form of these phases (the microstructure of the heat-treated steel) is determined by the cooling temperatures and rates.

Isothermal Heat Treatments

The solid-state eutectoid reaction resulting in the formation of ferrite (α) and cementite (Fe_3C) from austenite (γ) is rather slow, and the steel may cool below the equilibrium eutectoid temperature (A_1) before this transformation is complete. If the transformation occurs at temperatures below A_1, a nonequilibrium structure will form, and strengthening will probably result.

One convenient method of describing the nonequilibrium transformation of austenite during cooling is the isothermal transformation diagram, referred to more commonly as the time-temperature-transformation diagram, or TTT diagram

(Fig. 13.7(a)). This diagram shows the results of transforming austenite isothermally (i.e., at constant temperature) at temperatures below the upper critical temperature (the temperature at which the alloy under consideration is fully austenitic—A_3 for hypoeutectoid steels).

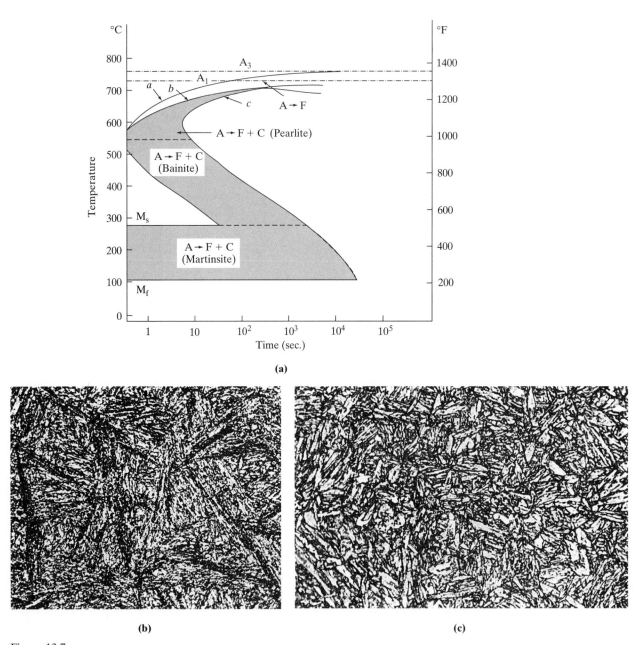

(a)

(b)   (c)

Figure 13.7

(a) Isothermal transformation (TTT) diagram for a hypoeutectoid steel (A, austenite; F, ferrite; and C, iron carbide [Fe_3C]): Transformation of austenite starts at curve *a*, and is complete at curve *c*; (b) photomicrograph showing bainitic structure in a hypoeutectoid steel (×750); (c) photomicrograph showing martensitic structure in tempered hypoeutectoid steel (×750) (P. A. Thornton and V. J. Colangelo, *Fundamentals of Engineering Materials,* Prentice-Hall, Inc., 1985. Photomicrographs courtesy of Theresa Brassard).

In Fig. 13.7(a), the shaded portion indicates the region where transformation is occurring as a function of time for a particular (constant) temperature. For example, at a temperature above 550°C (the nose of the TTT curve), transformation of the austenite to ferrite begins at curve *a* and continues isothermally until curve *b*. Austenite then begins to transform to pearlite and continues this reaction until curve *c* is reached. At *c* the austenite is completely transformed and a microstructure of ferrite and pearlite results.

At temperatures below the nose of the TTT curve, transformation of austenite produces a structure in which the ferrite and iron carbide are not lamellar, as in pearlite. This transformation product, called bainite, exhibits a fine, feathery or acicular (needlelike) structure (Fig. 13.7(b)). Depending on the actual transformation temperature, this nonequilibrium transformation product may have a higher strength and hardness than pearlitic structures while maintaining usable ductility and toughness.

Transformation of austenite at temperatures lower than those in the bainite range occurs as a function of temperature only; transformation is independent of time. This process involves a diffusionless transformation which starts at a temperature labeled M_s and produces a nonequilibrium or transition product called martensite, which has a fine grain size and is very hard and brittle (Fig. 13.7(c)). However, this phase is not stable and, given an opportunity, will proceed to form ferrite and iron carbide; as a result, we do not see martensite on the Fe—Fe$_3$C phase diagram. Normally, steels containing martensite are further heat treated (tempered) to produce more desirable microstructures and properties.

Continuous Cooling Heat Treatments

In practice, structural steels are strengthened through microstructural transformations which take place under continuous cooling conditions rather than isothermal conditions. In continuous cooling heat treatments, the material is cooled from above the A$_3$ temperature to room temperature continuously; the transformation from austenite to pearlite and/or martensite therefore occurs over a range of temperatures instead of at a single temperature, resulting in a more complex microstructure.

The products formed by continuous cooling procedures can be predicted using a continuous cooling diagram (Fig. 13.8), which differs from the isothermal transformation diagram in that the beginning and the end of the transformation are generally shifted to lower temperatures and longer times. Different rates of cooling, and hence different microstructures, are achieved through the use of different cooling media, as shown in Fig. 13.8. Essentially, as the cooling rate is increased, the hardness and the strength of the steel increase.

Rapidly cooling, or quenching, the austenite at a rate equal to or greater than the critical cooling rate (Fig. 13.8) will bypass the knee of the TTT curve and result in the formation of martensite. However, martensite is too hard and brittle for most purposes, and it is usually transformed to the equilibrium phases (α and Fe$_3$C) using a low-temperature heat treatment called tempering. The resulting microstructure is not lamellar, like that of pearlite, but contains many dispersed carbide particles. Tempering causes the strength and hardness of the martensite to decrease, while the ductility and impact properties are improved. By selecting the appropriate tempering temperature, a wide range of properties can be obtained.

Heat treatment processes involving slower cooling rates are primarily carried out to reduce or eliminate the undesirable effects of prior treatment of the steel. Two processes are in common use: normalizing and annealing (Fig. 13.9). In the

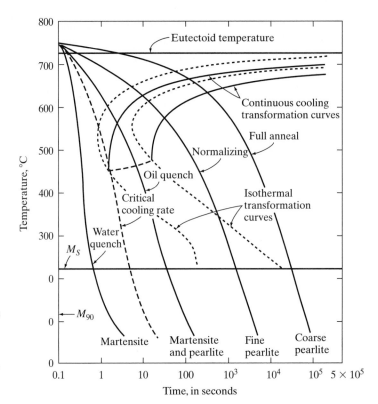

Figure 13.8
Variation of microstructure as a function of cooling rate for an eutectoid steel (after R. E. Reed-Hill, *Physical Metallurgy Principles*, D. Van Nostrand Co. Inc., 1964).

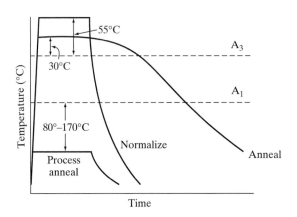

Figure 13.9
Schematic representation of heat treatment processes for a hypoeutectoid steel.

normalizing process, the material is heated to a temperature above the austenitizing temperature and, after a suitable holding time, cooled in still air, forming a fine pearlite microstructure. The principal effect of normalizing is that grain refinement occurs (i.e., grain size decreases), resulting in improved ductility and toughness. Furthermore, since recrystallization of the material occurs in this process, variations in the mechanical properties, caused by differential cooling rates following rolling cycles, will be reduced or eliminated. Thus, these properties will be more uniform, within a piece, and from piece to piece, than is the case with as-rolled steel.

In the full annealing process, the material is heated to a temperature above the austenitizing temperature and then slowly cooled in the furnace with the following effects: (1) The material is softened and its ductility is improved; (2) internal stresses induced, for example, by strain or work hardening, are relieved; and (3) recrystallization occurs. However, a partial annealing process, called process annealing or stress relieving, is more commonly used than full annealing: The steel is heated to a temperature below, but close to, the austenitizing temperature, followed by any desired rate of cooling. Process annealing is primarily carried out to relieve internal stresses resulting from work- or strain-hardening, but recrystallization may also occur.

13.4   MECHANICAL PROPERTIES

Various types of steel are available to meet specific service conditions (e.g., improved resistance to atmospheric corrosion). Selection of an appropriate grade of steel, or combination of type and strength level, has therefore become a primary consideration in the design of structural steel members.

Although there is some variability in their mechanical properties, structural steels are generally more uniform than most other construction materials. Variations in properties occur both within and between pieces because of differences in chemical composition (due to segregation of certain elements during the casting and solidification processes) and in mechanical and thermal treatment during the manufacturing processes (degree of reduction during rolling, cooling rates during heat treatment, etc.). The distribution of tensile properties for a particular structural steel (ASTM A2BS, grade C carbon steel plate) is shown in Figure 13.10. This type of variability has been recognized by the producers and users of structural steels and is taken into account in the design process.

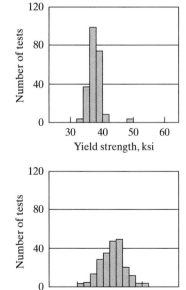

Figure 13.10 Distribution of tensile properties of ASTM A285, grade C carbon steel plate (Data represent all the as-rolled plate (224 heats for 6 steel mills) purchased to this specification by one Fabricator during a period of 8 years). (after Metals Handbook (10th ed.). Vol. 1 Properties and Selection: Irons, Steels, and High-Performance Alloys, ASM International, 1990, p. 196).

13.4.1 Stress-Strain Behavior

Under normal conditions, the response of structural steels to applied load or stress can be described as being ductile, in that substantial plastic deformation occurs before fracture, and independent of time. This type of behavior has been described in general terms in Chapter 5, but the particular aspects of the behavior of these materials are considered here.

The following characteristics of the behavior of a steel are shown by the stress-strain curve for uniaxial tensile loading conditions (Fig. 13.11):

1. *Elastic deformation.* The first section of the curve indicates that the initial behavior is linearly elastic (i.e., the relationship between stress and strain is linear and the strain is completely recovered when the stress is removed). This section is characterized by the following parameters: elastic modulus, the proportional limit (the highest value of stress at which the stress-strain relationship is linear), and the yield strength (the stress at which plastic or permanent strain, or yielding, is considered to begin). The yield strength is the most important criterion for engineering design, since it marks the normal design limit or useful engineering strength. However, for most materials, the stress at which yielding occurs is not readily apparent from the stress-strain curve, and several more or less arbitrary criteria or indicators of yield strength are commonly used (e.g., specified offset strain) (Fig. 13.11(a)). Plain carbon or mild steels have a double yield point (Fig. 13.11(b)), since in these steels higher stress is required to start yielding than to continue it once it has started.

2. *Uniform plastic deformation.* In this section of the curve, the strain in the specimen is permanent and its cross section is decreasing uniformly. In some materials, yielding does not begin gradually, but comes suddenly and results in a large plastic deformation at a more or less constant stress; this phenomenon is known as "discontinuous yielding" (Fig. 13.11(b)). If an ever-increasing stress is required to produce further yielding or strain, the process is known as strain hardening. This section of the curve is characterized by the ultimate strength, or maximum stress that the material can withstand.

3. *Localized plastic deformation (necking).* At the maximum load point, a segment of the specimen that is weaker than all other segments along its length deforms more and decreases in cross section more, leading to a condition of plastic instability. This condition is thus concentrated in a localized section of the specimen and is known as necking. It ends when fracture or failure of the specimen occurs.

13.4.2 Fracture Energy (Toughness)

As noted in Chapter 6, fracture energy (toughness) is the ability of the material to absorb energy, in particular under suddenly imposed stresses (impact conditions), by deforming plastically prior to fracture. Toughness is not normally considered directly in engineering design, but it is an important property in the selection of a structural steel for use under loading and/or environmental conditions which may result in brittle fracture.

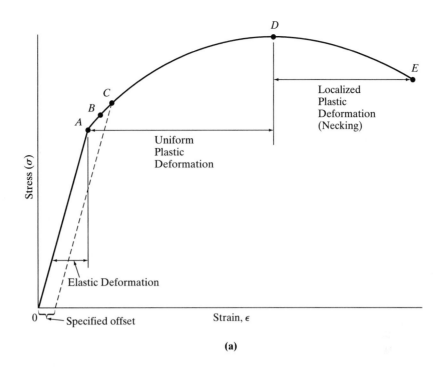

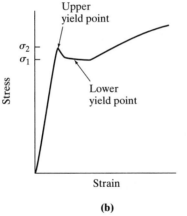

Figure 13.11
(a) Typical engineering stress-strain curve for a ductile material; (b) discontinuous yielding.

The toughness of a material can be evaluated in several ways:

1. As the area under the stress-strain curve in the uniaxial tension test. This method is not normally adequate, because it is not representative of material behavior under brittle fracture conditions.

2. Through the use of impact tests (i.e., Charpy or Izod notched-bar tests). This method is preferable to the preceding test because the test conditions are more representative of those resulting in brittle fracture, although the tests are essentially empirical in nature.

3. Through the use of fracture mechanics tests which provide a measure of the stress or energy required to propagate a sharp crack in the material.

In practice, the Charpy V-notch impact test (Fig. 6.15) is the method most commonly used to assess the toughness of structural steels. In particular, this test is used to determine the transition temperature where the failure behavior of a steel changes from ductile to brittle (Fig. 6.14).

If a steel is to resist brittle fracture under impact loading, it must have the ability to absorb considerable energy whenever stress concentrations occur. Brittle fracture can occur at stress levels below normal design levels when steel is subjected to impact loads at comparatively low temperatures. Under such conditions, steel undergoes a transition from ductile (high-energy absorption) fracture behavior at the higher temperatures to nonductile, or brittle (low-energy absorption) fracture behavior at lower temperatures. Different steels vary widely in the temperature at which the transition from ductile to brittle fracture occurs.

High energy-absorption values are associated with a ductile, shear-type, fibrous-appearing fracture with plastic deformation; low energy-absorption values are associated with a brittle, granular-appearing fracture; and in the transition zone, the fracture surface of the specimen exhibits a mixed mode (Fig. 13.12).

For design against brittle fracture, three factors must be considered simultaneously: (1) the presence of stress raisers, or notches, in the structural member; (2) the stress conditions, particularly tensile stresses, in the structure; and (3) the notch toughness of the steel at the operating temperatures. Good design can eliminate geometrical discontinuities and hence minimize the detrimental effect of notches or stress raisers and reduce tensile stresses that must be present for brittle fracture to

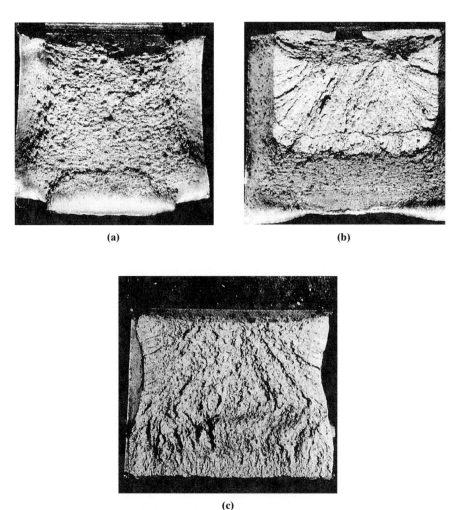

(a)   (b)

(c)

Figure 13.12
Typical Charpy V-notch fracture surfaces: (a) completely brittle fracture; (b) mixed mode fracture (part ductile, part brittle); (c) completely ductile fracture. (F. Larson, U.S. Army Materials Research Agency).

occur. Good notch-toughness behavior of the steel is promoted by (1) low carbon content, (2) absence of small voids, (3) grain size reduction, and (4) elimination of internal stresses by heat treatment.

13.4.3 Weldability

Welding is now the major fabrication method used to join sections or pieces of steel to form civil engineering structures. Fusion welding is the best and, hence, the most commonly used of the different welding processes available. In this process, part of the parent metal of each piece being joined is heated to a suitable temperature above the recrystallization or melting range, with or without the application of pressure and/or the introduction of filler metal, and solidified to form a joint. Fusion processes which involve heating by an electric arc are commonly used to fabricate and join structural members. In these processes, the filler electrode composition should be chosen to give a joint with satisfactory mechanical and corrosion properties.

During welding, a temperature gradient is established which varies from the fusion temperature at the weld metal to room temperature at some point in the parent metal away from the weld (Fig. 13.13). This produces changes in the metallurgical condition and properties of the parent metal in the vicinity of the weld, the so-called heat-affected-zone (HAZ). The microstructural changes in this region can be considerable and are generally accompanied by a deterioration of its mechanical properties. These changes depend on (1) the composition of the parent metal, (2) its original condition, and, possibly, (3) the cooling rate after the weld.

Weldability may be defined as the capacity of a metal to be joined by welding into a structure that can perform in a satisfactory manner for an intended service. The weldability of a steel is a complex property; it covers both the sensitivity to weld

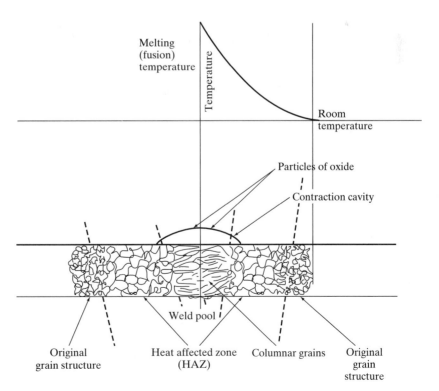

Figure 13.13
Schematic diagram of a fusion weld, showing temperature distribution and corresponding changes in grain structure in the heat-affected-zone (HAZ) in the parent metal (adapted from J. M. Illston, J. M. Dinwoodie, and A. A. Smith, *Concrete, Timber, and Metals, the Nature and Behaviour of Structural Materials,* Van Nostrand Reinhold Co. Ltd., 1979).

cracking and the toughness in both the weld and the HAZ required by service conditions and temperatures.

Weldability decreases as the carbon and alloy content of steel is increased. A convenient way to assess weldability and to evaluate the effect of alloying elements on weldability is to use a "carbon-equivalent" (CE) formula, such as

$$CE = \%C + (1/6)(\%Mn) + (1/5)(\%Cr + \%Mo + \%V) \qquad (13.1)$$
$$+ (1/15)(\%Cu + \%Ni),$$

where %Mn, etc., are the percentages of manganese, etc., in the steel being welded. The value of CE should not exceed about 0.25 for heavy structural steels; if it is larger than this, controlled cooling of the weld is necessary to avoid risks of embrittlement.

Welding has a very strong influence on fatigue strength due to associated stress concentrations which result from the design of the welded joint, surface effects, slag inclusions or voids in the weld, mechanical properties of the weld metal and/or heat-affected zone, and residual stresses. When different geometries and types of welded joints are compared on steels with different strength levels, it is usually found that the fatigue strength is more dependent on the geometry of the joint, the type of weld, and the freedom from notch effects than on the strength level of the steel.

13.5   CORROSION AND CORROSION PROTECTION

13.5.1 Corrosion Mechanism

Corrosion is the destructive attack, or deterioration, of a metal by chemical or electrochemical reaction with its environment. Corrosive attack of metals is an electrochemical process that is represented in Fig. 13.14.

In a galvanic cell, two dissimilar metals (e.g., iron and copper) are placed in electrical contact in the presence of oxygen and moisture. Separate chemical reactions take place at the surfaces of the two metals, creating a flow of electrons through the connecting wire. At the iron surface, or anode, oxidation of iron takes place in accordance with the following anodic reaction:

$$2Fe - 4 \text{ electrons} \rightarrow 2Fe^{++}. \qquad (13.2)$$

At the copper surface, or cathode, reduction of oxygen occurs in accordance with the following cathodic reaction:

$$O_2 + 2H_2O + 4 \text{ electrons} \rightarrow 4OH^-. \qquad (13.3)$$

The actual loss of metal involved in the process takes place at the anode, as indicated by Eq. 13.2. The iron atoms are transformed to ferrous ions (Fe^{++}) which dissolve in the solution around the anode. They may diffuse and combine with the hydroxyl ions (OH^-), with the precipitation of ferrous hydroxide [$Fe(OH)_2$] in accordance with the following net redox reaction:

$$2Fe + O_2 + 2H_2O \rightarrow 2Fe(OH)_2. \qquad (13.4)$$

The hydrous ferrous oxide formed according to Eq. 13.4 ($FeO \cdot H_2O$) is further oxidized to form hydrous ferric oxide ($Fe_2O_3 \cdot nH_2O$), which is rust.

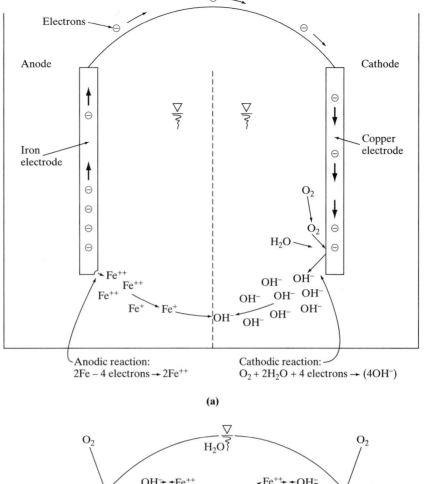

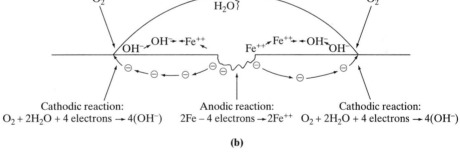

Figure 13.14
Schematic representation
of the corrosion process:
(a) simulation by a galvanic
cell, (b) corrosion on metal
surface exposed to a humid
environment.

Iron will corrode without the presence of a separate cathodic metal. Anode-cathode pairs can be set up on a steel surface where different sites have different electrochemical potentials or tendencies for oxidation. An electrical potential difference between possible anode and cathode sites can be the result of differences in composition, differences in residual strain, or differences in oxygen or electrolyte concentrations in contact with the surface.

13.5.2 Forms of Corrosion

1. *General corrosion:* General corrosion or rusting is the most familiar form of steel corrosion. It can be considered a uniform corrosion process in which numerous microcorrosion cells are activated at the corroded area. The cells could be minute grains where the boundary tends to be the anode, for example.

In atmospheric exposures, oxygen in the air is the usual oxidizing agent, and the water necessary for the reaction is readily available in the form of rain, condensation (dew, for example), or humidity (water vapor in the air). In the rusting of ordinary steel, the corrosion product (rust) does not form an effective barrier to further corrosion, but permits reactants to penetrate to the steel surface beneath and continue the rusting cycle.

2. *Pitting corrosion:* This is a nonuniform, highly localized form of corrosion that occurs at distinct spots where deep pits form. (A pit is a small electrochemical-corrosion cell, with the bottom of the pit acting as the anode.) Chloride-induced corrosion is of this type and can be seen frequently in structures exposed in coastal areas.

3. *Galvanic corrosion:* When two metals of different electrochemical potential are joined or coupled electrically in the presence of moisture or an aqueous solution, one will act as the anode and corrode; the corrosion of steel when it is in contact with copper is a familiar example. This principle is used to advantage when steel is protected by galvanic methods (for example, galvanized steel or the use of other surficial anodes).

4. *Stress-corrosion:* Under stress, corrosion processes proceed much faster and can lead to brittle failure as corrosion tends to be localized. Corrosion of this kind can occur in prestressing tendons in concrete.

5. *Crevice corrosion:* This form occurs when moisture and contaminants retained in crevices accelerate corrosion.

13.5.3 Corrosion Control

For most applications of structural steel, some form of corrosion control is essential, as discussed next.

Protective Coatings

Paint applied to steel functions as a barrier between the steel and the atmosphere, thereby preventing attack as long as the coating is intact. Epoxy coatings on reinforcing bars serve the same purpose, but may not perform as expected due to defects in the coating.

Galvanic Protection

Hot-dip galvanizing is a process in which an adherent, protective coating of zinc or zinc-iron compounds is developed on the surfaces of iron and steel products by immersing them in a bath of molten zinc, whereas metallizing involves the application of zinc onto the steel surface by means of a flame spray gun. The usefulness of zinc coatings as corrosion protection depends on (1) the barrier effect of zinc and its surface oxide film, (2) the relatively low rate of corrosion of zinc as compared with that of iron or steel, and (3) the electrolytic, or sacrificial, protection afforded to iron by zinc (i.e., the preferential oxidation of zinc at normal atmospheric temperatures, which acts as the anode relative to the steel).

Cathodic Protection

This method is used for structures located below ground or immersed in water, usually in conjunction with a protective coating. Because corrosion results from, or is accompanied by, a flow of electrical current between anodic and cathodic surfaces, it is possible to reduce or eliminate it by controlling the magnitude and direction of current flow. By reversing the current to the original anodic steel surface, the steel is made a cathode and does not corrode. A reverse-current flow is obtained either by electrically connecting the steel structure to a metal of higher electromotive energy (commonly zinc or magnesium, in the form of sacrificial anodes) or by artificially impressing a direct current from an outside source (for example, a power line and a rectifier). Effective protection is provided as long as the proper reverse-current flow is maintained. A protective coating, such as asphalt, tar, or an epoxy, is commonly applied to the structure to reduce power consumption.

Corrosion-resistant Steels

These steels contain a combination of alloying elements selected to provide a special type of oxide coating after prolonged exposure to the atmosphere (weathering). They usually contain copper and develop a resistance to atmospheric corrosion from four to eight times that of a plain-carbon steel. In addition to copper, phosphorus, chromium, nickel, and silicon are among the elements added (usually in combination) to achieve this special corrosion resistance. On exposure to the atmosphere, these steels gradually develop a tightly adhering oxide coating that acts as a barrier to moisture and oxygen and eventually almost prevents further corrosion. Furthermore, if this coating is damaged, it will heal itself.

13.6   CLASSIFICATION AND PROPERTIES OF STRUCTURAL STEELS

Steel is employed as the structural frame in various forms of construction (bridges, buildings, etc.), as well as in reinforced portland cement concrete structural members. The composition and engineering properties of the steels used in these diverse applications vary substantially. Although a detailed examination of all structural-quality steels is beyond the scope of this text, the main grades, types, and classes of these steels will be briefly described.

As noted previously, structural steels are classified on the basis of strength (class or grade) as well as composition (type). For example, the composition and selected properties of some (1) structural carbon steels (ASTM A-36M), (2) high-strength, low-alloy manganese vanadium steels (ASTM A-441/A-572M), and (3) high-yield-strength, quenched and tempered alloy steels (ASTM A-514M) are provided in Table 13.2 and Fig. 13.15. The differences in strength and ductility among the various types can be significant, and the selection of an appropriate steel should be based on the end use of the structural element.

Steels used as reinforcing elements in portland cement concrete structural members are available in several forms and with a variety of properties. Distinction must be made between (1) plain and deformed reinforcing bars (Fig. 13.16(a)), (2) wire for welded wire fabric (Fig. 13.16(b)), and (3) bar, wire, and strand (Fig. 13.16(c)), for prestressing. Gradings and selected properties are presented in Table 13.3, and typical stress-strain curves are shown in Fig. 13.17. The time-dependent behavior of steels is of interest in prestressed members, where relaxation of the steel reduces its

TABLE 13.2 Properties and Composition of Selected Structural Steels

Type	Chemical Composition, % wt.					Mechanical Properties		
	C (max)	Mn	P (max)	S (max)	Si	Yield Strength (min)(MPa)	Tensile Strengh (MPa)	Elongation (%)
Carbon[a]	0.25–0.29[d]	0.60–1.20[d]	0.04	0.05	0.15–0.40	250	400–550	20–23[d]
High strength low alloy[b]	0.21–0.26[d]	1.35–1.65[d]	0.04	0.05	0.15–0.40[d]	290–450[d]	415–530[d]	15–24[d]
Quenched & tempered alloy[c]	0.10–0.21[d]	0.40–1.50[d]	0.03	0.04	0.15–0.80[d]	620–690[d]	690–895[d]	18

[a] ASTM A-36M.
[b] ASTM A-572M.
[c] ASTM A-514M.
[d] Range of values, depending on shape and size of structural element.

TABLE 13.3 Properties of Selected Steels for Concrete Reinforcement

Type of Reinforcement	Grade	Yield Strength (min) MPa	Tensile Strength (min) MPa
Bars[a]	300	300	500
	400	400	600
	500	500	700
Steel wire for welded wire fabric	Smooth[b]	485	550
	Deformed[c]	485	550
Prestressing bars[d]	Plain	880	1035
	Deformed	825	1035
Seven-wire strand[e]	250		1725
	270		1860

[a] ASTM A-615M.
[b] ASTM A-82.
[c] ASTM A-496.
[d] ASTM A-722.
[e] ASTM A-416.

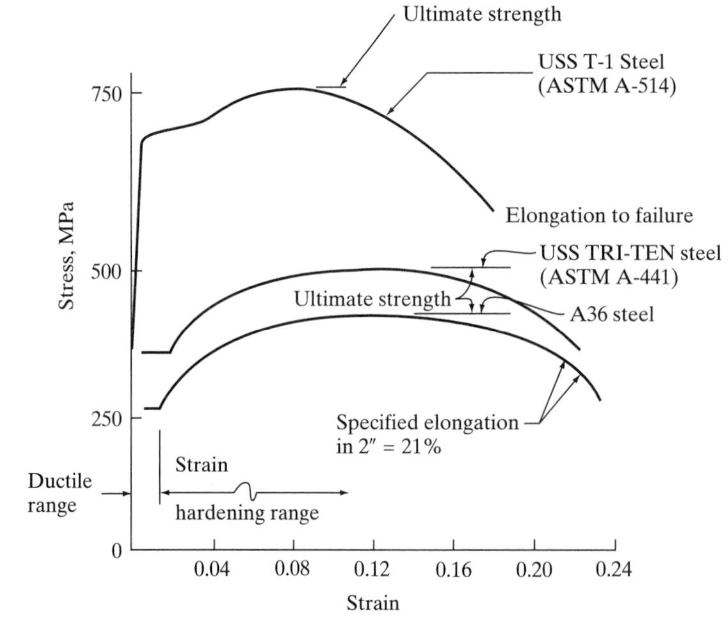

Figure 13.15
Stress-strain curves of selected structural steels (from S. E. Cooper, *Designing Steel Structures*, Prentice Hall, 1985, p. 63, adapted from *Steel Design Manual*, 1981, U.S. Steel Co.).

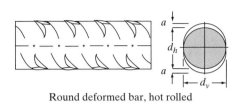

Round deformed bar, hot rolled

(a)

(b)

Figure 13.16
Shape of selected steels
used for concrete rein-
forcement: (a) deformed
bars, (b) welded wire fabric
(Courtesy of Tree Island
Industries, New Westmin-
ster, B.C.) (c) wire strand
for prestressing

(c)

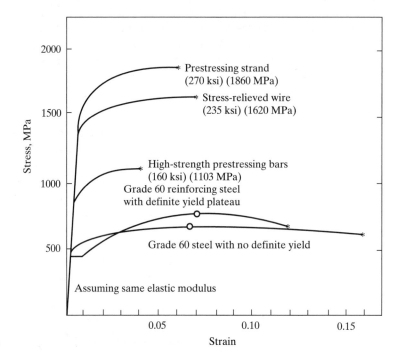

Figure 13.17
Stress-strain curves for selected steels for concrete reinforcement (after A. E. Naaman, *Prestressed Concrete Analysis and Design,* McGraw-Hill, 1982, p. 34).

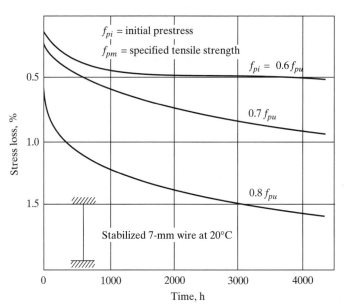

Figure 13.18
Stress relaxation in pretensioned steel for prestressing (after A.E. Naaman, *Prestressed Concrete Analysis and Design,* McGraw-Hill, 1982, p. 34).

stress level with time, as shown in Fig. 13.18. Note that the stress loss is greater for higher prestress levels (see Chapter 7).

Attention also should be given to the behavior of structural steels in service. Corrosion under general atmospheric conditions was discussed in Section 13.5, and in portland cement concrete in Section 12.5. In addition, performance at high temperatures with respect to fire safety is important. Above 300°C there is a tendency for some loss in strength (Fig. 13.19) and stiffness, which may become very signifi-

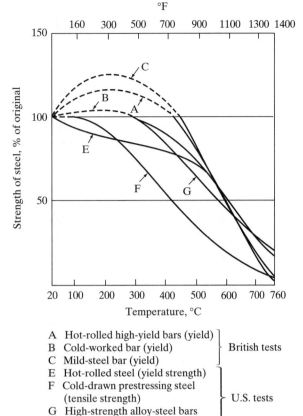

Figure 13.19
Effect of temperature on the strength of various types of steels (after B. K. Bardhan-Roy, "Fire Resistance-Design and Detailing," Chapter 14 in *Handbook of Structural Concrete,* F. K. Kong, R. H. Evans, E. Cohen, and F. Roll, Eds., Pitman Advanced Publishing Program, 1983, pp. 14–16)

A   Hot-rolled high-yield bars (yield)
B   Cold-worked bar (yield)   } British tests
C   Mild-steel bar (yield)

E   Hot-rolled steel (yield strength)
F   Cold-drawn prestressing steel
     (tensile strength)   } U.S. tests
G   High-strength alloy-steel bars
     (Tensile strength)

cant above 600°C. Although the strength loss may be recovered (depending on the level of temperature) on cooling, one should consider the load-carrying capacity of the steel in the high-temperature condition with respect to its ability to support the loads imposed on the structure without failing while the conditions of high temperature exist. Excessive deformation in a steel structure exposed to fire may be aggravated by the reduction in modulus of elasticity at high temperature.

BIBLIOGRAPHY

D. T. LLEWELLYN, *Steels: Metallurgy and Applications.* Butterworth-Heinemann Ltd., 1992

Manual of Steel Construction, (8th Edition), American Institute of Steel Construction, 1980

Metals Handbook Volume 1, Properties and Selection: Irons, Steels, and High-Performance Alloys, (10th Edition), ASM International, 1990.

PROBLEMS

1. What is a structural steel? What is the difference between a plain carbon and a low alloy structural steel?
2. What is the effect of an increase in carbon content from 0% to 1% on the (i) iron carbide (cementite, Fe_3C) content, (ii) strength and (iii) toughness of an annealed, plain carbon steel?
3. Identify and describe the two microstructural constituents which make up a typical structural steel. How is each of these microconstituents formed? How does each affect the mechanical properties of the steel?
4. What is a "killed" steel and how is it produced?
5. What is the mechanism by which carbon and nitrogen strengthen low alloy steels?
6. Briefly describe the strengthening mechanism resulting from work-hardening (strain-hardening). Why is this method of strengthening not commonly used for structural steels?
7. How are bainite and martensite formed? What is the difference in the properties of these two phases? Why are these phases not shown on the iron-iron carbide phase diagram?
8. What differences in microstructure and mechanical properties are produced by quenching, normalizing, or fully annealing a sample of structural steel?
9. Explain the differences between (i) elastic and plastic deformation, (ii) proportional limit and yield strength, and (iii) discontinuous yielding and strain-hardening.
10. How can the brittle fracture of structural steel members be prevented?
11. What is "weldability" and how is it affected by increasing carbon and alloy contents in a structural steel?
12. What conditions may result in the formation of a galvanic microcell on the surface of a structural steel?
13. Identify and briefly describe the most common form of steel corrosion.
14. What factors influence the usefulness of zinc coatings as a form of corrosion protection for structural steels?

14

Wood and Timber

14.1 INTRODUCTION

Wood is a naturally occurring, biological material. It is probably the world's oldest structural material, and because of the relative ease with which it can be produced and handled, it was until fairly recently by far the most widely used building material. Even now, the annual production of wood, about 10^9 tonnes (metric tons), is about equal to the world production of iron and steel and is comparable to the world production of concrete ($\sim 4 \times 10^9$ tonnes).

The fact that wood is relatively weak compared to structural metals or concrete can be misleading. If we compare the structural properties of wood and other materials on a *per unit weight* basis (i.e., *specific* properties), wood comes out very well, as shown in Table 14.1. Since wood is also considerably cheaper than other structural materials (and is to many people more aesthetically pleasing), its popularity as a structural material should come as no surprise. Wood is, of course, subject to fire and to decay by various forms of biological attack (insects, rot, etc.). Nonetheless, if properly maintained, it can last for a very long time.

Wood is probably more complex than the other materials discussed in this book (which are manufactured, rather than natural), for a number of reasons:

1. There are at least 30,000 species of trees, and this alone leads to a tremendous variation in properties.
2. Wood is a composite material, made up of a large variety of structural units of very different sizes and with distinct properties. Thus, the properties of wood (like those of portland cement concrete) must be examined at different levels: molecular, microscopic and macroscopic.

309

TABLE 14.1 Specific Properties of Some Structural Materials

Materials	E/ρ	$\sigma_{tensile}/\rho$	$\sigma_{compressive}/\rho$	K_{IC}/ρ
Wood	20–30	120–170	60–90	1–12
Mild steel	26	30	—	18
Aluminum alloys	25	180	130	8–16
Concrete	15	3	30	0.08

3. Because it is a naturally occurring material, wood contains many flaws or imperfections, of different sizes and degrees of severity, which tend to control its structural behavior.

4. Perhaps the greatest difference between wood and most other structural materials is the degree of *anisotropy* of wood. The other structural materials examined in this book are largely isotropic, particularly at the macroscopic level. That is, their bulk mechanical properties are more or less independent of specimen orientation. However, because of the way in which trees grow, wood develops a remarkably anisotropic structure. Its mechanical properties, measured in different orientations with respect to the direction of tree growth, may easily vary by a factor of at least 20.

The terms *wood* and *timber* are often used interchangeably. However, in this chapter, we will give them each specific meanings. *Wood* will refer to small, clear specimens, which are free of any macroscopic defects. That is, wood is the basic material which we obtain from trees. Wood specimens are used to study the relationships between microstructure and properties, and they are used for other fundamental investigations. Timber, on the other hand, will be used to describe the normal sawn structural members which you might buy in a lumberyard. Such timber will, in general, contain a bewildering number of macroscopic defects (cracks, knots, etc.), of different shapes, sizes, and orientations. The structural properties of timber are governed as much by these macroscopic features as by the underlying structure of the wood.

The combination of anisotropy and numerous large defects provides a limitation to the use of wood as a structural material. Many attempts have been made to overcome these problems, through the development of *manufactured wood products,* such as plywood, glulaminated members, and other such products.

In this chapter, we will first examine the structure and properties of wood. We will then turn our attention to timber, and finally to some of the manufactured wood products. Physical and mechanical properties and durability will be considered.

14.2 THE STRUCTURE OF WOOD

A tree may be considered to consist of three distinct parts, as shown in Fig. 14.1:

1. The *root system* anchors the tree to the ground and absorbs moisture and minerals from the ground for transfer through the trunk to the crown.

2. The *trunk* provides strength and rigidity and holds the crown above the ground. It serves to transport both moisture and minerals up from the roots and sap down from the crown.

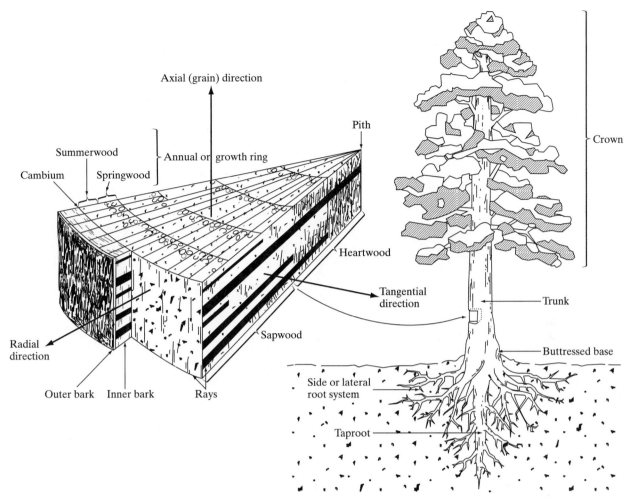

Figure 14.1
Schematic illustration of the parts of a tree and of the structure of wood.

3. The *crown,* consisting of leaves and their supporting branches, produces both food for the tree and seed. The leaves contain chlorophyll. In the photosynthetic process that occurs there, light breaks up CO_2 from the air, releasing O_2; the carbon combines with the sap to form sugar, cellulose, and a host of carbohydrates essential to the growth of the tree.

For engineering purposes, however, we will focus on the trunk of the tree, for it is the trunk that provides the structural timber.

14.2.1 Macrostructure of Wood

A cross section of a tree trunk is shown in Fig. 14.2. This may be seen to consist of a number of different components:

1. The *outer bark* is a dense, often rough, layer that protects the interior of the tree.

2. The *inner bark* transports the sap from the leaves to the growing parts of the tree.

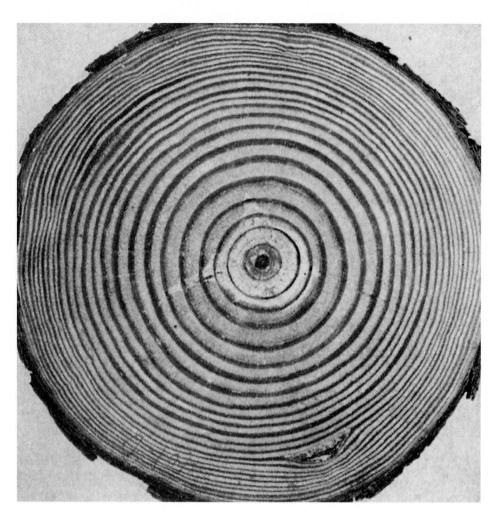

Figure 14.2
Cross section through the trunk of a tree. The annual growth rings, the darker heartwood, and the lighter sapwood can all be clearly seen (*Wood Handbook,* Forest Products Laboratory, U.S. Dept. of Agriculture. Agriculture Handbook No. 72, Revised 1974).

3. The *cambium* consists of a layer of tissue, only one cell thick, between the bark and the wood. It is the repeated subdivision of the cambium that forms both new wood and new bark.

4. The wood itself may be divided into sapwood and heartwood. *Sapwood* is the wood near the outside of a log; the width of this region is generally less than one-third the radius of the log. It is within the sapwood that moisture is conducted up from the roots and that food (carbohydrates) is stored until needed for further growth. *Heartwood* is the inner core of the trunk, composed entirely of nonliving cells. It is more resistant to decay than sapwood and is drier and harder.

5. The *annual rings* are the most distinct feature of the cross section of the trunk (Fig. 14.2). As the cells of the cambium grow and divide in the course of an annual growing season, they form (in cross section) a ring of cells, of diameters ranging from 0.02 to 0.5 mm, around the tree. In the spring, during the period of rapid growth, these cells are relatively large, with thin walls, and are referred to as *springwood,* or earlywood. Later in the growing season, *summerwood,* or latewood, is formed. It has smaller cavities and thicker walls than springwood and hence is harder and stronger mechanically.

6. Finally, at the center of the trunk is found the *pith*, the small cylinder of primary tissue around which the annual rings form.

It is also convenient here to distinguish between two different botanical groups of trees: *softwoods* and *hardwoods*. The terms *soft* and *hard* in this context do not refer to the actual hardness of the wood, since some softwoods are harder than some hardwoods. Rather, *softwoods* are trees that in most cases have needlelike or scale-like leaves (i.e., conifers such as Douglas fir, pine, or hemlock). *Hardwoods,* on the other hand, are trees that have broad leaves, such as oak, maple, walnut, or ash. Hardwoods tends to have a more complicated cell structure than do softwoods.

14.2.2 Microstructure of Wood

On a microstructural level, wood may be crudely modeled as being made up of a bundle of aligned thin-walled tubes (known as cellulose cells or fibers), glued together (Fig. 14.3). The walls of these cells are made up of several layers of mainly cellulose macromolecules, divided into the primary wall and secondary wall (to be discussed in Sec. 14.2.4).

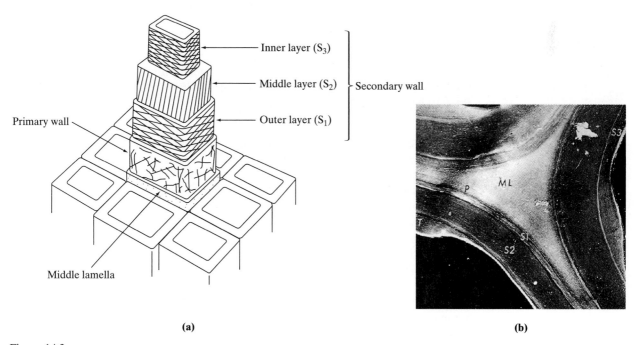

(a)   **(b)**

Figure 14.3
(a) Simplified model of the microstructure of wood showing bundles of aligned thin-walled tubes (known also as cellulose cells) and the structure of the cell wall (Building Research Establishment © Crown copyright); (b) Transmission electron micrograph of a cell wall cross-section, showing layered structure, including middle lamella (ML), primary wall (P), and the three layers of the secondary cell walls, S_1, S_2, S_3. (Courtesy of Dr. Simon Ellis, University of British Columbia).

In softwoods, about 90% of the volume of the wood consists of longitudinally oriented cells called *tracheids,* the remaining 10% is transversely oriented cells. The tracheids are typically about 2 to 5 mm long with an aspect ratio (ℓ/d ratio) of about 100. They are responsible for the mechanical support of the tree and for vertical conduction of water and sap. The transversely oriented *parenchyma* cells are typically 200 μm by 30 μm in size and are mostly located in rays (i.e., in strips of cells extending radially within a tree). They store the food for the tree and transport it

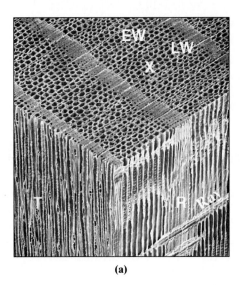 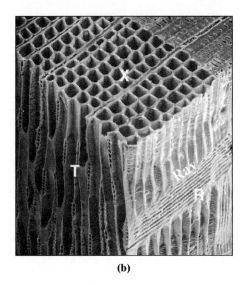

(a)   (b)

Figure 14.4   Microstructural characteristics of soft wood: (a) Softwood block showing three complete and part of two other growth rings in the cross-sectional view (X). Individual cells can be easily detected in the earlywood (EW), whereas the smaller latewood cells are difficult to distinguish in the latewood (LW). Note the abrupt change from earlywood to latewood. The two longitudinal surfaces (R—radial; T—tangential) are illustrated. Rays which consist of food-storing cells are evident on all three surfaces. (b) View of a softwood specimen with a broad latewood zone. Individual latewood cells with thick walls and small radial diameters can be detected. Note also the appearance of the wood rays in all three planes of study. X—cross-sectional plane; R—radial plane; T—tangential plane (courtesy of Dr. Simon Ellis, University of British Columbia).

horizontally. The typical microstructure of softwoods is shown in Fig. 14.4; the appearance of the tracheids is shown in Fig. 14.5.

The microstructure of hardwoods is more complex since they contain, in addition to tracheids and parenchyma, fibers and vessels (pores), as seen in detail in Fig. 14.6. The mechanical support of hardwood trees is due to the *fibers,* typically 0.7 to 3 mm long, with an aspect ratio of about 100. They have very tapered ends. Vertical conduction is carried out through *vessels* (or *pores*), typically of lengths of 0.2 to 1.2 mm and widths of about 0.5 mm (Fig. 14.6). The wall thickness of the closed-ended fibers is several times that of the open-ended vessels. However, in spite of the larger number of cell types in hardwoods, again about 90% of the cells are aligned vertically.

The function and properties of the different cell types for both hardwoods and softwoods are given in Table 14.2. The relative proportions and dimensions of these different cell types largely determine the densities of different woods and hence, as we will see later, many of their mechanical properties.

The cells are interconnected by *pits*—that is, discontinuities in the secondary cell walls in adjacent pairs of cells. Various type of pits can occur, but they all fulfil functions of providing a pathway for liquid movement between cells, both vertically and horizontally.

14.2.3 Molecular Structure of Wood

Wood is, of course, an *organic* material. On an elemental level (Table 14.3), the composition of all dry woods is approximately 50% C, 44% O, 6% H, and 0.1% N. On a molecular level (Table 14.3), wood is made up primarily of cellulose, hemicellulose, lignin, and extractives; minor quantities (< 0.3%) of inorganic materials are also present.

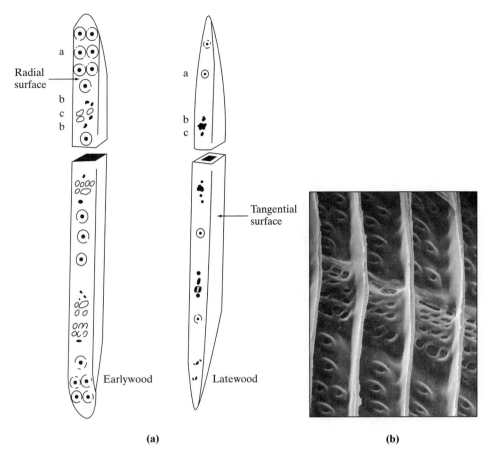

(a)   **(b)**

Figure 14.5

Cell structure of tracheid: (a) Isolated earlywood and latewood longitudinal tracheids. Note resemblance of tracheids to long cylindrical tubes. Tracheid lengths in this figure are considerably reduced because tracheids are normally about 100 times longer than wide. Note the rounded end of the earlywood tracheid in the radial view and the pointed end in the tangential view. Latewood tracheid ends tend to be pointed in both views: (a) bordered pits to adjacent longitudinal tracheids; (b and c) pits to adjacent ray cells (drawing from E. T. Howards and E. G. Manwiller, *Wood Science,* 2, pp. 77–86, 1969). (b) Internal cell walls of earlywood longitudinal tracheids. The circular, domelike structures are bordered pits which permit liquid flow between contiguous longitudinal tracheids. The smaller elliptical-shaped pits lead to adjacent transversely oriented ray cells (courtesy of Dr. Simon Ellis, University of British Columbia).

TABLE 14.2   The Functions and Dimensions of the Various Types of Cells Found in Softwoods and Hardwoods

Cells	Softwood	Hardwood	Function	Wall Thickness	Length	Width
Parenchyma	+	+	Storage		200 μm	30 μm
Tracheids	+	+	Support Conduction		2–5 mm	20–50 μm
Fibers		+	Support		0.7–3 mm	10–20 μm
Vessels (pores)		+	Conduction		0.2–1.2 mm	500 μm

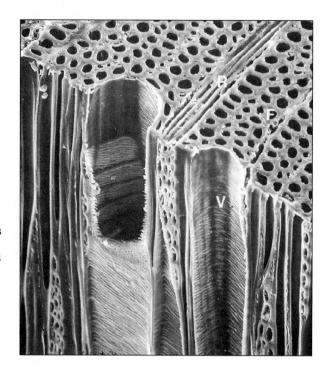

Figure 14.6 Hardwood showing a vessel (V), fibers (F), and ray cells (R). Notice the relative differences in size and shape of the cells. The ray shown in the tangential view is up to seven cells in width (courtesy of Dr. Simon Ellis, University of British Columbia).

TABLE 14.3 Composition of Wood

	% Composition	Polymeric Nature	Degree of Polymerization	Molecular Building Blocks	Role
Cellulose	45–50	Linear molecule, crystalline	5000–10,000	Glucose	Framework
Hemicellulose	20–25	Branched molecule, amorphous	150–200	Primarily non-glucose sugars	Matrix
Lignin	20–30	3-D molecule, amorphous	100–1,000	Phenolpropane	Matrix
Extractives	0–10	Polymeric	—	Polyphenols	Encrusting

Cellulose is the most important component of wood; it is a *linear polymer*, composed typically of several thousand glucose units. The glucose units are covalently bonded, both within and between units, and this strong bonding leads to the high tensile strength and stiffness of the cellulose molecules. The cellulose molecules are laterally bonded into linear bundles by a combination of hydrogen bonding and van der Waals bonding. The large amount of lateral hydrogen bonding leads to crystalline regions in the wood, separated by amorphous areas. The degree of crystallinity in wood lies in the range of 65% to 90%.

The hydroxyl (OH^-) groups in each unit of glucose making up the cellulose chain are important for two reasons associated with their ability to generate van der Waals and hydrogen bonding to adjacent molecules:

1. They attract the hydroxyl groups of adjacent molecules of cellulose, creating *microfibrils*.
2. They strongly attract water molecules, and hence are largely responsible for the swelling and shrinking that wood undergoes as it is wetted and dried.

Hemicelluloses, like cellulose, are also polymeric molecules. However, they are built up from a variety of different sugar molecules (rather than simply glucose); different species have different relative amounts of these various sugars in the hemicellulose chains. Hemicelluloses are important in the paper-making process since they provide much of the fiber-to-fiber bonding.

Lignins have complex, three-dimensional structures; they are built up from phenyl propane units, linked in a variety of ways. They permeate the matrix of cellulose microfibrils in the cell walls and fill the spaces between wood cells. Together, the lignins and the hemicelluloses surround the cellulose units, bonding them together. The lignins, in particular, impart rigidity and compressive strength to the cell walls. The structure of lignins is not completely understood.

The *extractives* do not form a part of the basic wood structure. They consist of a very wide variety of chemical substances and are in part responsible for such wood properties as color, odor, taste, resistance to decay, strength, flammability, and hygroscopicity.

Microfibrils are threadlike bundles of cellulose molecules that are arranged approximately parallel to each other, with lignins and hemicellulose between the cellulose molecules and helping to bond them together. The crystalline regions of the cellulose are not chemically reactive. The exact structure of microfibrils is not known. However, it is believed that they consist of a largely crystalline cellulose core, surrounded by partly crystalline hemicellulose and cellulose, and then by amorphous lignin. This sort of structure would lead to a high shear strength between microfibrils, which contributes to the tensile strength and toughness of wood. However, the covalent bonding forces along the length of the microfibrils are much stronger than the secondary forces binding them together laterally. This leads to the highly orthotropic behavior of wood, which will be discussed later in this chapter.

14.2.4 Cell Wall Structure in Wood

Wood is often considered to be fibrous composite material; this may be seen clearly if we consider the structure of cell walls. If one examines the cell wall structure at high magnifications (Fig. 14.3b), it is possible to distinguish a number of different layers with different orientations of the microfibrils in the different cell wall layers:

1. The *middle lamella* is a lignin-rich layer which joins together the neighboring wood cells.
2. The *primary wall* is formed first in the development of a wood cell. It is relatively thin, with a more or less random orientation of microfibrils.
3. The *secondary wall,* which is formed after the primary wall, itself consists of three layers:
   a. The S_1 layer is relatively thin, consisting of four to six layers of microfibrils. The microfibrils in successive layers (laminae) are arranged in alternating right- and left-handed helices, inclined about 50 to 70° from the vertical axis.

b. The S_2 layer is thick, making up most of the cell wall. The microfibrils are inclined only 10 to 30° from the vertical axis and all are arranged in a right-handed helix.

c. The S_3 layer is very thin and is similar to the S_1 layer.

The thickness of the various cell wall layers, and the orientation of the microfibrils in each layer, are summarized in Table 14.4.

TABLE 14.4   Thickness of Various Cell Wall Layers and Microfibril Angle within the Layers

Wall Layer	Relative Thickness (%)	Avg. Angle of Microfibrils
Primary Wall	±1	Random
S_1	10–22	50–70°
S_2	70–90	10–30°
S_3	2–8	0–90°

14.3   THE ENGINEERING PROPERTIES OF WOOD

14.3.1 Orthotropic Nature of Wood

As stated earlier, because of the way in which trees grow in nature, wood is a highly orthotropic material. Its properties in the directions of its three principal axes (shown in Fig. 14.7) are very different. The *longitudinal axis, L*, is parallel to the direction of the fiber growth (i.e. parallel to the grain). The *radial axis, R*, is normal to the growth rings and perpendicular to the grain. The *tangential axis, T*, is also perpendicular to the grain, but tangential to the growth rings. A number of terms are used to describe the direction of the grain (or fibers) in a piece of wood. The most important for our purposes are as follows:

1. *Slope of grain*: The angle between the direction of the grain and the axis of the specimen.
2. *Cross grain*: A pattern in which the fibers deviate from a line parallel to the sides of the specimen.
3. *Diagonal grain*: A grain in which the fibers form an angle with the axis of the specimen as a result of the sawing direction.
4. *Spiral grain*: An arrangement of the fibers which results from their growth in a spiral direction about the bole of a tree.

14.3.2 Effects of Relative Density

For all species of wood, the relative density (specific gravity) of the cell wall material itself is about 1.5. However, relative densities of different species range from about 0.04 for balsa wood to about 1.4 for lignum vitae. This implies that the differences in relative densities between species (and even within the same species) are due largely to differences in the void space (or *porosity*) associated with the geometry of the wood cells and their grouping together. For most materials, the relationship between mechanical properties and porosity can be written (empirically) as

$$S = S_0(1 - p)^n \tag{14.1}$$

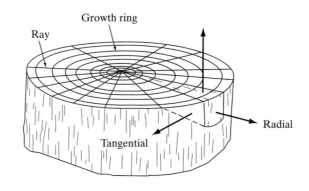

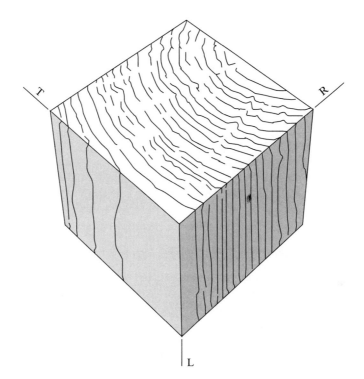

Figure 14.7
The anisotropic axes of
wood structure.

where S is the mechanical property in question, S_0 is the value of that property at zero porosity, p is the porosity, and n is an empirical constant, with a value for many materials of around 3. Thus, relatively small changes in porosity can lead to relatively large changes in mechanical properties.

14.3.3 Effects of Moisture Content

Wood can exchange moisture with the atmosphere relatively easily. Thus, given time, it will reach moisture equilibrium with its surroundings. Moisture can exist in wood in two states:

> **1.** As *free water* (or water vapor) within the cell cavities
> **2.** As *bound water,* physically adsorbed in the cell walls.

As green wood dries, the free water evaporates first. We may define the *fiber saturation point* as the condition in which all of the free water in the cell cavities has evaporated, but the cell walls are still fully saturated. This generally occurs at moisture contents in the range of 25% to 30%. As drying continues below the fiber saturation point, the removal of water from the cell walls allows a compaction of the molecular structure and the formation of additional hydrogen bonding. That is, the wood will both shrink and become stronger. This process is reversible.

Moisture Effects on Mechanical Properties

It has been found that above the fiber saturation point, changes in moisture content have little or no effect on the mechanical properties of wood. However, below the fiber saturation point, most mechanical properties can be related to the moisture content. An empirical relationship between moisture content and mechanical properties, written in a logarithmic form, is as follows:

$$\log P = \log P_{12} + [(M - 12)/(M_p - 12)] \log\left(\frac{P_g}{P_{12}}\right), \tag{14.2}$$

where P = property of interest,
P_{12} = value of the property at a moisture content of 12%,
M = moisture content,
M_p = moisture content below which property changes due to drying are first observed (this may be slightly less than the fiber saturation point and is often taken to be 25%), and
P_g = value of the property for moisture contents above M_p.

This formula does not adequately apply to large, structural members. In addition, it is not recommended for impact, bending, toughness, or tension perpendicular to the grain.

Shrinkage

Changes in moisture content above the fiber saturation point also have no effect on the dimensional stability of wood. Below the fiber saturation point, the *volumetric shrinkage* of wood is approximately proportional to the volume of water lost. However, because of the anisotropic structure of wood, the shrinkage is not in the same in all directions, as shown in Fig. 14.8a. The loss of water from the cell walls induces attractive forces between the microfibrils which cause them to bunch together. Since the microfibrils are oriented primarily parallel to the axis of the cells (i.e., to the longitudinal axis of the tree), most of the resulting decrease in volume is *perpendicular* to the axis of the cells; only a small amount of shrinkage takes place longitudinally. Moreover, depending on how the wood specimen was sawn with respect to the grain direction, this unequal shrinkage may also cause considerable *distortion* (warping, twisting, etc.) of the wood, as shown in Fig. 14.8b.

In general, as shown in Table 14.5, tangential shrinkage varies from about 5% to 12%, depending on the species; the radial shrinkage is about half as much. The longitudinal shrinkage is comparatively small, typically in the range of 0.1% to 0.2%.

For many species, the relationship between shrinkage and moisture content is known. If it is not, the change in dimension can be estimated by the empirical expression

$$S_{M_2-M_1} = \frac{M_2 - M_1}{30} S_{G-D} \tag{14.3}$$

where $S_{M_2-M_1}$ is the shrinkage between moisture contents M_2 and M_1, S_{G-D} is the total shrinkage from the green to the oven-dry state, available from tables, and 30 is the

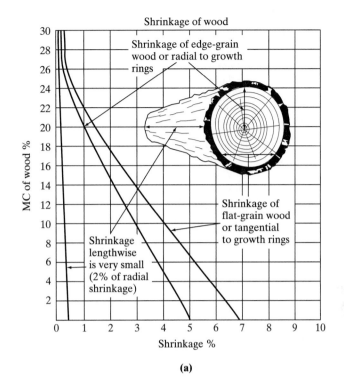

(a)

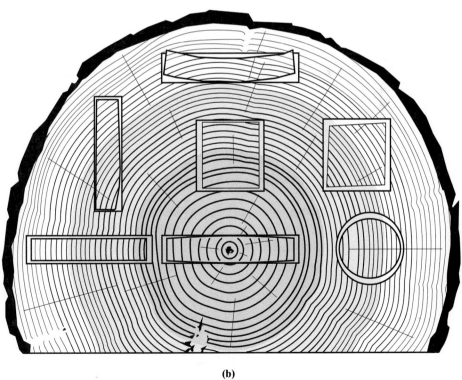

Figure 14.8
Characteristic shrinkage and distortion of flats, squares, and rounds, as affected by the direction of the annual rings. Tangential shrinkage is about twice as great as radial.

(b)

TABLE 14.5. Selected Data of Shrinkage Values of Wood, Relative to Dimensions when Green

| Species | Shrinkage from Green to Oven Dry, % of Green Size | | |
	Radial	Tangential	Volumetric
Softwoods			
Cedar, western red	2.4	5.0	6.8
Cedar, northern white	2.2	4.9	7.2
Douglas fir[a]	4.8	7.6	11.8
Firs, true (Hem-fir)[a]	4.5	9.2	13.0
Hemlock, eastern	3.0	6.8	9.7
Hemlock, western	4.2	7.8	12.4
Larch, western	4.5	9.1	14.0
Pine, eastern white	2.1	6.1	8.2
Pine, lodgepole	4.3	6.7	11.1
Pine, ponderosa	3.9	6.2	9.7
Pine, red	3.8	7.2	11.3
Pine, southern[a]	5.4	7.7	12.3
Pine, sugar	2.9	5.6	7.9
Pine, western (Idaho) white	4.1	7.4	11.8
Redwood, California	2.6	6.9	11.2
Spruce, eastern[a]	4.1	7.8	11.8
Spruce, Englemann	3.8	7.1	11.0
Spruce, Sitka	4.3	7.5	11.5
Hardwoods			
Alder, red	4.4	7.3	12.6
Ash[a]	5.0	8.1	15.2
Aspen[a]	3.5	7.9	11.8
Basswood	6.6	9.3	15.8
Beech	5.5	11.9	17.2
Birch, yellow	7.3	9.5	16.8
Cherry, black	3.7	7.1	11.5
Cottonwood, black	3.6	8.6	12.4
Hickory, pecan	4.9	8.9	13.6
Hickory, true[a]	7.7	12.6	19.2
Maple, sugar	4.8	9.9	14.7
Oak, white	5.6	10.5	16.3
Oak, red, northern	4.0	8.6	13.7
Oak, red, southern	4.7	11.3	16.1
Walnut, black	5.5	7.8	12.8

[a]These names represent several species or varieties, and values given are maximum shrinkage values for the group.

(Data from *Wood Handbook*, Forest Products Laboratory, U.S. Dept. of Agriculture, Agriculture Handbook No. 72, Revised 1974.)

assumed fiber saturation point (%). Both shrinkage and swelling on rewetting can be estimated in this way.

14.3.4 Mechanical Properties of Wood

In this section, we will deal only with the mechanical properties of *small, clear specimens* of wood. These are useful in trying to understand the relationships between the strength and the microstructure of wood. They also provide a reference point for the more practical studies on structural lumber containing macroscopic defects (knots, checks, cracks, etc.), which will be considered later in this chapter.

Due to the orthotropic nature of wood, the direction of measurement with respect to the structure of the wood must be specified for any mechanical property. In particular, we consider two primary directions:

1. *Parallel to the grain* refers to properties measured in the longitudinal direction of the wood (i.e., parallel to the orientation of the long axis of the majority of microfibrils).
2. *Perpendicular to the grain* refers to both the radial and tangential directions (i.e., the directions perpendicular to the growth rings and tangent to the growth rings, respectively, or perpendicular to the orientation of the majority of the microfibrils). It is a reasonable approximation to lump these two directions together; even though properties such as strength and elastic modulus are somewhat higher in the radial direction than in the tangential direction, in both cases they are much lower than the same properties measured in the longitudinal direction.

It must be emphasized that compared to the other materials discussed in this book, the properties of wood display a very high degree of variability. It is not uncommon to find coefficients of variation in excess of 30% for many properties. This variability must be taken into account when establishing design values, as will be discussed later. In this section, we will deal only with average values.

Modulus of Elasticity

The stress-strain curve for wood is nonlinear, even at low stresses. This may be seen particularly at very low stress rates, if sensitive equipment is used. However, at the more usual rates of loading and with normal laboratory equipment, the stress-strain curves do appear to have a linear region, as shown in Fig. 14.9; the wood then reaches its *proportional limit,* beyond which considerable nonlinear deformation can occur. For tension parallel to the grain, the proportional limit is about 60% of the ultimate load; in compression, it is about 20% to 50% of the ultimate load. The three elastic moduli of wood (in the different grain directions) are given in Table 14.6. As with strength properties, the elastic moduli decrease with increasing temperature and with increasing moisture content (up to the fiber saturation point).

Figure 14.9
Stress-strain curves for wood in tension and compression parallel to grain. Signs of stress and strain have been ignored so that both may be plotted together. Note particularly the difference in strength and in the extent of the nonlinear deformation prior to maximum stress.

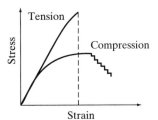

Tensile Strength

The tensile strength of wood in tension parallel to the grain is quite high, ranging from about 70 to 150 MPa for common structural species of wood. This is accompanied by relatively little deformation; failure strains are on the order of 1%. These properties can be explained in terms of the wood microstructure discussed previously. The microfibrils are oriented largely parallel to the grain. Failure appears to occur either in the S_1 layer, or as shear failure between the S_1 and S_2 layers. In any

TABLE 14.6   Elastic Constants of Selected Timbers

Species	Density (kg/m³)	Moisture content (%)	E_L	E_R	E_T
Hardwoods					
Balsa	200	9	6300	300	106
Khaya	440	11	10200	1130	510
Walnut	590	11	11200	1190	630
Birch	620	9	16300	1110	620
Ash	670	9	15800	1510	800
Beech	750	11	13700	2240	1140
Softwoods					
Norway Spruce	390	12	10700	710	430
Sitka Spruce	390	12	11600	900	500
Scots Pine	550	10	16300	1100	570
Douglas Fir[a]	590	9	16400	1300	900

Note: E is the modulus of elasticity in a direction indicated by the subscript (in MPa).
[a]Listed in the original as Oregon pine.

event, failure in tension parallel to the grain involves the breaking of strong, primary bonds.

In tension perpendicular to the grain, tensile strengths are much lower, on the order of 2 to 9 MPa. The loading is perpendicular to the microfibrils, and so failure involves mostly the breaking of weak secondary bonds. However, because of the ease of distortion of the "hollow tubes" in this form of loading, relatively large distortions may occur before failure, and the σ-ε curve is much more nonlinear.

Compressive Strength

To understand the behavior of wood in compression, it is helpful to look at the idealized model of wood structure shown in Fig. 14.3. In compression parallel to the grain (Fig. 14.10a), the failure mechanism involves the development of small "kinks" within the microfibrils, eventually leading to localized buckling of the cell walls. This manifests itself to the naked eye as a line of wrinkles on the surface of the specimen (Fig. 14.11), horizontal on the LR face, and inclined on the LT face. The strength in compression parallel to the grain is only about one-half that in tension, in the range of about 25 to 60 MPa for wood. There is also a greater strain to failure.

Figure 14.10
Compressive loading of wood in different directions: (a) when wood is loaded along the grain most of the cell walls are compressed axially; (b, c) when loaded across the grain, the cell walls bend.

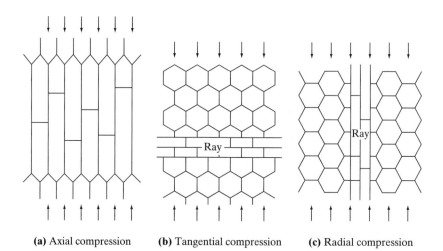

(a) Axial compression   **(b)** Tangential compression   **(c)** Radial compression

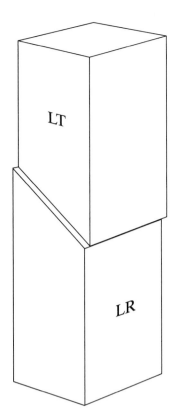

Figure 14.11 Failure under longitudinal compression at the macroscopic level. On the longitudinal radial plane the crease (shear line) runs horizontally, while on the longitudinal tangential plane the crease is inclined at 65° to the vertical axis.

In compression perpendicular to the grain, the hollow wood cells simply collapse, or flatten. This process of failure can clearly lead to very large deformations, with no well-defined maximum load. Indeed, once the cells have collapsed, the stress may continue to increase to very high values. The stresses at which the cells begin to collapse (marked by the beginning of extensive nonlinearity in the σ-ε curve) are in the range of about 3 to 10 MPa.

Flexural Strength

In flexure, there will be tensile stresses acting at the bottom of the beam and compressive stresses acting at the top; shear stresses may also be present. However, the compressive strength of wood (parallel to the grain) is only about one-half the tensile strength. Therefore, failure will generally begin as compression failure in the upper part of the beam. As the stress increases, the neutral axis will move downward. Eventually, final failure will occur when the stresses in the bottom fibers exceed the tensile strength of wood. Typical flexural strengths are in the range of 40 to 100 MPa (i.e., in between the compressive and tensile strengths parallel to the grain).

Shear Strength

As shown in Fig. 14.12, there are six principal modes of shear in wood, which can be divided into three groups of two: shear perpendicular to the grain, shear parallel to the grain, and rolling shear. The strongest of these is *shear perpendicular to the grain,* which would require the breaking largely of primary bonds. In fact, this shear strength is so high that it is rarely reached, since other modes of failure will occur first. *Shear parallel to the grain* involves the relative sliding of microfibrils parallel to their long axes, requiring the breaking mostly of secondary bonds. This shear

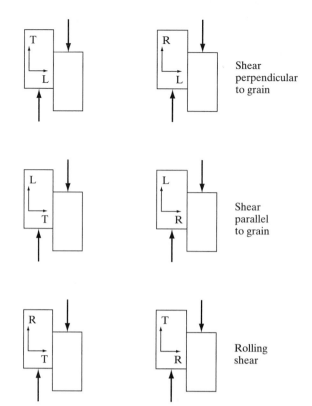

Figure 14.12
Schematic representation of the six principal modes of shear failure in wood, divided into three groups, as shown.

Shear perpendicular to grain

Shear parallel to grain

Rolling shear

strength is on the order of 5 to 15 MPa. In *rolling shear,* the direction of sliding is also perpendicular to the grain, but the mode of failure involves the tracheids "rolling over" each other—again, the "bundle of straws" analogy may be helpful. Rolling shear strengths are on the order of one-quarter the values of shear parallel to the grain.

Effects of Temperature

There are two different types of temperature effects on the mechanical properties of wood. In the long term, prolonged exposure to high temperatures will degrade wood in an irreversible manner; this effect will be examined later when we consider durability. The other temperature effect, which will be dealt with here, is the short-term response to temperature, which is reversible. Although not all properties are affected to the same degree, there is, in general, a decrease in mechanical properties with an increase in temperature. As a crude rule of thumb, in the range −18°C to 66°C (0°F to 150°F), a 1°C change in temperature will change the mechanical property by 0.6% to 1.0%. Dry wood is less sensitive than green wood to temperature effects.

Effects of Rates and Duration of Loading

Wood is a viscoelastic material, and thus some of its mechanical properties are strongly time dependent. Here, we will examine the four major phenomena in which this time dependency is manifested: rate of loading, duration of load, fatigue, and creep.

RATE-OF-LOADING EFFECTS

As we noted in Chapter 6, most materials exhibit an increase in their apparent strength as the rate of loading increases. Wood also exhibits this behavior. Although the physical mechanisms responsible are not yet understood, it may be that some type of critical strain phenomenon is involved. As a rough estimate, a change in the loading rate by a factor of 10 will change the apparent strength by about 8%.

DURATION OF LOAD

Closely related to the rate of loading effect is the problem of duration of load. That is, how will the apparent strength of wood change if a load is applied continuously for a long period of time? The "standard" curve is that published by Wood in 1951, though the curve later developed by Pearson (Fig. 14.13) is probably a better representation of what occurs. These data suggest that wood which is to carry a dead load for 50 years may be stressed to only about one-half of the value obtained from short-term strength tests. It must be emphasized, however, that the data of Fig. 14.13 can only be applied to wood under laboratory conditions. More realistic tests on timber indicate that Wood's curve is conservative for times less than about three years, but it underestimates the duration of load effect for longer times.

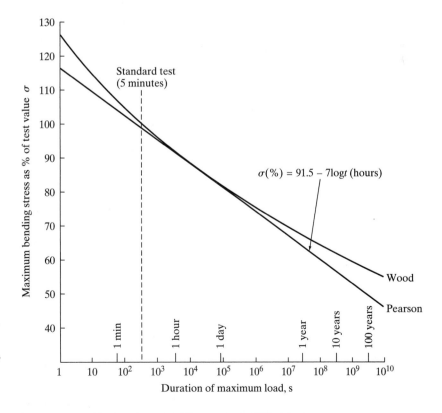

Figure 14.13
The effect of duration of load on the bending strength of timber (after L.W. Wood, "Relation of strength of wood to duration of stress," *U.S. Forest Products Laboratory,* Madison, Wisconsin, Report No. R-1916, 1951, and R. G. Pearson, "The effect of duration of load on the bending strength of wood," *Holzforschund,* 26, No. 4, pp. 153–158, 1972).

$\sigma(\%) = 91.5 - 7\log t$ (hours)

FATIGUE

There is not an extensive literature on the fatigue properties of wood. It would appear that, for small clear specimens, the fatigue strength at 2×10^6 cycles of loading is about 60% of the static strength. In structural timber containing defects, the reduction in strength under fatigue loading is more severe.

CREEP

When wood is first loaded, it undergoes elastic deformations; if the load is maintained, additional deformations occur over time. These additional deformations are termed *creep;* the phenomenon of creep was described in general terms in Chapter 7. Creep will occur in wood even at very low stresses and will continue more or less indefinitely, though at ever-decreasing rates. For the ordinary loads used in practice, creep deformation may be on the same order of magnitude as the initial elastic deformation. Upon unloading, it is found that typically about one-half of the total creep is irrecoverable. The magnitude of the creep is increased both at higher temperatures and at higher moisture contents. In addition, *cycling* either the temperature or the moisture content under constant load will further increase the creep deformations.

Having discussed the properties of wood (i.e., small, clear specimens), we can now turn our attention to timber, the material actually used in construction. *Timber* is wood which contains a variety of macroscopic defects. In many cases, these flaws have a much larger effect on the behavior of timber than do the properties of the wood material itself.

14.4   DEFECTS AND OTHER NONUNIFORMITIES IN WOOD

The mechanical properties measured on carefully selected small, clear, straight-grained wood are modified by natural defects and abnormalities in timber, which occur as a function of the way in which trees grow. In general, these defects all have a tendency to reduce the strength measured on small, clear specimens. The defects which are commonly encountered include the following:

1. *Fiber and ring orientation:* In structural timber, depending on how a log was sawn and on how uniform the grain was in the original tree, the grain direction may not coincide with the axes of the board, as shown in Figs. 14.14a to 14.14e. The effects of this on strength were described in Sec. 14.3.

2. *Knots:* A knot is that portion of a limb that has been surrounded by subsequent growth of wood. A *spike knot* (Fig. 14.14f) is a knot sawn in a nearly lengthwise direction. A *loose knot* (Fig. 14.14g) is a knot which is not held firmly in place by the surrounding wood, while a *sound tight knot* (Fig. 14.19h) is firmly fixed in place in the piece of timber. The effect of knots on structural properties depends on the size, type, frequency, and location of the knots.

3. *Checks:* A check (Fig. 14.14j) is a lengthwise separation of the wood which usually extends across the growth rings; it commonly results from the drying process.

4. *Wanes:* The wane (Fig. 14.14k) is the lack of wood on the face of a piece, for any reason at all.

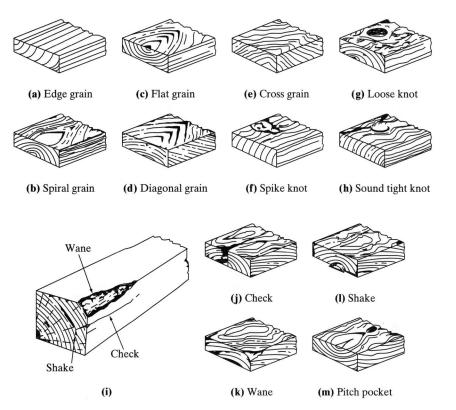

Figure 14.14
Some common defects in wood.

(a) Edge grain   **(c)** Flat grain   **(e)** Cross grain   **(g)** Loose knot

(b) Spiral grain   **(d)** Diagonal grain   **(f)** Spike knot   **(h)** Sound tight knot

Wane

Check

Shake

(i)

(j) Check   **(l)** Shake

(k) Wane   **(m)** Pitch pocket

5. *Shakes:* A shake (Fig. 14.14l) is a separation along the grain, between the annual growth rings.

6. *Pitch Pocket:* A pitch pocket (Fig. 14.14m) is an opening between growth rings containing resins or bark.

In addition to these defects, there are other inhomogeneities in wood. There are differences in properties between earlywood and latewood, and between heartwood and sapwood. Moreover, the density of a log varies considerably with location, as shown in Fig. 14.15. Thus, even within a single species or for specimens cut from a single tree, the variability will be very great. These variabilities come on top of those induced by changes in the slope of grain.

Of course, a number of the properties of wood, such as thermal properties, electrical properties, acoustic properties, creep, and shrinkage, are not much affected by flaws and will not be considered further. We will focus primarily on the effects of flaws on the strength properties of timber.

14.5   EFFECTS OF FLAWS ON MECHANICAL PROPERTIES OF TIMBER

The effect of most flaws is both to decrease the strength of timber compared to the strength of small, clear specimens of wood and to change the mode of failure. Clearly, flaws such as checks and shakes are essentially just cracks, and they will weaken timber in the same way as artificially introduced cracks. Indeed, the presence of such natural cracks makes fracture mechanics (see Chapter 6) an attractive tool for analyzing the failure of timber. Similarly, the effect of grain orientation has been discussed earlier in this chapter.

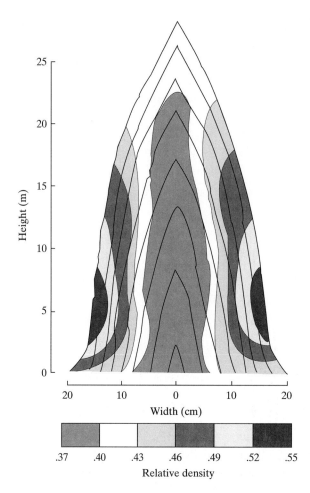

Figure 14.15
Diagrammatic representation of the density distribution in a spruce stem. From Kellman and Côté, *Principles of Wood Science and Technology, I-Solid Wood.* Springer-Verlag, 1968.

The effects of knots are more complex, depending on the type of knot, its size and location in the member, and the type of loading. In the general case, knots decrease the strength of wood (except for compression and shear perpendicular to the grain). This strength reduction is roughly proportional to the area of the cross section occupied by knots. Although there is a very wide scatter in experimental results, ASTM D-245 prescribes linear strength reduction equations, two of which are shown in Fig. 14.16.

Knots act to reduce strength in several ways:

1. Most significantly, they distort the fibers around the knot. This often leads to the occurrence of localized tensile stresses *perpendicular to the grain,* regardless of whether the wood is loaded in tension, bending, or even compression. Since wood is particularly weak in tension perpendicular to the grain, failures will often initiate in such areas.

2. The discontinuities in fiber growth around knots also lead to the presence of *stress concentrations,* which will reduce the load-carrying capacity of a member (Chapter 6).

3. In the drying process, *checking* often occurs around knots.

The effects of knots in determining the failure pattern in timber beams in bending are shown in Fig. 14.17, both for knots on the bottom (tension) face of the beam and on the top (compression) face.

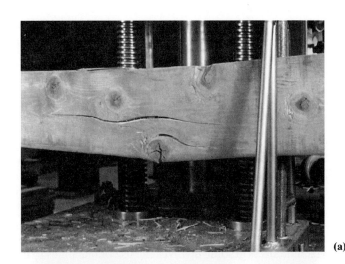

Figure 14.16
Knot area displacement versus strength ratio (SR). (From R. J. Hoyle, Jr., in *Wood as a Structural Material* (A. G. H. Dietz, E. L. Shaffer and D. S. Gromala, eds.) Educational Modules for Materials Science and Engineering, The Pennsylvania State University, 1980, 1981, 1982.)

For bending SR of knots on narrow face or on centerline of wide face for compression SR of any knots

For bending SR of knots on edge of wide face

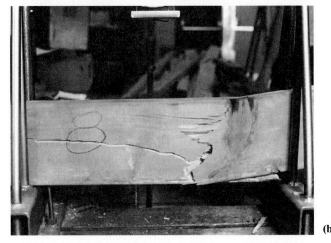

Figure 14.17
Effects of knots on the failure pattern in timber beams: (a) knot at bottom of beam; and (b) knot at top of beam.

The flaws described previously are not uniformly distributed within a tree, and trees themselves will be different from one another. Every piece of structural timber will have a different flaw distribution from every other piece, even those cut from the same log. These differences will lead to a large variability in strength. Therefore, as a matter of convenience for design purposes, lumber is *graded* into different classes, according to strength and appearance. These are, in general, *stress grades* (i.e., the various grades are associated with a particular set of mechanical properties).

The details of any grading practice vary from country to country, and the grading rules can be complex. However, all grading practices share some common features. There are, basically, two ways of grading timber: *visually* and *mechanically*.

14.6.1 Visual Grading

By far the oldest method of grading timber is by visual stress grading; probably 95% of lumber is graded in this way. As the name implies, this method involves assessing the properties of any particular piece by its visual appearance. The most important criteria are as follows:

1. Density—related to the rate of growth and the percentage of latewood
2. Decay
3. Heartwood or sapwood—may be important for durability considerations, but not for strength
4. Knots—type, size, location
5. Slope of grain—often referred to as cross grain
6. Shakes, checks, and splits
7. Wane
8. Pitch pockets.

Unfortunately, even though graders are trained carefully, there is a great deal of subjectivity in the quality of grading. Therefore, in recent years, there has been a move toward a more objective form of grading—that is, machine stress grading.

14.6.2 Mechanical Grading

Machine stress grading involves the use of machines to measure the *stiffness* of a piece of timber; the stiffness is, of course, related to the modulus of elasticity. Although there is no *theoretical* relationship between the modulus of elasticity and strength properties, an *empirical* relationship has been observed. This method is not yet widely used and, of course, it too is not perfectly reliable. It is, however, slowly growing in popularity, particularly as better stress grading machines are now being developed.

14.6.3 Description of Visual Stress Grades

As indicated earlier, grading practices differ somewhat from country to country, and even within a country (say, the United States or Canada), a number of different authorities are responsible for grading different categories of wood. In Canada, the lumber categories and grades are as given in Table 14.7. Identical categories exist in the United States, except that the additional grades, No. 1 Dense, No. 2 Dense, and No. 3 Dense, are sometimes used to indicate denser (and hence stronger) boards within the same visual stress grade.

TABLE 14.7 Lumber Categories, Grades, and Uses in Canada

Lumber Product	Grade Category	Nominal Sizes (inches)*	Grades	Principal Uses
Dimension lumber	Light framing	2" to 4" thick 2" to 4" wide	Construction	Widely used for general framing purposes.
			Standard	Pieces are of good appearance but graded primarily for strength and serviceability.
			Utility	Used in non-load-bearing walls where economical construction is desired for such purposes as studding, blocking, plates, and bracing.
			Economy*	Temporary or low-cost construction where strength and appearance are not important.
	Structural light framing	2" to 4" thick 2" to 4" wide	Select Structural	Intended primarily for use where high strength, stiffness, and good appearance are desired, such as trusses.
			No. 1	
			No. 2	For most general construction uses.
			No. 3	Appropriate for use in general construction where appearance is not a factor, such as studs in non-load-bearing walls.
	Stud	2" to 4" thick 2" to 6" wide	Stud	Special-purpose grade intended for all stud uses.
			Economy stud*	Temporary or low-cost construction where strength and appearance are not important.
	Structural joists and planks	2" to 4" thick 5" and wider	Select structural	Intended primarily for use where high strength, stiffness, and good appearance are desired.
			No. 1	
			No. 2	For most general construction uses.
			No. 3	Appropriate for use in general construction where appearance is not a factor.
			Economy*	Temporary or low-cost construction where strength and appearance are not important.
	Appearance	2" to 4" thick 2" and wider	Appearance	Intended for use in general housing and light construction where lumber permitting knots but of high strength and fine appearance is desired.
Decking	Decking	2" to 4" thick 6" and wider	Select	For roof and floor decking where strength and fine appearance are required.

*Note: 1" = 25.4 mm

(Continued)

TABLE 14.7   Continued

Lumber Product	Grade Category	Nominal Sizes (inches)	Grades	Principal Uses
			Commercial	For roof and floor decking where strength is required but appearance is not so important.
Timber	Beams and Stringers	5" and thicker, width more than 2" greater than thickness	Select structural	For use as heavy beams in buildings, bridges, docks, warehouses, and heavy construction where superior strength is required.
			No. 1	
			Standard* Utility*	For use in rough, general construction.
	Posts and Timbers	5" × 5" and larger, width not more than 2" greater than thickness	Select structural	For use as columns, posts, and struts in heavy construction, such as warehouses, docks and other large structures where superior strength is required.
			No. 1	
			Standard* Utility*	For use in rough, general construction.

Note: All grades are stress graded, meaning that working stresses have been assigned (and span tables calculated for dimension lumber) except those marked by an asterisk.

14.7   DESIGN PROPERTIES

Of course, the purpose of all of the grading rules is to permit the designer to choose appropriate mechanical properties for structural purposes. Obviously, different *species* of timber have different mechanical properties. In Canada, all of the commercial species are divided into five groups; within each group, the various species are assumed to have identical properties. In the United States, however, allowable stresses are specified for each of the commercial species separately.

As indicated earlier, most testing is carried out on small, clear specimens, while timber, which contains flaws, is considerably weaker. The problem is to go from small, clear values to those of structural timber. There are a number of factors involved.

Clear wood itself is a highly variable material, and timber even more so. If *average* values were used for design purposes, about one-half of the material would fall below these values, and an unacceptably high proportion of failures would occur. Therefore, the statistical variability in wood properties is taken into account explicitly in the design codes. Assuming that wood properties obey the *normal distribution,* strength properties are chosen so that only 5% of the specimens tested would be expected to fall below the specified strength. This is referred to as the *5% exclusion limit* and may be determined from the properties of the normal distribution as

$$5\% \text{ exclusion limit} = \sigma_m - 1.645s, \tag{14.4}$$

where σ_m is the mean strength and s is the standard deviation. The exceptions to this are the modulus of elasticity and the strength in compression perpendicular to the grain, where average values may be used. This is justified on the basis that failure is unlikely to occur if these two design values are exceeded.

As stated previously, the strength values so obtained refer only to small, clear specimens. To go from those to structural timber, the adjustment factors shown in Table 14.8 must first be applied (by dividing the small, clear values by the appropriate adjustment factor). Then a second set of adjustment factors must be applied to take into account the particular *grade* of lumber. Typical minimum *strength ratios* for bending are shown in Table 14.9, to indicate the sorts of differences in quality that are expected for the different grades.

Table 14.8 Adjustment Factors to be Applied to the Clear-Wood Properties

	Softwoods	Hardwoods
Bending strength	2.1	2.31
Modulus of elasticity in bending	0.94	0.94
Tensile strength parallel to grain	2.1	2.3
Compressive strength parallel to grain	1.9	2.1
Horizontal shear strength	4.1	4.5
Proportional limit in compression perpendicular to grain	1.5	1.5

TABLE 14.9 Visual Grades Described in the National Grading Rule[a]

Lumber Classification	Grade Name	Bending Strength Ratio (%)
Light framing (2–4 inch thick, 4 inch wide)[b]	Construction	34
	Standard	19
	Utility	9
Structural light framing (2–4 inch thick, 2–4 inch wide)	Select structural	67
	1	55
	2	45
	3	26
Studs (2–4 inch thick, 2–4 inch wide)	Stud	26
Structural joists and planks (2–4 inch thick, 6 inch and wider)	Select structural	65
	1	55
	2	45
	3	26
Appearance framing (2–4 inch thick, 2–4 inch wide)	Appearance	55

[a]Sizes shown are nominal.
[b]Widths narrower than 100 mm (4 inch) may have different strength ratio than shown. Contact rules-writing agencies for additional information.

Finally, there is another series of stress modification factors to be applied. These include the following:

Load sharing: If a system consists of three or more parallel members at 600 mm (24 inch) centers or less, arranged and connected so that they mutually support the load (e.g., roof or floor joists), the allowable bending stresses may be increased by 15% according to United States practice (or 10% in Canada).

Duration of load: As discussed earlier, the strength of wood is highly dependent on the rate of loading and on the duration of load (Fig. 14.13). Based on these data, allowable stresses are assigned on the assumption of a 10-year-long maximum load duration. If the load is applied for longer or shorter than the normal 10 years, a further adjustment is made, as shown in Table 14.10.

Treatment factor: When timber is treated with chemicals for fire retardation, allowable stresses are reduced by 10%.

Size factor: Large specimens are apparently weaker than smaller ones; a small correction factor is specified for beams greater than 300 mm (12 inch) in depth.

TABLE 14.10   Load Duration Factors (K_D)

Duration of Loading	Definition	Typical Application	K_D
Continuous	Structure subjected to more or less continuous full design load throughout its life, such as dead loads or dead loads plus live loads imposed continuously.	Tanks and bins containing fluids or. granular material; retaining walls; floors continuously subjected to full design load; support of dead loads; loads due to machinery.	0.90
Normal	Structure subjected to loads less than full design load frequently but to full design load only occasionally, such that cumulative period of full design load does not exceed 10 years throughout life of structure and the structure is stressed to not more than 90% of allowable for remainder of life of structure.	Dead loads plus live loads for most floors of residential, business, commercial, and assembly occupancies; wheel loads on bridges carrying normal to heavy traffic; miscellaneous structures; some retaining walls.	1.00
Two months	Structure subjected to full design load for no more than two months, continuously or cumulatively, throughout the life of the structure, or less than full loads for equivalent longer durations.	Structures subjected to snow loads plus dead loads; some temporary structures; some bridges having light to moderate traffic.	1.15
Seven days	Structure subjected to full design load for no more than seven days, continuously or cumulatively, throughout the life of the structure, or less than full design loads for longer durations.	Dead loads plus live loads for some concrete formwork and falsework.	1.25
One day	Structure subjected to full design load for no more than one day, continuously or cumulatively, throughout the life of the structure, or less than full design load for longer durations.	Dead loads and live loads combined with wind loads and earthquake loads; some concrete formwork.	1.33
Instantaneous	Structure subjected to full design load for no more than one minute, cumulatively, throughout the life of the structure.	Dead loads plus live loads increased appropriately for impact, such as wheel loads on bridges and cranes, or elevators; since none exceed 100% (K_D of 2.00), impact neglected for timber.	2.00

Note: (1) Modification factors apply to all types of allowable unit stress and to allowable loads on fastenings, but not to modulus of elasticity.
(2) Typical applications are not mandatory but are included for guidance of the designer.

Service condition: Since the mechanical properties of wood are decreased as the moisture content increases, an adjustment must be made depending on the service conditions. A moisture content less than 15% is defined as dry; above this, the timber is considered to be wet. Typical modification factors are shown in Table 14.11.

Temperature: The mechanical properties of wood vary somewhat with temperature. Thus, adjustments similar to those shown in Table 14.12 are often applied. Below 65°C, changes in mechanical properties due to temperature are reversible; if wood is held above 65°C for a long period of time, however, there will be a permanent loss in mechanical properties.

Thus, calculating allowable timber properties is relatively simple; the various adjustment factors described previously are merely applied simultaneously:

$$\text{Allowable property} = (5\% \text{ exclusion limit}) \times (\text{adjustment factor}) \times (\text{strength ratio})$$
$$\times (\text{other modification factors, as applicable}).$$

14.8   WOOD-BASED COMPOSITES

Structural timber does have several natural limitations: (1) The size of sawn timbers, in terms of both length and cross-sectional dimensions, is limited by the size of a tree; and (2) the presence of knots and other imperfections, particularly in larger timbers, puts a severe limit on the allowable stress levels. Therefore, to improve the utilization of wood in construction, a wide variety of wood-based composites have been produced (e.g., plywood, glued-laminated timber, particleboard, and so on), with the particular aims of

1. Producing different sizes and shapes, ranging from panel products to very large beams, or curved beams
2. Producing materials with better mechanical properties than those of structural timber
3. For economic considerations, finding a use for more of the volume of the tree and for wood wastes and scrapped wood.

There is insufficient space here to describe all of the different manufactured wood products that have been developed over the years. We will describe only three materials: plywood, glued-laminated timber, and a new class of products made by pressing and gluing very small pieces of wood into structural sizes.

14.8.1 Plywood

The oldest, and still the most common, form of panel or sheet material made from wood is *plywood.* Plywood is produced by first peeling a debarked log on a special lathe to produce thin veneers. These veneers are then glued together in layers (or plies), with the grains of the successive plies at right angles to each other. This *cross-laminated* material has a number of advantages over wood, apart from the fact that it can be produced in large sheets (typically, 4 × 8 ft, or 1200 × 2400 mm in North America). The principal advantages compared to sawn lumber are as follows:

1. Because of its cross-laminated structure, plywood is very split resistant.
2. Its properties along the length of the sheet are similar to its properties across the width.

TABLE 14.11 Service Condition Factor, K_S

| Service Condition | Nominal Lumber Thickness (inches) | Bending | | Compression | | | Tension Parallel to Grain (K_{St}) | Modulus of Elasticity (K_{SE}) |
		Stress at Extreme Fiber (K_{Sb})	Longitudinal Shear[a] (K_{Sv})	Parallel to Grain (K_{Sc})	Perpendicular to Grain (K_{ScL})			
Dry manufacture and dry service[b]	2"–4"	1.08	1.05	1.17	1.00		1.08	1.05
Dry service	All thicknesses	1.00	1.00	1.00	1.00		1.00	1.00
Wet service	4" or less	0.84	0.96	0.69	0.67		0.84	0.94
	Over 4"	1.00	1.00	0.91	0.67		1.00	1.00

[a]Dry manufacture and dry service applies where the moisture content does not exceed 15% at the time of manufacture or at any time in service.
[b]For lumber manufactured at more than 19% moisture content and used under either wet or dry service conditions, the modification factor for "wet" service conditions shall be used for longitudinal shear.

TABLE 14.12 Percent Increase or Decrease in Design Values for Each 1°C Decrease or Increase in Temperature (%)[a]

Property	Moisture Content	Cooling Below 68°F (Minimum −300°F)	Heating above 68°F (Maximum 150°F)
Modulus of elasticity	0	+0.07	−0.07
	12	+0.27	−0.38
Other properties	0	+0.31	−0.31
	12	+0.58	−0.88

[a]*Source:* National Design Specification for Wood Construction.

3. Because knots extend only through 1 ply, they are not particularly detrimental to strength.

4. Shrinkage and swelling are minimized, because the tendency of an individual ply to swell or shrink perpendicular to the grain is restrained by the resistance to movement of the adjacent plies at right angles to the ply in question.

14.8.2 Glued-laminated Timber

Glued-laminated timber (which will be referred to here as *glulam*) is manufactured by gluing together a large number of relatively short pieces of dimension lumber (each up to ~50 mm thick) to build up timbers, which may be up to at least 40 m (130 ft) in length and up to over 2 m deep; they may be straight or curved. The pieces are glued together so that the grain directions in all of them are essentially parallel. The mechanical properties are thus not as limited by knots or other imperfections as they would be in sawn timber, and hence the design stresses are actually somewhat higher even than those of the select structural grade of sawn timber. Although glulam members are generally more expensive than sawn timbers because of the cost of the laminating and gluing process, they have a number of advantages over sawn timber:

1. It is easy to produce much larger sections than can be sawn as a single member, and the sections may be curved.

2. Because the thin laminations are well seasoned before the glulam member is fabricated, checking and other seasoning defects are minimized.

3. It is possible to use material of a lower grade in those sections of a member which are less highly stressed.

4. This type of material lends itself to innovative architectural effects.

14.8.3 Manufactured Wood Products

Research is currently being carried out in a number of places to try to develop other high-performance wood-based materials. One such product, which has been developed by McMillan-Bloedel in Canada, is called *Parallam.*® This material is made by producing veneer (as for plywood) and then chopping the veneer into small strips. These small strips are then dried, oriented in the same grain directions, and then glued and pressed into large panels. The panels are subsequently sawn into dimension lumber.

The appearance of Parallam (and of the other similar materials under development) is shown in Fig. 14.18. Because of this form of composite structure, where the individual strips of veneer are small, knots and other flaws are relatively

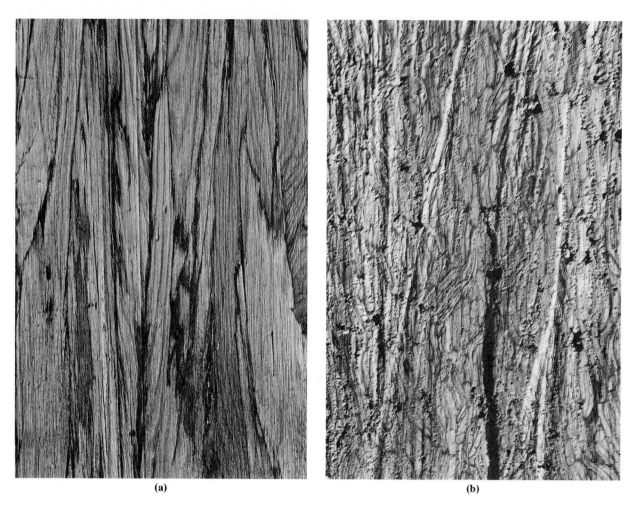

Figure 14.18
Parallel strand lumber (Parallam): (a) side view, and (b) end view.

ineffective. In addition, because of the close quality control of the manufacturing process, the variability in these materials is substantially less than that of ordinary sawn lumber. As a result, manufactured wood products such as Parallam have design stresses somewhat higher than those of select structural-grade sawn lumber.

14.9 DURABILITY

Although wood can be a very durable material, it is, like other natural organic materials, subject to deterioration over time. The major causes of deterioration which we will discuss in this section are as follows:

1. Fire
2. Decay
3. Termites
4. Marine borers (teredos).

14.9.1 Fire

Wood is, obviously, a highly flammable material; it has been used as a fuel since time immemorial. However, because of the particular burning characteristics of wood, timber structures can have a remarkable degree of fire resistance; indeed, timber structures may be able to retain strength in fires even better than steel structures.

The fire resistance of wood (at least in large sections) is due to the way in which wood burns, which is strongly dependent on the conditions when wood is exposed to high temperatures: temperature, rate of heating, moisture content of the wood, and geometry of the section. The degradation of wood on exposure to high temperature is summarized in Table 14.13. As wood is heated above 100°C, it begins to emit volatile gases. These gases will, typically, ignite at about 250°C in the

TABLE 14.13   Thermal Degradation of Untreated Wood upon Exposure to Heat

Zone	Temp (°C)	Pyrolytic Degradation	Flaming Combustion
D	500°	-Secondary pyrolysis continues to form combustible gases. -Fast heating = little charcoal, much tar and highly flammable gases, methane, carbon monoxide. -Slow heating = much charcoal, little tar, less gases in which water vapor and carbon dioxide are most important.	-Oxidation reduction occurs at more rapid rate as a result of heat released in Zone C; process exothermic. -Oxygen available from air. -Virtually same as pyrolysis, but reactions take place sooner because of oxidation. -Charcoal glows with little or no flame
C	500°	-Only charcoal residue remains. -Air (oxygen) lacking; limited combustion; pyrolysis vigorous and becomes exothermic. -Combustible gases = hydrogen, methane, carbon monoxide. -Tars = smoke is combustible. -Residue = charcoal (c) acts as catalyst. -Primary charcoal has crystalline structure with C—C bonds; carbonization complete by 390°. -Most important zone; Zones A and B moved inward.	and is consumed. Ashes result. -Heat liberated as a result of combustion; also it comes from glowing combustion of charcoal. -Secondary pyrolysis yields combustible gases which mix with air; flame ignites them. Thus, after first vigorous flaming, diminution occurs until sufficient heat passes through charcoal and pyrolizes deeper portion of wood.
B	280°	-Charring heavier. -Endothermic reaction moves Zone A deeper in wood. -Volatile products same as Zone A; less water vapor; carbon monoxide formed. -Hemicelluloses decompose first; cellulose pyrolyzes second at 240° to 350°; lignin at 280° to 500°.	-Volatile gases emitted, but not readily ignitable. -Exothermic reactions reached at lower temperature than in pyrolysis, which occurs out of contact with air. -Wood can be reduced to charcoal without flame. -Exothermic reaction occurs without ignition.
A	200°	-Endothermic process starts. -H_2O evaporates from wood. -Hemicelluloses degrade first to methane, formic acid, acetic acid, carbon dioxide. -Outside surface of wood absorbs heat and chars slightly.	-Sound wood does not ignite in Zone A. -Wood starts to char in Zone A rather than Zone B. -Oxidation reactions are exothermic in Zone A; completely exothermic by time Zone B is reached. -Gases released not ignitable.
	70°		

presence of an open flame, but they will self-ignite only at about 500°C. As the burning continues, a layer of *char* develops on the surface (Fig. 14.19), which then slowly progresses into the wood. At the base of the char layer, the temperature is about 290°C. However, wood is an excellent insulator; 6 mm in from the char base, the temperature is only about 180°C, and 123 mm in the temperature is only about 93°C (in Douglas fir). Thus, the charring rate is relatively slow, on average only about 38 mm/h for dry Douglas fir; this permits wood to retain its strength for relatively long periods in a fire, as long as the timbers are large enough in cross section. Of course, light structural members will burn and lose strength much more rapidly.

Char layer

Char base ($T = 288°$ C)

Pyrolysis zone

Pyrolysis zone base

Normal wood

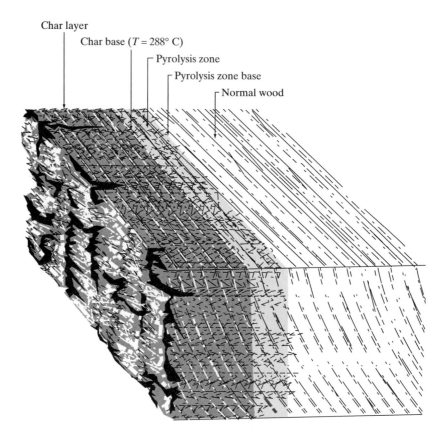

Figure 14.19
Degradation zones in a wood section exposed to fire on one surface. From D. E. Lyon in *Wood as a Structural Material* (A. G. H. Dietz, E. L. Shaffer and D. S. Gromala, eds.) Educational Modules for Materials Science and Engineering, The Pennsylvania State University, 1980, 1981, 1982.)

Fire-retardment treatment of wood improves its performance in fires in two ways: (1) by reducing the amount of flammable gases released; and (2) by reducing the amount of heat released in the initial stages of the fire. There are two primary methods of wood treatment to improve its fire resistance:

1. *Pressure impregnation with water-soluble salts,* such as zinc chloride, boric acids, sodium tetraborate, and mono- or di-ammonium phosphates. To provide such protection, at least 35 to 70 kg of salt per cubic meter of wood must be retained by the wood. This technique is used primarily for new timber construction.

2. *Fire-retardant chemicals painted on the surface* are less effective and are used mainly to improve the fire resistance of existing structures. When they are exposed to high temperatures, they tend to react in such a way as to help insulate the wood, and they also restrict the flaming of the combustible gases from the wood.

14.9.2 Decay

Decay in wood is caused by a class of organisms (*wood-destroying fungi*) which attack the cellulose and lignin, eventually causing the wood to *rot*. They do this by producing spores which secrete an enzyme that de-polymerizes the long-chain molecules in wood. Eventually, the wood becomes very soft and weak; the fracture of a piece of timber affected by rot is shown in Fig. 14.20.

Figure 14.20 Failure surface of wood severely affected by rot.

To thrive, fungi require food, oxygen, warmth, and moisture. The food comes from the wood itself: the cellulose, the lignin, and the cell contents. There is generally enough oxygen within wood to promote the growth of fungi; the prime exception to this is submerged wood. Fungi grow best in temperatures between about 20°C and 35°C. Above this temperature, most fungi will die upon prolonged exposure; at low temperatures, fungi will remain inactive but will not die.

The effects of moisture are more complicated. At moisture contents below 20%, fungi will either die, or at least remain dormant; they are most active when the moisture content is above the fiber saturation point (25% to 30%). Since fully air-dried wood has a moisture content well below 20%, it should be resistant to rot. The term *dry rot* is a misnomer, since wood *must* be damp to rot. This term refers to the action of certain types of fungi that have water-conducting strands capable of transporting water (usually from the soil) to the wood in question, so that the wood becomes moist locally and subject to rot.

Different species of wood have different degrees of resistance to rot, though they will all rot under the appropriate conditions.

14.9.3 Termites

There are several insects that attack wood: Termites, carpenter ants, and powder-post beetles are the most common, and of these termites are by far the most important. There are two classes of termites: subterranean and non-subterranean. They are found throughout most of South America and parts of North America (excluding most of Canada) and in tropical and subtropical zones throughout the world.

Perhaps 95% of the damage due to termites is caused by subterranean termites. They generally nest in soil and require moist, warm conditions. Their principal food is cellulose obtained from wood. To obtain enough moisture, if the wood itself is not moist enough, they can build "shelter tubes," or covered runways, to maintain direct contact between the wood and the moist soil. Indeed, the best means of preventing attack by subterranean termites is to block access from the soil to the wood through appropriate construction techniques.

Non-subterranean termites can live in dry wood or damp wood, depending on the species of termite, without direct contact with the soil. They are not as destructive as subterranean termites, but over time they too can do a great deal of damage.

14.9.4 Marine Borers

Various types of marine borers will attack wood in saltwater or in brackish water worldwide, though they are most destructive when the water temperature is around 10°C (50°F).

Shipworms are wormlike molluscs, related to clams and oysters, the most destructive of which are *teredos*. They burrow into the wood and live on wood borings and on the organic matter contained in the sea water. *Wood lice* are crustaceans related to crabs and lobsters which also bore into wood, though they confine themselves to the regions just below the surface. No wood species are truly resistant to attack by marine borers. Some form of preservation is required to minimize attack by these organisms.

14.9.5 Preservative Treatments

Over the years, a great many preservatives have been developed to protect wood from attack by rot, insects, and marine borers.

Creosote is obtained from the distillation of coal tar. It may be used as is, or it may be mixed with petroleum oil, coal tar, pentachlorophenol, or copper naphthenate. Creosote is one of the most effective preservatives against attack by decay, insects, and marine organisms. It is often used on heavy timbers (railroad ties, bridge timbers, and so on). Its major drawbacks are an unpleasant odor and the fact that creosote-treated woods cannot be painted.

Pentachlorophenol, in petroleum solvents such as No. 2 fuel oil or liquefied natural gas, is similar to creosote in its behavior but it is not effective against marine borers. It too cannot be painted.

A number of *waterborne salts* are also used as preservatives. They include acid copper chromate, ammoniacal copper arsenite, chromated copper arsenate, and chromated zinc chloride. They are less effective than creosote but are used when the treated wood is to be painted.

BIBLIOGRAPHY

ASTM, Annual Book of ASTM Standards, Vol. 04.09, *Wood,* American Society for Testing and Materials, Philadelphia, 1997.

BARRETT, J. D. and FOSCHI, R. O., (eds.), *Proceedings of the First International Conference on Wood Fracture,* Banff, Alberta, Canada, 1979.

BODIG J. and JAYNE, B.A., *Mechanics of Wood and Wood Composites,* Van Nostrand Reinhold, 1982.

HEARMON, R. F. S., Elasticity of Wood and Plywood. *Special Report on Forest Products Research, London,* No. 7, 1948.

ILLSTON, J. M., (ed.), *Construction Materials,* 2nd ed., E&FN SPON, 1994.

MADSEN, B., *Structural Behavior of Timber,* Timber Engineering Ltd., North Vancouver, British Columbia, Canada, 1992.

PROBLEMS

14.1. What are the principal differences between *wood* and *timber?*

14.2. How can we relate the properties of wood to the properties of structural timber?

14.3. Why is timber much stronger in the longitudinal direction than it is in either the radial or tangential directions?

14.4. Describe the factors that control the *durability* of timber.

14.5. Why do there have to be *grading rules* for timber? What are the essential differences between visual grading and machine grading of timber?

14.6. What are the advantages and disadvantages of in-grade testing?

14.7. How does the moisture content affect the properties of timber?

14.8. How much does the microstructure of *wood* affect the properties of *timber?* Discuss.

14.9. How do defects and other nonuniformities affect the mechanical properties of timber?

14.10. Discuss the factors which determine the properties of manufactured wood products.

14.11. Why do we *cure* timber? What happens to the timber during the curing process?

14.12. Why does timber *shrink* much more in the radial and tangential directions than in the longitudal direction?

14.13. What is the significance of the *grain* in timber?

15

Polymers and Plastics

15.1 INTRODUCTION

Polymeric materials are used extensively in the construction industry, mainly for nonstructural applications, such as components in the envelope of buildings, sealants, adhesives, repair, restoration, and interior finishing. Their main advantage is in the variety of properties that can be tailored for various applications. The basic structure and general behavior of these materials were described in Chapter 2. In this chapter this topic will be further developed with the object of providing the general background required from the viewpoint of using these materials in the construction industry.

15.2 CLASSIFICATION AND PROPERTIES

The variety of compositions and properties in this class of materials is an advantage from the viewpoint of end products, but at the same time it requires from the user extensive knowledge which is beyond the scope of this book. Therefore, in this section a simplified treatment will be given based on the concepts described in Chapter 2.

Generally, it is useful to classify materials within a certain class on the basis of chemical composition. In the case of polymeric materials, nomenclatures can be based on the name of the monomer which makes up the backbone of the polymer. This is more applicable to thermoplastic polymers. The more common polymers are based on vinyl-type monomers, which have the following general composition:

$$-\left[CH_2 - \underset{\underset{R_2}{|}}{\overset{\overset{R_1}{|}}{C}}\right]-$$

R_1 and R_2 are substitutional groups, and common examples are given in Table 15.1, along with the name of the polymer derived from these basic monomer units. Some of these names will be familiar to the reader because they represent materials which are widely used.

TABLE 15.1   Thermoplastic Polymers with Carbon Backbone

Name of Polymer	Composition of R_1	Composition of R_2
Polyethylene	H	H
Polypropylene	H	CH_3
Polyvinyl chloride (PVC)	H	Cl
Polystyrene	H	C_6H_5
Polymethylmethacrylate	CH_3	$COOCH_3$

Polymers having the same composition as defined previously may be different in their properties due to differences in the characteristics of the structure of the chain, such as chain length and branching. For example, polyethylene may be in the form of branched or linear chains; the linear material would be characterized by higher transition temperature and strength due to its ability to pack more densely, leading to higher intensity of van der Waals forces between the chains and a greater tendency for crystallization. These differences show up in the density of the two types of polyethylene, which is higher in the linear material (~ 960 versus 920 kg/m^3) as well as in the mechanical properties (Fig. 15.1). Thus, the polyethylenes are classified

Figure 15.1
Effect of crystallinity on the stress-strain curve of polyethylene: high-density polymer of high degree of crystallinity and low-density polymer of lower crystallinity (after J. A. Sauer and K. D. Pae, "Mechanical Properties" of "High Polymers," in H. S. Kaufman and J. J. Falcetta, eds., *Introduction to Polymer Science and Technology,* Wiley, 1977, p. 397).

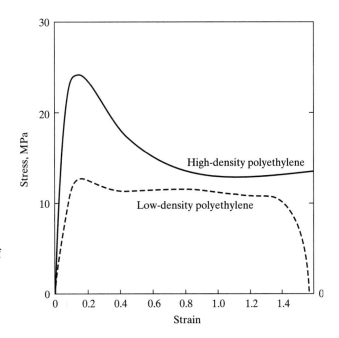

into high- and low-density products, where the high-density product can be used to manufacture rigid pipes for hot water flow and the low-density is used for flexible packaging material or sheets which serve as moisture barriers in structures. Further modifications in properties can be obtained when the polymer is combined with fillers, as discussed in Sec. 15.3. Variations of the nature just described can be achieved in many of the compositions outlined in Table 15.1, thus indicating that knowing the composition of the polymer is far from being sufficient to characterize its properties. Therefore, relying on the composition or chemical notation of the polymer to characterize its suitability for a particular engineering application can be misleading. This is different from with other materials, such as steels and cements. A powerful tool to characterize the performance of a thermoplastic material is by its response to temperature, evaluated by stress-strain curves (Fig. 15.2). Modulus of elasticity versus temperature curves are often used to classify the polymers on the basis of their mechanical performance. An example is presented in Fig. 15.3, for polymers of similar overall chemical composition but different chain structure. The crystalline polymer below the melting transition temperature, T_m, and the amorphous polymer below the glass transition temperature, T_g (i.e., glassy polymer) are both solidlike materials, with the crystalline being more rigid (i.e., higher modulus of elasticity), stronger, and more ductile. The greater ductility of the crystalline material is due to the ability of the crystallites to be opened up and extend their chains, as seen schematically in Fig. 15.4. In the glassy polymer, no such mechanism is available. Above T_m, the materials are essentially fluids of high viscosity. Above T_g, the amorphous polymer will be elastomeric in nature if it is lightly crosslinked. Without the crosslinks, the polymer molecules would easily slip one over the other, resulting in a viscous behavior. A schematic description of the stress-strain curves of these different forms of the thermoplastic materials is given in Fig. 15.5. The mechanisms for the different behaviors were discussed generally in Chapter 2 and are summarized schematically in Fig. 15.6. A typical range of mechanical and physical properties is provided in Table 15.2.

The value of the transition temperatures relative to the temperature of the environment can be a useful guide for the application of a particular polymer, or

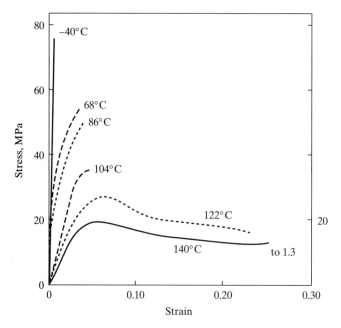

Figure 15.2
Effect of temperature on the tensile stress-strain curve of polymethylmethacrylate; a brittle to ductile transition occurs between 86 *to* 104°C, which is in the range of the glass transition temperature (after T. Alfrey, *Mechanical Behaviour of Polymers*, Wiley Interscience, 1967).

Chap. 15   Polymers and Plastics

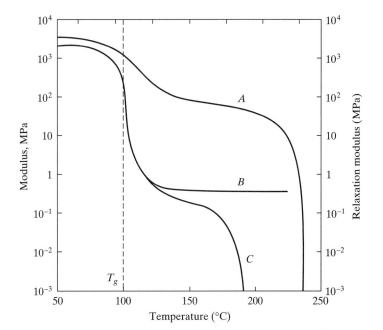

Figure 15.3
The modulus of elasticity versus temperature curves for crystalline isotactic polystyrene (a), slightly cross-linked atactic polystyrene (b), and linear amorphous polystyrene (c), (after A. V. Tobolsky and H. F. Mark, *Polymer Science and Materials,* Wiley Interscience, 1971).

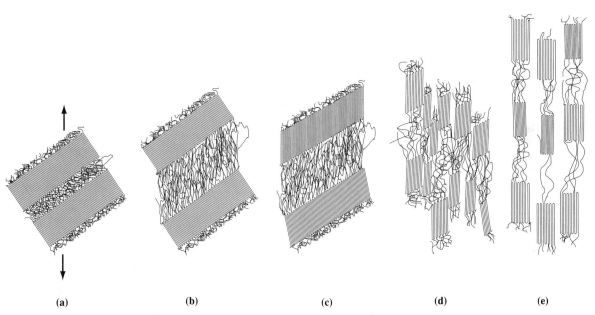

(a)   **(b)**   **(c)**   **(d)**   **(e)**

Figure 15.4
Changes in the structure of a crystalline polymer upon loading, showing opening of the crystallites during yielding: (a) two adjacent chain-folded lamellae and interlamellar amorphous material before deformation, (b) elongation of amorphous tie chains during the first stage of deformation, (c) tilting of lamellar chain folds during the second stage, (d) separation of crystalline block segments during the third stage, (e) orientation of block segments and tie chains with the tensile axis in the final deformation stage (after J. M. Schultz, *Polymer Materials Science,* Prentice Hall, 1974, pp. 500–501).

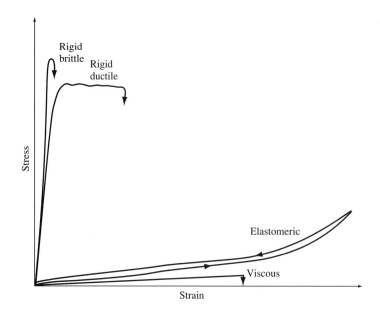

Figure 15.5
Schematic description of the stress-strain curve of various types of thermoplastic polymers.

Polymer type	Description	Shape and structure of polymer chains		
		Before failure	After failure/extensive deformation	
Viscous	Amorphous $T > T_g$			Chain uncoiling + slippage
Elastomer	Amorphous lightly cross-linked $T > T_g$			Chain uncoiling
Glassy	Amorphous $T < T_g$			← Crack
Crystalline	$T < T_m$			Separation of crystalline block

Figure 15.6
Schematic description of the response to stress of different types of thermoplastic polymers.

alternatively one may specify the transition temperature required for a specific use. If the melting transition temperature is significantly higher than room temperature, the polymer can be considered as a solid material and can be used for making products such as pipes and rigid polymeric foams. If below the glass transition temperature the polymer is amorphous (i.e., glassy polymer at room temperature), the rigid polymer tends to be brittle and transparent and can be used for glazing. Although more brittle than a crystalline polymer, it is more ductile than silicate glass. In practice we can obtain a fully glassy polymer at room temperature (i.e., completely amorphous), but rarely is it possible to obtain a fully crystallized polymer (see Chapter 2). Thus, for polymers with crystalline and glass transition temperatures above room temperature, the solid obtained at room temperature would have properties in between those of the crystalline and glassy state. In such systems it is possible to get solids of a wide range of properties (strength, modulus of elasticity, ductility) by adjusting the contents of the two phases and their distribution. As a result of all of these factors, the properties of thermoplastic polymers of similar composition can vary over a wide range, as seen in Table 15.2.

TABLE 15.2 Some Properties of Thermoplastic, Elastomeric, and Thermosetting Polymers

Polymer	Tensile Strength (MPa)	Modulus of Elasticity (GPa)	Elongation at Failure (%)	Density (kg/m³)	Heat Deflection (Temp. 66 psi (°C))	Examples of Application
Thermoplastics						
Polyethylene[a]						Packing films,
Low density	8.5–22	0.1–0.3	50–800	920	42	pipes
High density	22–40	0.4–1.3	15–130	960	85	
Polypropylene[a]	28–42	1–1.5	10–700	900	115	Tanks, fibers
Polyvinylchloride (PVC)[a]	35–64	2.1–4.3	2–100	1400		Pipes, floor tiles
Polystyrene[a]	23–57	2.7–3.2	1–60	1060	82	Insulating foams, lighting panels
Polymethyl-methacrylate[a] (PMMA) (Perspex)	43–86	0.14–0.36	300–450	1220	93	Windows, windshields
Polytetrafluoroethylene[a] (Teflon) (PTFE)	14–50	0.4–0.55	100–400	2170	120	Seals, coatings
Thermosettings						
Epoxies[b]	28–107	2.8–3.6	0–6	125-	—	Adhesives
Polyurethanes[b]	35–71	3–6	1300	—		Coatings, insulating foams
Polyesters[b]	43–93	2.1–4.6	0–3	1280	—	Matrix in FRP, laminates
Elastomers						
Polychloroprene[c] (Neoprene)	25	—	800	1240	—	Hoses
Butadiene-styrene[d] (SBR)	4.3–2.1	—	600–2000	1000	—	Coatings
Silicones[e]	2.5–7	—	100–700	1500		Sealants

[a]Composition given in Tables 2.5 and 15.1.
[b]Composition given in Sec. 15.2.
[cde]Composition same as isoprene (Table 2.5) with Cl replacing CH_3 is a copolymer of styrene and butadiene based on Si-O-Si backbone with organic groups attached.

Transition temperatures are usually estimated by standard tests, in which deflection of a loaded specimen is measured as a function of temperature. The temperature at which deflection becomes excessive is defined as the heat deflection temperature (or softening temperature), which is close to the transition temperature. As seen in Table 15.2, even for solid thermoplastics the transition temperatures and modulus of elasticity are considerably lower than those of conventional construction materials, such as steel and concrete; this is related to the coiling of the polymeric chain, which is not entirely avoided even below T_g. This also results in a solid, whose mechanical properties change readily with a small increase in temperature (Fig. 15.2) and which is much more sensitive to time-dependent deformations, as outlined in the discussion of viscoelastic behavior in Chapter 7. On the other hand, the strength of the polymeric materials can be high (Table 15.2). Crosslinking in thermosetting polymers improves the thermal stability but does not enhance the modulus of elasticity to the range of concrete or steel.

Many of the polymers contain along their chain backbone some functional groups that are molecules with a composition different from the monomer that makes up the backbone of the polymer chain. The functional groups are usually more polar and more reactive than the typical monomer unit. As such, they may have two types of influences: (1) Due to their polarity, they may enhance the secondary van der Waals bonds between neighboring chains and may even lead to formation of stronger hydrogen bonding, resulting in a polymer with a higher transition temperature and higher tendency to crystallize, all of which should result in improved properties; and (2) due to their higher activity, these functional groups may be activated to form chemical bonding between neighboring chains, leading to the formation of a crosslinked polymer (thermost), which is usually stronger and more temperature resistant than the thermoplastic polymer.

Some common functional groups of this kind are as follows:

$$\text{Epoxy:} \quad \overset{\displaystyle O}{\overset{\diagup\ \diagdown}{C-C}} \qquad\qquad \text{Ester:} \quad \overset{\displaystyle O}{\overset{\|}{C-C}} \qquad\qquad \text{Urethane:} \quad \overset{\displaystyle O}{\overset{|}{NH-C-O-}}$$

These functional groups can be placed along the polymer chain backbone at different densities. Since these groups have a major impact on intermolecular physical and chemical bonding, their nature and density will determine to a large extent the mechanical and physical performance of the polymer. Thus, the names of many of the polymers containing these functional groups are derived from the name of the group; polymers named epoxies, polyesters, and polyurethanes are very familiar. This nomenclature is limited in the sense that each of these names represents a family of materials that can be different in properties, depending on the density of the functional group and the nature of the chain backbone between the functional groups. For example, polyurethane can be in the form of a very hard and strong material if the urethane groups are densely located; it can also be in the form of a flexible material if the urethane groups are spaced apart and the polymer backbone between them is sufficiently free to behave like a lightly crosslinked elastomeric material.

Many of the polymers that we are using consist of non-crosslinked, relatively short polymer chains (prepolymers) having active functional groups along their backbone. The prepolymer is in a thin or thick fluid state; its functional groups can be activated later to form crosslinks and, as a result, lead to hardening of the material. The activation can be obtained by mixing the prepolymer with a reactive component, or it can be activated when in contact with species existing in normal environments, such as moisture and oxygen. These characteristics are the base for many known materials and processes in the construction industry, such as epoxies and polyesters used to produce fiber-reinforced plastics (fiber mats are impregnated with a fluid prepolymer which later hardens) and adhesives and sealants, which are applied in a fluid state on the surfaces of components at the joints and harden later. The polar functional groups are also effective in bonding (physically and chemically) to surfaces in contact with the polymer and are thus effective in promoting adhesion. Epoxy is one of the noted functional groups of this kind.

The range of compositions of the polymer offers the possibility to engineer materials with a wide variety of properties. However, this is not always sufficient to satisfy the end user's need, and therefore additional measures are needed based on combining the polymers with various additives and fillers. Several classes of such materials can be identified:

1. *Plasticizers:* These are low-molecular-weight (MW) molecules that reduce internal bending between polymer chains, thereby lowering the glass transition temperatures.

2. *Lubricants:* These additives are intended to reduce external friction in the processing stage of the manufacturing of products by methods such as extrusion. They also improve the ability of the polymer in the processing state to flow and wet and mix efficiently with other additives and fillers. The lubricants are usually materials of low molecular weight.

3. *Stabilizers:* These additives are intended to improve the durability performance of the polymer by enhancing its resistance to effects of heat, radiation, and oxidation (for additional discussion, see Sec. 15.4.3).

4. *Fire retarders:* A major disadvantage of polymers is that they may release toxic gases on combustion (see Sec. 15.4.2). Fire retarders are used to inhibit burning.

5. *Fillers:* Fillers are usually inert substances whose content can be in the range of 5 to 60%. They are intended to enhance mechanical properties, such as hardness, abrasion resistance, and strength. Sometimes fillers are used as low-cost extenders to lower the price of the polymer material. Common fillers are materials such as carbon black and calcium carbonate.

6. *Reinforcement:* The properties of the polymers are not always adequate for structural and semistructural applications. Therefore, for these applications, they are always reinforced with fibers to yield composites which can be of high performance. This topic of fiber-reinforced plastics (FRPs) will be discussed in Chapter 16, which deals with fiber-reinforced composites.

In practice, most of the polymeric products used contain more than one additive or filler to optimize the properties for the end use. The process of modifying the polymer with additives and fillers is commonly referred to as compounding. This stage can be done on a small scale by producers who obtain the basic polymer (sometimes referred to as resin) from a larger manufacturer. The compounding stage is not a simple one because the enhancement of one property using a certain additive or filler may be accompanied by a deleterious effect on other properties. Thus, a considerable amount of know-how is required for proper compounding. In view of the differences in performance between the basic polymer and the compounded materials, it has been suggested by some to refer to the first one as a polymer and to the second as plastic.

15.4 PROPERTIES FOR CIVIL ENGINEERING APPLICATIONS

The range of properties of the polymers and plastics is wide. Even materials that have common compositional characteristics and are labeled by the same name (polyethylenes, epoxies, polyesters, etc.) vary in properties to such an extent that they each must be considered to represent a family rather than a material of unique properties. Therefore, when selecting a polymer or plastic for a certain end use, one must not rely on the technical name of the product but assess it on the basis of the required properties that will assure adequate performance. For example, polyester resins are usually suitable for application as a matrix for fiber-reinforced plastic composites. Yet commercial products may be different in properties, such as long-term durability and fire resistance. Because the general classification of a polymer to the polyester group is not sufficiently informative, it is essential to have information regarding the actual performance of the material, in particular, with regard to the relevant properties for the end use. The relevant standards offer a variety of test methods and specification regarding short- and long-term properties.

Despite the wide range of properties, there are certain common characteristics of the polymers which distinguish them from other construction materials and require special considerations if one is to take advantage of the many superior properties that are existent in various polymers, such as high strength, extensibility and flexibility, adhesion, and resistance to aggressive solutions (e.g., acids). The properties that are needed to be addressed are the modulus of elasticity, viscoelastic characteristics, and sensitivity to temperature, fire, and other environmental influences.

15.4.1 Mechanical Performance

In terms of mechanical behavior, the polymeric materials are characterized by a much lower modulus of elasticity and higher viscoelastic response (i.e., higher creep and stress relaxation) than conventional construction materials, like concrete and steel. The difference can be by orders of magnitude (e.g., moduli of elasticity values of 210, 20–35, and less than 5 GPa for steel, concrete, and polymers, respectively).[1] The higher viscoelastic response (for more details, see Chapter 7) is associated with the relatively low transition temperature of the polymers (70 to 250°C). This implies that structures prepared from polymers may undergo large deflections, which may be a limiting factor in civil engineering applications. To overcome this deficiency, polymeric materials are usually applied as part of a fiber-reinforced plastic composite in spatial structures where the geometry of the structure compensates for the low modulus of elasticity. Thus, polymeric materials may be used to produce pipes and containers which, due to their circular shape, have sufficient rigidity in spite of the low modulus of elasticity of the polymer.

15.4.2 Thermal and Fire Performance

Polymeric materials are more sensitive to the effect of heat and fire than other conventional construction materials. With regard to thermal effects, attention should be given to the higher coefficient of thermal expansion, which is much greater in polymers than in conventional construction materials. The coefficient of thermal expansion for polymers can be in the range of $50–300 \times 10^{-6}°C^{-1}$ compared to about $10 \times 10^{-6}°C^{-1}$ for steel and concrete. Thus, in structures where these materials are incorporated together, special considerations should be given to the overall design and detailing. For example, gutters from polymeric material installed on a concrete structure must be connected with flexible devices to allow for differential movement between the two materials.

---

[1]Modern molecular engineering has created crystalline with rigid backbones that can have a modulus of elasticity similar to steel (e.g., Kevlar). However, these polymers are only produced as fibers and are used in composites where high strength to weight ratios are required, as in aerospace applications.

Performance in fire requires special attention because during heating the polymer can soften more readily due to the lower transition temperature (see the values of softening temperature in Table 15.2), and some compositions may be more sensitive to ignition or may release toxic fumes. These may occur even before actual combustion, as the material starts to decompose due to the effect of heat. The behavior from this viewpoint can be different for various polymer compositions, some of them performing considerably better than others due to control of the polymeric structure itself and proper compounding. Thus, the performance of the polymeric material exposed to fire should be assessed on the basis of several criteria, such as ignition, flame spread, smoke, toxicity of fumes, and softening and dripping. Tests to characterize such properties and grade the various materials according to their fire performance are outlined in various standards, such as ASTM E-906 and ASTM E-1321. The physical structure of the material can play an important role; the high surface area of plastic foams may lead to greater sensitivity. In judging the fire performance, there is also a need to address the whole component and not just the material; this is done in fire endurance tests such as ASTM E-119.

15.4.3 Weathering and Durability

Polymeric materials usually have excellent resistance to various chemicals, such as acids, alkalies, and salt solutions, and as such are used extensively for coating and resurfacing of components exposed to aggressive chemical environments. Some of the polymers (in particular, the thermoplastics) are sensitive to organic fluids such as fuels and solvents. These fluids consist of low-molecular-weight organic molecules, which can readily penetrate in between the polymer chains and reduce the van der Waals bonding, resulting in effects such as softening and swelling.

In civil engineering applications, attention should be given to the outdoor weathering of the polymers because exposure to solar radiation, moisture, and heat can lead to processes that may result in loss of strength, toughness (embrittlement), cracking (crazing), and changes in appearance (color, opacity, and gloss). These effects, when they occur, are usually the result of complex chemical and physical processes which take place at the same time, such as the following.

Photodegradation processes resulting from the absorbed solar radiation. The ultraviolet (UV) light wavelength (290–410 nm) has a quantum energy on the same order of magnitude as the bonds in the polymer chains (99–70 kcal/mol); thus, if such radiation is absorbed in the polymer and cannot be dissipated, it may cause dissociation of bonds. This event can lead to one or more of the following changes: breakage (cleavage) of the polymer molecule into smaller units, crosslinking, formation of double bonds, depolymerization, and hydrolysis. These processes are more likely to affect thermoplastic polymers and lead to a variety of mechanical and physical changes, as outlined previously.

Oxidation reactions, which are usually initiated by the photodegradation process and are therefore referred to as photooxidation. The processes described previously, with regard to the changes in bonding due to photodegradation, may involve oxidation, and this may determine the outcome of the process (whether it is chain scission, crosslinking, etc.)

Leaching of plasticizer by volatization or oxidative degradation.

Changes in dimensions due to temperature and humidity cycling, leading to cracking and crazing, which may enhance the chemical processes discussed previously.

Moisture effects, which cause swelling when moisture penetrates into the polymeric material, as well leaching of some soluble products that were generated in the photodegradation process. Water may participate in secondary photochemical reactions.

For many practical purposes, the weathering performance is assessed by accelerated tests, in which the conditions to accelerate the processes discussed previously are induced. These tests involve cycles of radiation, heating-cooling, and wetting-drying, in instruments which are collectively called "weather-o-meters." Such instruments and the conditions for running them are outlined in ASTM G-23. The weathering performance is evaluated by comparing the specified property before and after the test. Empirical relations have been developed between the time in the accelerated test and the natural exposure time, leading to similar changes in properties. Natural exposure is usually carried out in testing and research laboratories which operate natural exposure sites, as outlined in ASTM D-1435.

15.4.5 Adhesion

Most of the applications of polymeric materials in civil engineering are for nonstructural building products, where maximum advantage can be taken of their special properties (with minimum of interference from some of the disadvantages just described). These include foams for insulation, window and door assemblies, finishing materials, and adhesives and sealants. Of particular interest are the latter two applications, because they are in many cases of prime importance and consideration in modern construction systems. Here, a much greater input of the engineer is required to match the adhesive or the sealant to the performance needed from the structure. The more modern adhesives and sealants are based on systems of prepolymers, which harden on site by a crosslinking reaction and at the same time can bond well with the substrate, as described previously. These materials may be given an overall characterization by addressing at the same time their peel and shear/tensile bond strength (Fig. 15.7), as outlined in standards such as ASTM C-794, ASTM C-906, ASTM C-961, and ASTM C-1135. The various types of bonds can relate to the actual performance of such systems and to their hardness, which is more a function of the degree of crosslinking (Fig. 15.8). High peel bond strength is obtained with more flexible adhesives, which can accommodate large local deformations during peeling. High shear/tensile bond strength are obtained with stronger compositions, where higher densities of crosslinking are responsible for the enhanced strength. Yet crosslinking that is too dense may lead to reduction in shear/tensile bond strength, because with such compositions the polymer is too rigid and may

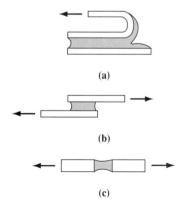

Figure 15.7
Schematic description of bond tests: (a) peel bond test; (b) shear bond test; and (c) tensile bond test.

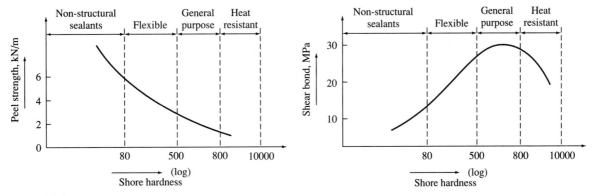

Figure 15.8
Classification of adhesives and sealants based on their bond strength values and hardness.

enhance stress concentration in the bonded joint. These influences are readily accountable for the curves in Fig. 15.8. Low hardness (i.e., smaller crosslinking and higher flexibility) provides higher peel bond strength and is more suitable for materials serving as sealants; intermediate hardness results in higher shear/tensile bond strength due to the increased crosslinking and is suitable for general-purpose adhesives; very high hardness is obtained with excessive crosslinking, resulting in a more brittle adhesive, where shear/tensile bond is reduced due to stress concentrations. However, the additional crosslinking provides greater stability for high-temperature performance and thus forms the basis for high-temperature adhesives.

BIBLIOGRAPHY

C. HALL, *Polymer Materials,* second edition, Wiley, New York, 1989.

N. C. MCCRUM, C. P. BUCKLEY, and C. B. BUCKNALL, *Principles of Polymer Engineering,* Oxford University Press, Oxford, 1989.

PROBLEMS

15.1. What are the requirements from the structure of a polymer to ensure that it behaves as an elastomer.

15.2. A given polymer is characterized by the following transition temperatures:

$$T_s = -10°C \qquad T_m = 180°C$$

The polymer can be obtained in two forms: (a) with a high degree of crystallinity of 80 percent and (b) with a low degree of crystallinity of 20 percent.
What are the differences or similarities between these two polymers when the service temperatures are:
(i) 280°C
(ii) +20°C

15.3. Polyethylene can be obtained in the form of a flexible and soft material which can be used for packaging or as a rigid solid which can be applied to produce pipes. What are the differences in the structure of two forms of polyethylene which lead to these properties.

15.4. What are the possible applications of the three different polymers in a service temperature in the range of 10 to 30°C:

$$\text{polymer (a): } T_g = 100°C$$
$$\text{polymer (b): } T_g = 20°C$$
$$\text{polymer (c): } T_g = -50°C$$

15.5. Explain the possible reasons for the existence of polyurethanes which can be significantly different in their mechanical properties.

$$\text{The polyurethane group:} \quad -N-\overset{\overset{\displaystyle O}{\|}}{C}-O-O-$$
$$\underset{H}{|}$$

15.6. Amorphous polyethylene usually has a transition temperature which is lower than that of amorphous P.V.C.
Explain the reason for this difference.

$$
\text{Polyethylene} -\overset{\overset{\displaystyle H}{|}}{\underset{\underset{\displaystyle H}{|}}{C}}-\overset{\overset{\displaystyle H}{|}}{\underset{\underset{\displaystyle H}{|}}{C}}-\overset{\overset{\displaystyle H}{|}}{\underset{\underset{\displaystyle H}{|}}{C}}-\overset{\overset{\displaystyle H}{|}}{\underset{\underset{\displaystyle H}{|}}{C}}-\overset{\overset{\displaystyle H}{|}}{\underset{\underset{\displaystyle H}{|}}{C}}-; \quad \text{PVC} -\overset{\overset{\displaystyle H}{|}}{\underset{\underset{\displaystyle H}{|}}{C}}-\overset{\overset{\displaystyle Cl}{|}}{\underset{\underset{\displaystyle H}{|}}{C}}-\overset{\overset{\displaystyle H}{|}}{\underset{\underset{\displaystyle H}{|}}{C}}-\overset{\overset{\displaystyle Cl}{|}}{\underset{\underset{\displaystyle H}{|}}{C}}-\overset{\overset{\displaystyle H}{|}}{\underset{\underset{\displaystyle H}{|}}{C}}-
$$

15.7. Discuss the advantages and limitations of polymer materials in civil engineering applications.

15.8. What are the changes that can be made in the structure of a given polymer to increase its glass transition temperature?

15.9. Discuss the limitations of characterization of the performance of the adhesion characteristics of a given polymer adhesive by the bond strength value obtained by one particular test method.

16

Fiber-Reinforced Composites

16.1 INTRODUCTION

A general description of fiber-reinforced composites was given in Chapter 2 (see Fig. 2.46). In civil engineering structures, a variety of such materials are applied that have different types of matrices and fibers. In fact, the first fiber-reinforced composite material developed and used extensively in modern times was asbestos cement, which consists of a cement matrix reinforced with dispersed asbestos fibers. Products of this kind are manufactured as thin sheet materials, and their main application is for cladding, roofing, and pipes. Currently there is a trend toward diversification, and components of this kind are made with different combinations of human-made and natural fibers (e.g., cellulose) and with matrices which are either cementitious or polymeric. The geometries of the fiber reinforcement include several types shown in Fig. 16.1. Two-dimensional reinforcement is characteristic of thin sheet products, and three-dimensional is the distribution in fiber-reinforced concretes, where the length of the fibers dispersed in the matrix is smaller than the cross section (~25 mm long fiber in 50 to 150 mm cross-section concrete).

The concept underlying fiber-reinforced composites is the ability to produce materials with a fiber whose strength approaches the theoretical value (Chapter 1). The small cross section of the fiber results in reduction in the size of flaws which are responsible for the low strength of bulk material (relative to the theoretical strength), as is demonstrated by the concepts of fracture mechanics discussed in Chapter 6. In addition, in some materials, such as polymeric ones, the process of drawing the bulk material into a longitudinal fiber is accompanied by alignment of the chain molecules, thus providing higher density of aligned primary bonds. Properties and dimensions of some fibers which are used more frequently in civil engineering are provided in Table 16.1.

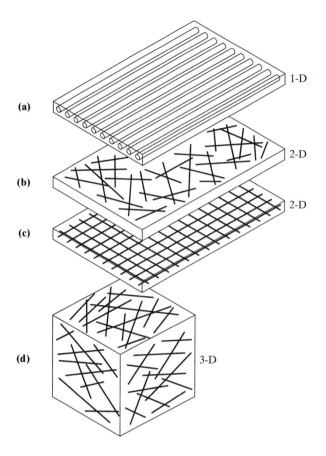

Figure 16.1
Geometries of fiber-reinforced materials: (a) one-dimensional arrangement; (b) two-dimensional arrangement, random; (c) two-dimensional arrangement, oriented; (d) three-dimensional, random arrangement.

TABLE 16.1   Typical Properties of Fibers

Fiber	Diameter (μm)	Relative Density[a]	Modulus of Elasticity (GPa)	Tensile Strength (GPa)	Elongation at Break (%)
Steel	5–500	7.84	200	0.5–2.0	0.5–3.5
Glass	9–15	2.60	70–80	2–4	2–3.5
Asbestos					
Crocidolite	0.02–0.4	3.4	196	3.5	2.0–3.0
Chrysotile	0.02–0.4	2.6	164	3.1	2.0–3.0
Fibrillated polypropylene	20–200	0.9	5–77	0.5–0.75	8.0
Aramid (Kevlar)	10	1.45	65–133	3.6	2.1–4.0
Carbon (high strength)	9	1.90	230	2.6	1.0
Nylon	—	1.1	4.0	0.9	13.0–15.0
Cellulose	—	1.2	10	0.3–0.5	—
Acrylic	18	1.18	14–19.5	0.4–1.0	3
Polyethylene	—	0.95	0.3	0.7×10^{-3}	10
Wood Fiber	—	1.5	71.0	0.9	—
Sisal	10–50	1.50	—	0.8	3.0

[a]Formerly specific gravity.

    The production of such composites can be carried out by various technologies, some of them simple and enabling even on-site production, while others are more complex and limited to carefully controlled industrial processes (Fig. 16.2). Some of these production methods can be applied for both polymeric and cementitious matrices, while others are limited to one of the matrices only. Generally, the cemen-

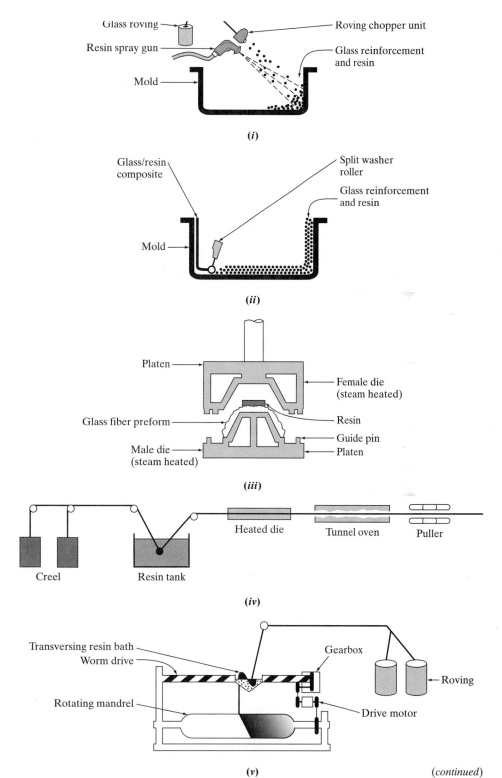

Figure 16.2
Production methods of fiber reinforced composites: (a) FRP—(i) spray up, (ii) hand lay up, (iii) hot press molding, (iv) pultrusion, (v) filament winding (after L. Hollaway, *Glass Reinforced Plastics in Construction,* Surrey University Press, 1978);

(*continued*)

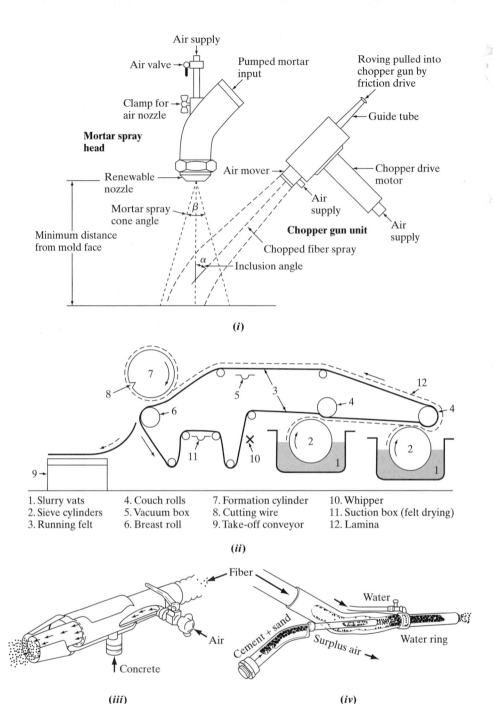

(i)

1. Slurry vats
2. Sieve cylinders
3. Running felt
4. Couch rolls
5. Vacuum box
6. Breast roll
7. Formation cylinder
8. Cutting wire
9. Take-off conveyor
10. Whipper
11. Suction box (felt drying)
12. Lamina

(ii)

(iii)   **(iv)**

Figure 16.2 (*continued*) (b) FRC—(i) spray gun of glass fiber-reinforced cement (after G. True, *GRC Production and Use,* Palladian Publications Ltd., 1986), (ii) Hatschek process for production of asbestos and cellulose reinforced cement (after J. E. Williden, *A Guide to the Art of Asbestos Cement,* J. E. Williden Publishers, 1986), (iii) wet fiber-reinforced concrete shotcreting, (iv) dry-fiber reinforced concrete shotcreting

titious matrix, which has a higher viscosity, does not mix as readily with fibers, whereas polymeric matrices can be produced with larger contents of fibers because they can be much more fluid in their fresh state prior to hardening. Thus, the two types of composites differ considerably in their matrix properties (the polymeric matrix is usually much more ductile) and in the content of the fibers, which can be as large as 50% to 80% in the polymeric composites, less than ~15% in the cementitious composites, and usually less than 2% for fiber-reinforced concretes, where fibers are simply introduced into the concrete mixer.

16.2.1 Overall Mechanical Behavior

In this section the basic concepts of the mechanics of fiber-reinforced composites will be presented, highlighting the similarities and differences between fiber-reinforced polymers (FRPs) and fiber reinforced cements (FRCs), which result from the different properties of the matrices and the contents of fibers that can practically be incorporated. The FRP composite can usually be treated as a ductile matrix composite, whereas FRC is a brittle matrix composite.

Let us first consider composites reinforced with continuous and aligned fibers that have the stress-strain curves shown in Fig. 16.3 for the FRP and FRC materials. In both systems the fiber is stronger and more rigid than the matrix. However, in the FRP composite the matrix is more ductile than the fiber, whereas with FRC it is just the opposite. Assuming a perfect bond between the two components in each of the composites, the rule of mixture applies (Chapter 10):

$$E_c = E_m \cdot V_m + E_f \cdot V_f, \tag{16.1}$$

where E_c, E_m, and E_f are the initial moduli of elasticity of the composite, matrix, and fiber, respectively, and V_m and V_f are the volume fraction contents of the matrix and the fiber, respectively ($V_m + V_f = 1$). The rule of mixtures shows a strong dependency

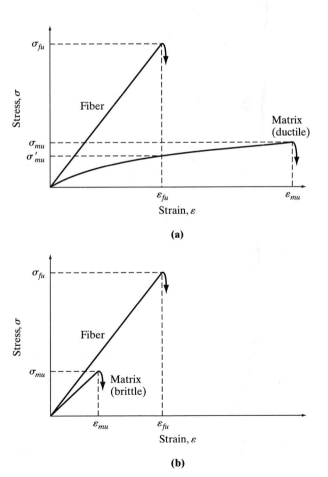

Figure 16.3
Stress-strain curves of components in fiber-reinforced composites: (a) fiber-ductile matrix systems (characteristic of FRP), and (b) fiber-brittle matrix systems (characteristic of FRC).

of the composite performance on the fiber content, in particular when the fiber property is considerably higher than that of the matrix. This is the case for the moduli of elasticity of the components of FRP. Thus, composites of this kind provide an efficient means to overcome the limitation of the low modulus of elasticity of polymers discussed in Chapter 15. The moduli of such composites, as well as their strengths, would depend to a large extent on the method of production, which controls the amount of fiber that can be added (Fig. 16.4).

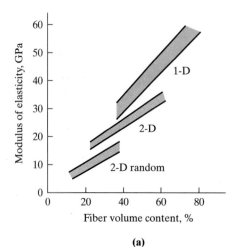

(a)

(b)

Figure 16.4
Mechanical properties of glass fiber-reinforced plastic as a function of the fiber content and geometry (1-D: typical of filament winding, 2-D: typical of hand lay-up, 2-D random: typical of spray up): (a) modulus of elasticity, (b) tensile strength.

The stress that can be supported by the composites up to the cracking of the fiber in the case of FRP (ε_{fu} in Fig. 16.3a) and the cracking of the matrix in the case of FRC (ε_{mu} in Fig. 16.3b) also follows the rule of mixtures:

$$\sigma_c = \sigma_m \cdot V_m + \sigma_f \cdot V_f, \tag{16.2}$$

where σ_c, σ_m, and σ_f are the stresses supported by the composite, matrix, and fiber, respectively, in the range below the cracking strains. In the range above the cracking strains, the equations governing the stress supported by the two types of composites will be different: In the FRP the only active component will be the matrix, whereas in the FRC it will only be the fiber:

Stress supported by FRP at $\varepsilon > \varepsilon_{fu}$:

$$\sigma_c = \sigma_m \cdot V_m = \sigma_m(1 - V_f) \tag{16.3}$$

Stress supported by FRC at $\varepsilon > \varepsilon_{mu}$:

$$\sigma_c = \sigma_f \cdot V_f \tag{16.4}$$

In each of the ranges in Eqs. 16.2, 16.3, and 16.4, the maximum stress-bearing capacity of the composite (i.e., composite strength) can be calculated by replacing the appropriate stress values of the components by the values at the ultimate strains. This will be discussed separately for FRP and FRC.

In the FRP system, if V_f is sufficiently large, the composite will be

$$\sigma_{cu} = \sigma'_{mu} \cdot (1 - V_f) + \sigma_{fu} \cdot V_f, \tag{16.5}$$

where the addition of subscript u indicates an ultimate value.

The first term in this equation is the contribution of the matrix, and the second one is the contribution of the fibers. If the fiber volume in the FRP is smaller than a critical value, then a situation may arise where the strength of the matrix alone, [i.e., $\sigma_{mu} (1 - V_f)$, is greater than the strength of the combined matrix and fiber:

$$\sigma_{mu}(1 - V_f) > \sigma'_{mu}(1 - V_f) + \sigma_{fu}V_f. \tag{16.6}$$

In this case, the strength will be controlled by the matrix:

$$\sigma_{cu} = \sigma_{mu}(1 - V_f). \tag{16.7}$$

These relations are presented in Fig. 16.5a, as curves of strength versus fiber volume content, showing the intersections between the two relations.

In the FRC system, if V_f is sufficiently large, the composite strength will be

$$\sigma_{cu} = \sigma_{fu}V_f. \tag{16.8}$$

If, however, the fiber volume in the FRC composite is low, smaller than a certain critical value, a situation may arise where the strength of the matrix alone [i.e., $\sigma_{mu}(1 - V_f)$] is greater than that of the fiber:

$$\sigma_{mu}(1 - V_f) > \sigma_{fu}V_f. \tag{16.9}$$

In this case, the strength will be controlled by the matrix, as in Eq. 16.7. These relations (Eq. 16.8 and 16.9) are presented in Fig. 16.5b, showing the intersection between the two sets of equations.

Assuming that the strength of the fiber, σ_{fu}, is much greater than the strength of the matrix, σ_{mu}, which is the case in the actual systems used, it can be shown readily that the critical fiber volume content, V_{crit}, is

$$V_{\text{crit}} = \frac{\sigma_{mu}}{\sigma_{fu}}. \tag{16.9}$$

Introducing typical values to Eq. 16.9 (FRP–glass fibers in polyester matrix, where $\sigma_{fu} = 2100$ MPa and $\sigma_{mu} = 65$ MPa; FRC–glass fibers in cement matrix, where $\sigma_{fu} = 2100$ MPa and $\sigma_{mu} = 5$ MPa), we obtain the critical fiber volume contents of about 3% and 0.3% for the FRP and FRC, respectively.

In the range above V_{crit}, the mode of failure of FRC and FRP will be different. In the FRP, no cracking will occur prior to the composite failure, while with FRC multiple cracking will occur as the matrix ultimate tensile strain is exceeded. This is shown schematically in Fig. 16.6.

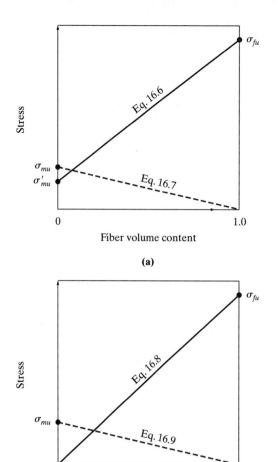

(a)

Figure 16.5
Composite strength-fiber
volume content relations:
(a) fiber-ductile matrix systems (characteristic of
FRP); (b) fiber-brittle matrix systems (characteristic
of FRC).

(b)

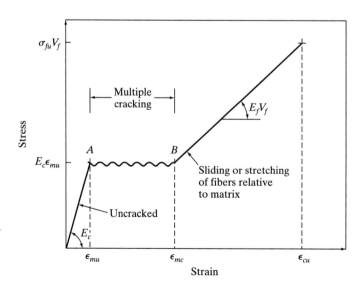

Figure 16.6
Ideal stress-strain curve for
FRC composite according
to the Aveston-Cooper-
Kelly (ACK) model.

16.2.2 Bonding

In practice, many of the fiber-reinforced composites consist of short fibers, and therefore the stress that they carry may be smaller than their strength. Their load-bearing capacity in the composite depends on the stress transfer across the fiber-matrix interface, which is a function of the shear bond stress that may develop across the interface and the length of the fibers. These characteristics can be modeled and tested by the pullout test shown in Fig. 16.7a. Assuming that the stress transfer mechanism across the interface is a frictional one (i.e., the interfacial shear bond strength is a constant value along the fiber), the tensile stress distribution along the fiber is a triangular one (Fig. 16.7b). Under such conditions the fiber will pull out of the matrix at a load P, and the interfacial shear bond strength, τ_{fu}, can be calculated as

$$\tau_{fu} = \frac{P}{\pi d l'}, \tag{16.10}$$

where d is the fiber diameter and l' the embedded length.

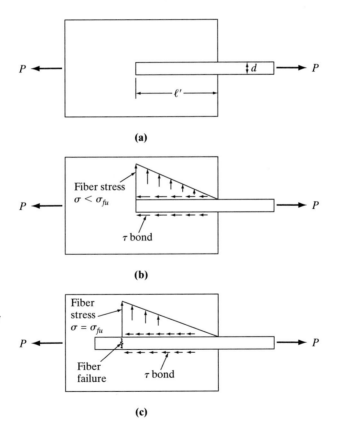

Figure 16.7
Pullout behaviour of a fiber. (a) General geometry of pullout. (b) Stress distribution in a fiber which pulls out. (c) Stress distribution in a fiber that fractures during the test, assuming frictional stress transfer.

If the fiber-embedded length is sufficiently long, the stress which is built up in the fiber will be as large as its strength (Fig. 16.7c) and the fiber will break rather than pull out. This situation will occur if the fiber-embedded length, l', is greater than

$$l' > \frac{\sigma_{fu} \cdot d}{4 \cdot \tau_{fu}}. \tag{16.11}$$

If we consider an actual composite, the minimum length of the fiber required for the buildup of a stress which is equal to its strength is called the critical fiber length, l_{cr}. It is a function of its geometry and interfacial shear bond strength (Fig. 16.8):

$$l_{cr} = \frac{\sigma_{fu} \cdot d}{2 \cdot \tau_{fu}}.$$

(16.12)

Accordingly, the tensile stress distribution in the fiber along its axis is a function of its length, as shown in Fig. 16.9. On this basis, it is possible to define an efficiency factor for the fiber, η_l, which is the average stress along the fiber, σ_{ave}, relative to its strength, σ_{fu} (i.e., σ_{ave}/σ_{fu})

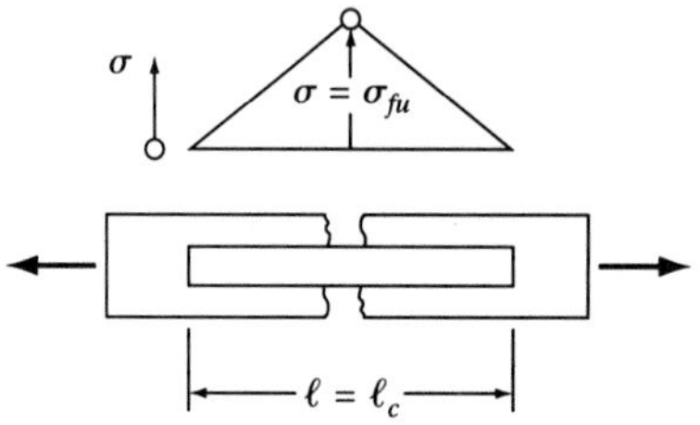

Figure 16.8
Definition of critical length, assuming frictional stress transfer.

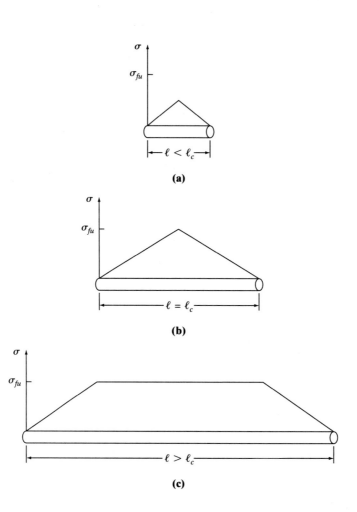

Figure 16.9
Stress profile along a fiber in a matrix as a function of fiber length.

Chap. 16   Fiber-Reinforced Composites

For $l > l_{cr}$,

$$\eta_l = 1 - \frac{l_{cr}}{2l} \qquad (16.13)$$

For $l < l_{cr}$,

$$\eta_l = \frac{\tau_{fu}}{\sigma_{fu}} \cdot \frac{l}{d} \qquad (16.14)$$

where l is the length of the fiber.

16.2.3 Influence of Bonding on Composite Behavior

The bonding characteristics will affect the strength of the composite and the shape of its stress-strain curve. For strength, it should be considered that the contribution of the fiber is smaller than $\sigma_{fu} \cdot V_f$ in Eqs. 16.5 and 16.8. The reduction in contribution is expressed in terms of efficiency coefficients for length (Eqs. 16.13 and 16.14) and efficiency coefficients for orientation, η_θ, both of which are in the range of 0 to 1. The values for η_θ will be discussed in Sec. 16.2.4. Thus, Eqs. 16.5 and 16.8 can be rewritten as

For FRP,

$$\sigma_{cu} = \sigma'_{mu} \cdot (1 - V_f) + \eta_l \cdot \eta_\theta \cdot \sigma_{fu} \cdot V_f \qquad (16.15)$$

For FRC,

$$\sigma_{cu} = \eta_l \cdot \eta_\theta \cdot \sigma_{fu} \cdot V_f. \qquad (16.16)$$

Since in most FRC composites failure is by pullout of fibers [i.e., $\eta_l = (\tau_{fu}/\sigma_{fu}) \cdot (l)/d$, according to Eq. 16.14], Eq. 16.16 can be rewritten as

$$\sigma_{cu} = \eta_\theta \cdot \tau_{fu} \cdot V_f \cdot \frac{l}{d}. \qquad (16.17)$$

The bonding characteristics have a significant influence on the shape of the stress-strain curve of the brittle matrix composite (in particular, in the multiple cracking zone). The nature of cracking at this stage and the spacing between the cracks is a function of the interfacial bond strength and fiber modulus of elasticity. Beyond the multiple cracking zone, additional load will result in slippage or straining of the fibers; in the case of fiber stretching, the slope of the curve will be

$$E_c = E_f \cdot V_f. \qquad (16.18)$$

The ideal stress-strain curve is given in Fig. 16.6. Example of curves of composites with fibers of different bond and modulus of elasticity are provided in Fig. 16.10. This indicates that too good a bond, as with asbestos fibers, gives high strength but low ductility; a very poor bond, as with polypropylene fibers, gives high ductility, but with no increase in strength and with the development of wide matrix cracks; glass (or steel) fibers give intermediate behavior.

16.2.4 Effect of Fiber Orientation

In the actual composite, the fibers are rarely oriented in one direction. They are either dispersed in 2-D (fibers in thin-sheet FRC or FRP), or in 3-D (fibers dispersed in concrete, where the cross section of the composite is significantly larger than the

Figure 16.10
Stress-strain behaviour of three types of FRC: (a) Asbestos fibers: strong bond, many fine cracks closely spaced (x' and w are small). Strong but limited ductility and ability to withstand impact. (b) Glass fibers: moderate bond, multiple cracking but greater spacing, more pullout of fibers. (c) Polypropylene fibers: low bond and low E; and thus cracks are very widely spaced. Fibers debond with considerable pullout. High ductility and resistance to impact (adapted from D. J. Hannant, *Fiber Cements and Fiber Concretes*, Wiley, 1978).

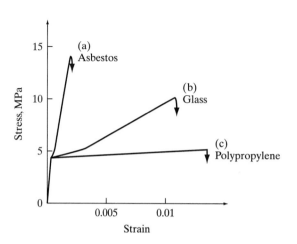

length of the fibers), or are arranged as laminates. Laminates are stacks of individual plates (laminae) with unidirectional fiber reinforcement in each lamina. The laminae may be oriented at right angles (cross-ply) or at various angles (angle-ply) (see Fig. 16.11). Plywood is a common example of a cross-ply laminate, while filament wound glass fiber tubes are angle-ply laminates. It is not possible to consider the response of laminates to stress in detail in this book. Suffice it to say that the approach to laminate behavior is to consider the effect of the applied stress on each of the lamina orientations. When failure criteria are reached for one orientation, all laminae in that orientation are considered to fail at the same time and the stress is

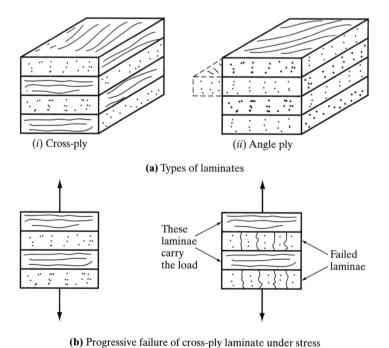

Figure 16.11
Schematics of laminate construction and response to load: (a) types of laminates, (b) progressive failure of cross-ply laminate under stress.

(i) Cross-ply   (ii) Angle ply

(a) Types of laminates

These laminae carry the load

Failed laminae

(b) Progressive failure of cross-ply laminate under stress transferred to the remaining laminae. The unbroken laminae continue to support the redistributed load until they too fail.

A general and simplified method to assess the effect of orientation on the efficiency of the fibers can be based on equations developed by Krenchel:

$$\eta_\theta = \sum a_\theta \cdot \cos^4 \theta, \tag{16.19}$$

where a_θ is the proportion of fibers oriented at an angle θ. For 2-D and 3-D randomly oriented fiber distribution, the η_θ efficiency coefficients are approximately 1/3 and 1/6, respectively.

16.3   FIBERS AND MATRICES

A wide variety of fibers are used for reinforcement of cementitious and polymeric matrices. A list of some of the more common ones and their properties was provided in Table 16.1, while matrix properties are summarized in Table 16.2. In reinforcement of concretes, use is made of fibers with relatively large diameters, greater than 0.1 mm, which include steel and polypropylene. The interfacial bond that can be achieved in such systems is relatively low because of the particulate nature of the cementitious matrix. Therefore, the steel fibers used are usually deformed into various shapes to provide mechanical anchoring. In the case of fiber reinforcement of concrete, the content of the fibers is equal to or smaller than the critical volume (due to mixing limitations) and their main task is to improve the toughness of the matrix by providing postcracking load-bearing capacity with only limited strength enhancement (Fig. 16.12).

Smaller-diameter fibers, mostly in the range of 10 μm, are used for reinforcement of polymeric matrices (FRP) and thin-sheet cementitious composites (FRC). The fibers in this category are usually high-performance, human made fibers, such as glass and carbon. Natural fibers, like those derived from wood pulping (cellulose fibers), and those found as natural minerals (asbestos fibers), are used extensively in various construction materials, mainly as reinforcement for thin-sheet products used for cladding and roofing, such as shingles, flat sheets, and corrugated sheets. In these

Figure 16.12
Schematic stress-strain curves for low and high fiber volume FRC (after A. Bentur and S. Mindess, *Fibre Reinforced Cementitious Composites,* Elsevier Applied Science, UK,1990).

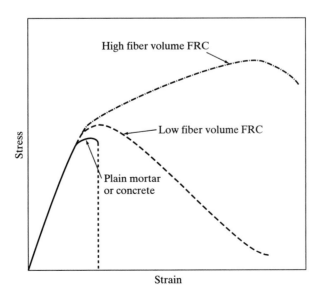

TABLE 16.2 Properties of Some Matrices Used in Fiber-Reinforced Composites

	Tensile Strength (MPa)	Compressive Strength (MPa)	Modulus of Elasticity (GPa)	Elongation at Fracture (%)	Density (kg/m³)
Epoxy	35–100	100–200	3–6	1–6	1100–1400
Polyester	40–90	90–250	2–4.5	2	1200–1500
Cement paste concrete	2–4	30–70	20–40	0.01	1800–2400

products the fiber content exceeds the critical volume content, and this shows up in the stress-strain curve of such composites, which exhibits an ascending branch beyond the first crack stress (Fig. 16.12).

BIBLIOGRAPHY

ASHBEE, K. H. G. *Fundamental Principles of Fiber Reinforced Composites,* second edition, Technomic Publication, Lancaster & Basel, 1993.

BENTUR, A. and MINDESS, S., *Fibre Reinforced Cementitious Composites,* Elsevier Applied Science Publishers, London, 1990.

HOLLAWAY, L., *Polymer Composites for Civil and Structural Engineering,* Blackie, London, 1993.

HOLLAWAY, L. (editor), *Polymers and Polymer Composites in Construction,* Thomas Telford, London, 1990.

HULL, D., *An Introduction to Composite Materials,* Cambridge University Press, Cambridge, 1992.

PROBLEMS

16.1 Calculate the tensile strength and initial modulus of elasticity of a thin composite material containing 30 percent by volume of uniformly dispersed fibers in 2 dimensions.

> Properties of the fibers:
> tensile strength—1500 MPa
> modulus of elasticity—70000 MPa
> diameter—20 micrometer

> Properties of the matrix:
> tensile strength at the ultimate strain of the fiber—20 MPa
> initial modulus of elasticity—2000 MPa

> Properties of the fiber-matrix interface:
> bond strength—0.7 MPa

16.2 Glass fibers are being used to reinforce a brittle and a ductile matrix. Both matrices have the same strength. Determine which of the composites will have higher strength. The fibers are continuous and are positioned parallel to the loading direction of the composite. Their content is 20 percent by volume.

> Properties of the fibers:
> Tensile strength—1300 MPa
> modulus of elasticity—70000 MPa
> stress-strain behavior—linear up to failure

Properties of the ductile matrix:
tensile strength—100 MPa

Properties of the brittle matrix:
tensile strength—100 MPa
stress-strain behavior—liner elastic up to failure with an ultimate strain of 0.5 percent

16.3 Calculate the stress-strain curves of the two composites in question 16.2 above.

16.4 Calculate the critical fiber volume of a composite consisting of the following components:

Properties of the fibers:
strength—1000 MPa
diameter—0.2 mm
length—20 mm
bond strength of the matrix—0.7 MPa
dispersion—case (a): random 2-dimension dispersion; case (b); random 3-dimension dispersion

Properties of the matrix:
strength—50 MPa
Stress-strain behavior—brittle, with ultimate strain smaller than that of the fibers.

16.5 Calculate the strength of the two composites with the properties outlined below, in which the fibers where surface treated to increase the bond strength from 1.5 MPa to 3.0 MPa.

Properties of the fibers:
tensile strength—3000 MPa
diameter—0.1 mm
length—50 mm

Properties of the matrix:
tensile strength—150 MPa
stress-strain behavior—brittle, with ultimate strain lower than that of the fibers.

16.6 What would be the difference in the failure mode of a brittle matrix reinforced with short fibers whose length is slightly smaller than the critical length and with fibers whose length is much greater than the critical length?

16.7 A composite material is made of 6 layers of continuous fibers which are unidirectional. Each layer is placed at an angle of 30° clockwise relative to the previous one.

The tensile strength of the fibers is 2500 MPa and their total volume content is 36%. Calculate the strength of the composite in a direction parallel to one of the fiber layers (i.e. 0°) and at an intermediate direction between layers (i.e. 15°).

Assume that the contribution of the matrix to the strength is negligible.

16.8 You are requested to design a composite in which the fiber efficiency will reach 90 percent. The fibers are carbon fibers with a strength of 2000 MPa and a diameter of 7 micrometer.

For production purposes of the composite it is required to use short fibers with a length which is not greater than 10 mm. What is the bond strength needed to achieve the required efficiency.

If the production process would be modified to enable the use of longer fibers of 15 mm length, what would then be the required bond?

Solutions to Numerical Problems

2.4 LiF: R = 0.44   octahedral
KCl: R = 0.74   cubic
KI: R = 0.61   octahedral
RbBr: R = 0.76   cubic
CsI: R = 0.78   cubic
(see Table 2.2)

2.5 Using Eq. 2.2
(i) 6×10^{-9} cm²/s; 3.7×10^{-8} cm²/s; 3.9×10^{-6} cm²/s
(ii) 0.4×10^{-20} cm²/s; 1.3×10^{-18} cm²/s; 5.7×10^{-12} cm²/s
(iii) Carbon is a much smaller atom

CHAPTER 3

3.2 (a) 1310°C
(b) W_L = 53% Cu, 47%Ni
(c) Amount of liquid $X_L = \dfrac{X_S - X}{X_S - X_L} = \dfrac{60 - 50}{60 - 47}$

$$= \frac{10}{13} = 77\%$$

$$X_S = 23\%$$

(d) 1325°C

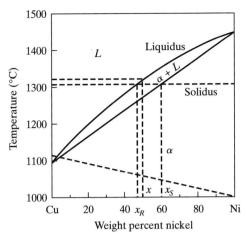

3.3 (a) 780°C

(b) Fraction of solid B = $0.5 = \dfrac{RA - PA}{AB - RA}$

Solve for RA = 77.5%

B = 77.5%, A = 22.5%

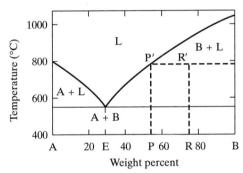

(c) Compostion of Eutectic (E) = 30%A, 70%B

CHAPTER 4

4.1 8 MPa. Bubble pressure exceeds atmosphere pressure; will be unstable. (Use Eq. 4.2). Pressure is reduced to 0.56 MPa; bubble will be stable. (28 dynes. cm^{-1} = 28 mJ.m^{-2}).

4.5 Concrete: 68 m when r = 500 nm
Brick: 1.4 m when r = 25 μm
(Eq. 4.13b)

4.6 In (p/p_o) = relative humidity = 87% at 25°C.

CHAPTER 9

9.1 (a) Parallel: E_c = 47 GPa; (b) Series: E_c = 28 GPa: (c) Hirsch: E_c = 32 GPa, (d) Aggregate: E_c = 34 GPa.

9.2 (a) Parallel: E_c = 60 GPa; (b) Series: E_c = 14 GPa; (c) Hirsch : E_c = 20 GPa; (d) Aggregate: E_c = 24 GPa.

10.1 3.6% (Eq. 10.3)

10.2 1217g (Eq.10.3)

10.3 (i) Moisture content of sand = 7.5% (Eq. 10.6).
(ii) This is greater than adsorption hence sand is wet.
(i) Moisture content of gravel = 1.2%.
(ii) This is less than adsorption, hence gravel is air dry.

10.4 F.M = 2.78

10.5 Max. Size = 1 in.

CHAPTER 11

11.2 (a) Type I; (b) Type V; (c) Type II; (d) Type III.

11.3 (a) Type II (or Type IPM); (b) Type I (or Type III; if low overnight temperatures are forecast; blended cement not recommended); (c) Type I (blended cement not recommended); (d) Type V (or Type IP, Type IS).

11.4 (a) (i) 0.27 cm³/g cement = 32% capillary porosity
(ii) gel porosity = 13%, thus total porosity = 55%
(b) (ii) 23% porosity; (ii) 36% total porosity.
(c) (i) 25% porosity; (ii) 41% total porosity.

11.5 (a) 26 MPa (Eq. 11.17)   31 MPa (porosity from Q.4)
(b) 40 MPa   46 MPa
(c) 39 MPa   42 MPa

(a) 24 GPa (Eqs. 11.17 and 11.21)
(b) 30 GPa
(c) 30 GPa

Index